Sustainable Development and Application of Renewable Chemicals from Biomass and Waste

Sustainable Development and Application of Renewable Chemicals from Biomass and Waste

Editors

Mohamad Nasir Mohamad Ibrahim
Patricia Graciela Vázquez
Mohd. Hazwan Hussin

MDPI • Basel • Beijing • Wuhan • Barcelona • Belgrade • Manchester • Tokyo • Cluj • Tianjin

Editors

Mohamad Nasir Mohamad
Ibrahim
School of Chemical Sciences
Universiti Sains Malaysia
Penang
Malaysia

Patricia Graciela Vázquez
Faculty of Sciences Exacts
University of La Plata
and Principal Researcher
of CONICET
La Plata
Argentina

Mohd. Hazwan Hussin
School of Chemical Sciences
Universiti Sains Malaysia
Penang
Malaysia

Editorial Office
MDPI
St. Alban-Anlage 66
4052 Basel, Switzerland

This is a reprint of articles from the Special Issue published online in the open access journal *Molecules* (ISSN 1420-3049) (available at: www.mdpi.com/journal/molecules/special_issues/bio_waste).

For citation purposes, cite each article independently as indicated on the article page online and as indicated below:

LastName, A.A.; LastName, B.B.; LastName, C.C. Article Title. *Journal Name* **Year**, *Volume Number*, Page Range.

ISBN 978-3-0365-6433-3 (Hbk)
ISBN 978-3-0365-6432-6 (PDF)

Contents

About the Editors

Mohamad Nasir Mohamad Ibrahim

Mohamad Nasir Mohamad Ibrahim obtained his B.Sc. (1994), M.Sc. (1997) and PhD (1999) from Missouri S&T (formerly known as University of Missouri-Rolla, USA). He is currently served as an Associate Professor in the School of Chemical Sciences, Universiti Sains Malaysia (USM). He has published more than 120 journal articles, 17 book chapters and 5 academic books throughout his twenty-year career in USM including "Graphene: A Versatile Advanced Material". Currently, his Scopus h-index is 35 with 4026 citations and he had been granted with ten patents for his R&D products/processes, where five of them are at an international level. More than thirty research grants (1 international grant and 5 industry grants) were secured and utilized to support his team's research activities. He has received fourteen international awards for his research outputs and currently serves as a Guest Editor for Frontiers in Chemistry. He had supervised more than 20 graduate students. His main research areas are Lignin and Lignocellulosic Materials, Nanoparticles, Graphene and Biomaterials and Petroleum Engineering. Recently, he is busy working in the microbial fuel cells topic especially in developing a novel electrode and has published several papers in well reputed journals such as Chemical Engineering Journal, Journal of Cleaner Production, etc. At the moment, he is the pioneer on MFCs research topic in the School of Chemical Sciences, USM. He enjoys sharing his industrial experience, where he spent two and half years working as a R&D Manager at NSE Resources Corporation Sdn Bhd., in his classes.

Patricia Graciela Vázquez

Patricia Vázquez obtained her PhD (1984) from University of San Luis (Argentina). She currently serves as an Associate Professor in the Faculty of Sciences Exacts, University of La Plata and Principal Researcher of CONICET (Argentina). She has published more than 110 journal articles, 10 book chapters and 3 academic books throughout in 30-year career in CONICET including IUPAC titular member of Green Chemistry Commission.

Currently, her Scopus h-index is 30. More than thirty research grants (international and national grants) were secured and utilized to support her team's research activities. She has received ten national awards for her research outputs and currently serves as a Guest Editor for BMC Nature, among others. She has supervised more than 15 graduate students and fellowships, nationals and internationals. Her main research areas are Nanoparticles, Biomaterials and Advanced Materials for Catalysts. Recently, she has been busy working in the antimicrobial fabric topic, especially in developing a novel additives. At the moment, she is the pioneer on Circular Economy in Argentina.

Mohd. Hazwan Hussin

Mohd. Hazwan Hussin received his B.Sc (Hons.) in Chemistry and M.Sc degrees from Universiti Sains Malaysia, respectively. He was awarded double-PhD degree program with USM and Universite de Lorraine, France. His research area involves lignocellulosic materials, lignin, cellulose and corrosion protections (inhibition and coatings). He was listed as Top 2% Scientist in the world (under the category of 2019 and 2020 citation) by Stanford University. He is also a Research Fellow for CEGEOTECH, UNIMAP and has been appointed as Visiting Professor at the Universite de Lorraine, France. Dr. Mohd Hazwan is currently served as the Deputy Director for Research Creativity and Management Office (RCMO) USM, Coordinator for USM-UL Centre and Point of Contact for Malaysia-France University Centre (MFUC).

Article

Phenol Liquefaction of Waste Sawdust Pretreated by Sodium Hydroxide: Optimization of Parameters Using Response Surface Methodology

Shihao Lv, Xiaoli Lin, Zhenzhong Gao, Xianfeng Hou, Haiyang Zhou * and Jin Sun *

College of Materials and Energy, South China Agricultural University, Guangzhou 510642, China
* Correspondence: hyzhou@scau.edu.cn (H.Z.); sunjin@scau.edu.cn (J.S.)

Abstract: In this study, a two-step method was used to realize the liquefaction of waste sawdust under atmospheric pressure, and to achieve a high liquefaction rate. Specifically, waste sawdust was pretreated with NaOH, followed by liquefaction using phenol. The relative optimum condition for alkali–heat pretreatment was a 1:1 mass ratio of NaOH to sawdust at 140 °C. The reaction parameters including the mass ratio of phenol to pretreated sawdust, liquefaction temperature, and liquefaction time were optimized by response surface methodology. The optimal conditions for phenol liquefaction of pretreated sawdust were a 4.21 mass ratio of phenol to sawdust, a liquefaction temperature of 173.58 °C, and a liquefaction time of 2.24 h, resulting in corresponding liquefied residues of 6.35%. The liquefaction rate reached 93.65%. Finally, scanning electron microscopy (SEM), Fourier transform infrared spectroscopy (FT-IR), and X-ray diffraction (XRD) were used to analyze untreated waste sawdust, pretreated sawdust, liquefied residues, and liquefied liquid. SEM results showed that the alkali–heat pretreatment and liquefaction reactions destroyed the intact, dense, and homogeneous sample structures. FT-IR results showed that liquefied residues contain aromatic compounds with different substituents, including mainly lignin and its derivatives, while the liquefied liquid contains a large number of aromatic phenolic compounds. XRD showed that alkali–heat pretreatment and phenol liquefaction destroyed most of the crystalline regions, greatly reduced the crystallinity and changed the crystal type of cellulose in the sawdust.

Keywords: waste sawdust; alkali–heat pretreatment; liquefaction; response surface methodology; residual content

Citation: Lv, S.; Lin, X.; Gao, Z.; Hou, X.; Zhou, H.; Sun, J. Phenol Liquefaction of Waste Sawdust Pretreated by Sodium Hydroxide: Optimization of Parameters Using Response Surface Methodology. *Molecules* **2022**, *27*, 7880. https://doi.org/10.3390/molecules27227880

Academic Editors: Mohamad Nasir Mohamad Ibrahim, Patricia Graciela Vázquez and Mohd. Hazwan Hussin

Received: 15 October 2022
Accepted: 11 November 2022
Published: 15 November 2022

Publisher's Note: MDPI stays neutral with regard to jurisdictional claims in published maps and institutional affiliations.

1. Introduction

Fossil energy is an important raw material utilized for the production of chemical products [1]. However, the utilization of fossil energy generates a number of greenhouse gases including sulfur oxide, nitrogen oxide, carbon oxide and carbon dioxide, causing air pollution and climate change [2]. Therefore, in order to alleviate energy crises and environmental pollution, researchers need to urgently search for cheap, clean and renewable energy resources [3]. As a natural renewable material with a huge storage capacity, lignocellulosic biomass has attracted increased research attention. It has a rich variety, including crop waste, wood processing residues, and wood product recycling waste, which is considered to be one of the most promising and sustainable alternatives to petroleum for the production of energy, materials and chemicals in the future [4,5].

Several treatment methods such as pyrolysis, liquefaction and gasification have been used to process lignocellulosic biomass to produce biofuels and other valuable chemicals [6,7]. Among them, liquefaction is an effective method for the integrated utilization of lignocellulosic biomass. Liquefaction technologies can be divided into two broad categories, hydrothermal liquefaction and solvent liquefaction [8]. Hydrothermal liquefaction is often carried out under harsh conditions, including high temperatures (200 °C to 400 °C), high pressures (5 MPa to 20 MPa), and a closed environment. Compared with hydrothermal liquefaction, solvent liquefaction can be carried out under mild reaction conditions and is an effective method for the integrated utilization of biomass [9]. Solvent liquefaction usually uses polyols or phenols to liquefy biomass at atmospheric pressure and relatively low temperature. Most often, acid is used as the catalyst and phenol as the solvent in solvent liquefaction, which reflects a high biological conversion rate [10]. The liquefied products are rich in phenolic compounds, which can be used to prepare phenolic resin adhesives after condensation with formaldehyde [11]. However, the liquefied products contain unreacted solvents and acidic catalysts that require recycling or alkali neutralization before preparing adhesives [12].

Further, during the liquefaction, the cellulose and hemicellulose of lignocellulosic biomass are wrapped in highly polymerized polyphenolic structured lignin, which makes it difficult to transform, thus hindering the high-value utilization of lignocellulosic biomass. Targeted pretreatment of lignocellulosic biomass before liquefaction can change the chemical structure and composition of lignocellulose, soften the raw materials, and induce depolymerization and chain breaks [13]. More importantly, the alkaline or acidic conditions of the pretreatment stage can be utilized to meet economic and environmental criteria during the liquefaction, and the pretreated lignocellulosic biomass exhibits better liquefaction performance [14].

Response surface methodology (RSM) is highly effective for the optimal design of a regression model, which is used to resolve problems related to nonlinear data processing [15]. By fitting the regression and plotting the model, the effect of the variables and their interactions on the response variables can be easily evaluated, and the optimal value of the response and the corresponding experimental conditions can be determined. The model obtained by RSM is continuous and can be analyzed continuously for each experimental level during the search for the optimizing experimental conditions, so we can better understand the experimental process [16]. In addition, RSM allows process optimization in a limited number of experimental runs, thus significantly reducing experimental time and costs [17].

In this study, a two-step process was used to achieve the effective liquefaction of waste sawdust. The first step entailed the pretreatment of waste sawdust with hot alkali under atmospheric pressure. In the second step, the pretreated sawdust was liquefied using phenol as the liquefying agent. The effects of the mass ratio of sawdust to NaOH on the liquefaction were studied. Based on RSM, the effects of the three experimental variables, namely the mass ratio of phenol to pretreated sawdust, liquefaction temperature, and liquefaction time, and their interactions on the liquefaction effect of waste sawdust were systematically investigated. At the same time, the specific residual content model was established using the design software, which was analyzed in detail to determine the optimal liquefaction conditions. The validity of the model was verified by repeated experiments. Finally, the untreated waste sawdust, pretreated sawdust, liquefied residues and liquefied liquid were characterized and analyzed by scanning electron microscopy (SEM), Fourier transform infrared (FT-IR) spectroscopy, and X-ray diffraction (XRD) to facilitate the high-value utilization of waste sawdust.

2. Results and Discussion

2.1. Alkali–Heat Pretreatment—Screening and Optimization Tests

The effects of the mass ratio of sawdust to NaOH at 140 °C on the liquefied residue yield were studied (Figure 1). It could be seen that the mass ratio of sawdust to NaOH had a significant effect on the liquefied residue yield. The liquefied residue yield decreased initially and then increased as the sawdust to NaOH mass ratio increased. The liquefied residue yield was 12.5% at a mass ratio of 1:1.5. The liquefied residue yield of sawdust without alkali–heat pretreatment under the same liquefaction conditions was 80.4%, which was really high, indicating that the alkali–heat treatment destroyed the internal structure of sawdust, which was conducive to liquefaction. In a certain range, the liquefied residue yield decreased with the reduction in the sawdust to NaOH mass ratio. This was explained by as the amount of NaOH increased, additional wood components were decomposed. Under alkali–heat pretreatment, the carbohydrates are degraded via the peeling reaction of the reducing end groups. Since hemicellulose has a substantially lower molecular weight than cellulose, it is also degraded via hydrolysis in hot alkali, increasing the number of reducing end groups and enhancing the erosion of polymer chains [18,19]. In addition, the presence of NaOH separated the bonds between lignin and carbohydrates, and disrupted the ether bonds between lignin polymers, resulting in the depolymerization and degradation of lignin [20,21]. However, when the mass ratio of sawdust to NaOH was 1:2, the yield of the liquefied residue was slightly increased, indicating that the high proportion of NaOH enhanced the repolymerization of degradation products or inhibited the depolymerization of lignocellulose [22].

At a sawdust to NaOH mass ratio of 1:1, the yield of liquefied residue was relatively low. However, when the mass ratio decreased to 1:1.5, the yield of liquefied residue changed slowly. In terms of energy conservation, a satisfactory liquefied residue yield was obtained at a sawdust to NaOH mass ratio of 1:1. Meanwhile, it should be noted that some of the adsorbed NaOH crystal on sawdust during the pretreatment was dissolved during the subsequent liquefaction, which induced cellulose swelling and increased the porosity and specific surface area of the sawdust, thus increasing the accessibility of phenol to sawdust and promoting the degradation of the sawdust [23–25]. According to the literature, the liquefied residue of sawdust has a significant impact on the performance of the subsequent fabrication of bio-based phenolic resins [26]. Therefore, it is important to optimize the pretreatment conditions and thus reduce the residual content of liquefaction.

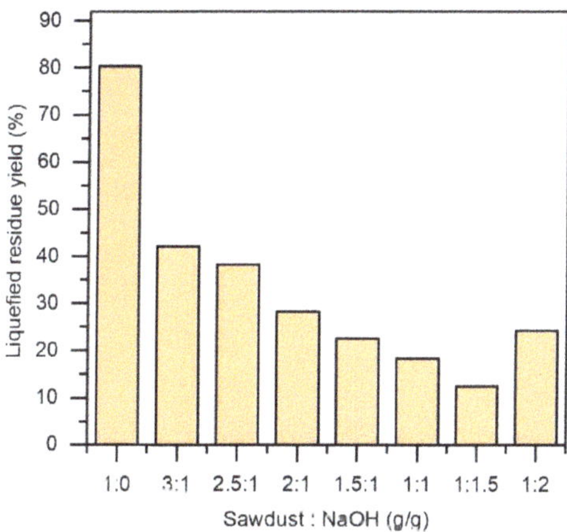

Figure 1. The effect of alkali–heat pretreatment on the liquefied residue yield. The phenol liquefaction temperature was 150 °C, the mass ratio of phenol to pretreated sawdust was 5:1, and the liquefaction time was 2 h.

2.2. Optimization of Phenol Liquefaction of Waste Sawdust

2.2.1. Model Fitting

The results of the whole experimental runs are summarized in Table 1. In general, the residual content after phenol liquefaction of waste sawdust was relatively low. Under different liquefaction conditions, the residual content varied from 6.82% to 17.13%, suggesting that phenol liquefaction is an effective and feasible biomass conversion method. The liquefied products can further react with formaldehyde to produce valuable biobased wood adhesive [27]. The lowest residual content was observed in a phenol-to-pretreated sawdust mass ratio of 4.21, a liquefaction temperature of 173.58 °C, and a liquefaction time of 2.24 h. The fitting quadratic multiple regression equation after the exclusion of the insignificant terms for the residual content of waste sawdust is determined based on these data, as shown in Equation (1).

$$R = 162.74864 - 24.02153A - 1.03750B - 14.186057C - 0.134375BC$$
$$+2.15511A^2 + 0.003566B^2 + 7.64545C^2 \tag{1}$$

where R is the residual content (%), A denotes the mass ratio of phenol to pretreated sawdust (P/S), B refers to the liquefaction temperature (°C), and C is the liquefaction time (h).

Table 1. Liquefaction variables and levels.

Order	Factors			
	Mass Ratio of Phenol to Pretreated Sawdust, P/S	**Liquefaction Temperature, T (°C)**	**Liquefaction Time, t (h)**	**Residual Content, R (%)**
1	3	140	1.5	17.13
2	5	140	1.5	12.55
3	3	180	1.5	15.44
4	5	180	1.5	12.61
5	3	140	2.5	16.62
6	5	140	2.5	13.36
7	3	180	2.5	9.38
8	5	180	2.5	8.22
9	2	160	2	17.8
10	6	160	2	12.23
11	4	120	2	14.11
12	4	200	2	10.09
13	4	160	1	15.72
14	4	160	3	12.36
15	4	160	2	7.58
16	4	160	2	7.35
17	4	160	2	7.81
18	4	160	2	6.28
19	4	160	2	5.78
20	4	160	2	6.12

In order to further verify the adequacy of the quadratic model, an analysis of variance (ANOVA) was carried out and the results are shown in Table 2. *p*-values less than 0.05 indicate the model variables are statistically significant, and the smaller the *p*-value, the more significant the corresponding coefficient and contribution to the response variable [17]. In this study, the *p*-value of the model was less than 0.0001 and the adjusted R^2 value was 0.9189. This indicated that the response regression model was highly significant and sufficient for the response variables tested. Further, the "lack of fit f-value" of 2.43 implied that the lack of fit was insignificant relative to the pure error. There was a 17.65% chance of a lack of fit f-value due to noise. The insignificant lack of fit was good, which indicated that the proposed model fitted the data well. Based on the *p*-values, it can be seen that the variables of A, B, C, the interaction term of BC, and the quadratic terms of A^2, B^2, C^2 were significant, indicating that the variables of the reaction were interactive and complex. The significant effect of every single variable on the residual content decreased as follows: A > B > C. The significant effect of interacting variables on residual content decreased in the order of BC > AB > AC. The magnitude of the f-value reflects the importance of each test variable on the index, and the larger f-value indicates the greater importance of the test index [28]. The importance of the influence of the test variables on residual content was as follows: A > B > C, which was identical to the significant effect of every single factor on residual content. In addition, the R^2 of the selected model was 0.9573, and the fitting degree was more than 95%, indicating that the model effectively reflected the changes in response values.

Table 2. Results of the analysis of variance for residual content.

Source	Sum of Squares	df	Mean Square	F Value	p Value	
Model	284.38	9	31.60	24.91	<0.0001	significant
A	32.98	1	32.98	26.00	0.0005	
B	30.39	1	30.39	23.96	0.0006	
C	17.79	1	17.79	14.02	0.0038	
AB	1.85	1	1.85	1.46	0.2546	
AC	1.12	1	1.12	0.8811	0.3700	
BC	14.45	1	14.45	11.39	0.0071	
A^2	116.78	1	116.78	92.07	<0.0001	
B^2	51.15	1	51.15	40.33	<0.0001	
C^2	91.85	1	91.85	72.42	<0.0001	
Residual	12.68	10	1.27			
Lack of fit	8.98	5	1.80	2.43	0.1765	not significant
Pure error	3.70	5	0.7404			
Cor total	297.06	19				

The comparison between the actual residual content obtained in the experiment and the predicted residual content based on the quadratic model is shown in Figure 2. Generally, each experimental data point should be approximately close to the regression line of the prediction data, which suggests that the estimated effect is true and effective [29]. In this study, the 20 groups of actual experimental values were distributed near the predicted value, which implied a strong correlation between the experimental responses and predicted responses.

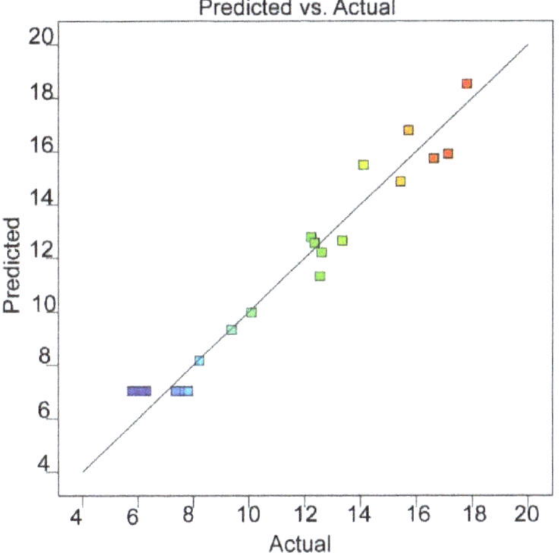

Figure 2. Comparison between the predicted (straight line) and actual response values (points) obtained from the model for the response of residual content.

2.2.2. Main Response Surface Plots and Optimization

The three-dimensional response surface plots and contour plots were used to delineate the effect of independent variables and their interactions on residual content based on the obtained regression equation. The response surface plots of residual content were a function of two specific variables, while the other variable remained at a fixed value. The curve shapes of all response surfaces were upward concave, and all the center and edge points were within the studied range, which indicated that there was an optimal response for the content of liquefied residues.

The effect of the P/S ratio and the liquefaction temperature on the residual content is shown in Figure 3a,b. It can be seen that P/S and liquefaction temperature significantly affected the residual content. In a specific range, the residual content decreased with the increase in P/S and liquefaction temperature and then increased with the further increased P/S and temperature after reaching the optimal critical point (P/S and temperature of 4.21 and 173.58 °C, respectively). Similar phenomena were observed in other woody materials [30]. The increase in P/S in the range of 2 to 4.21 reduced the residual content, indicating that the P/S values in this range facilitated the liquefaction of sawdust. However, when the P/S increased to a higher value, it negatively affected the sawdust liquefaction. These results suggest that lower P/S values lead to the recondensation of the low-molecular-weight compounds into insoluble residues [31]. An insufficient amount of phenol in the reaction system increases the viscosity of the liquefied products [32]. The P/S value of 4.21 is reasonable to ensure minimum residual content of waste sawdust for adequate liquefaction.

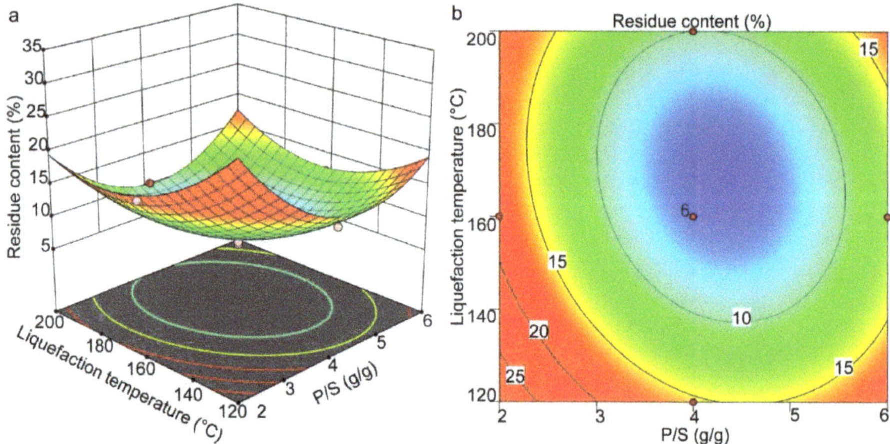

Figure 3. The response surface (**a**) and contour plots (**b**) showing the effects of P/S and liquefaction temperature on the residual content.

Further, the effect of liquefaction temperature on the residual content showed a similar trend compared with the P/S value. With the increase in reaction temperature, the residual content decreased first and then increased above 173.58 °C which is explained by the incomplete bond cleavage of different components of waste sawdust at lower temperatures due to less violent reaction conditions. In addition, reactions, such as hydrolyzation and depolymerization, resulting in smaller molecules could not be completed [33]. As the temperature increases, the glycosidic bonds break, leading to dehydration and decarboxylation, resulting in the cleavage of large molecules into smaller fragments [34], and a decrease in the residual content. However, with the further increase in temperature (>173.58 °C), the residual content increased gradually. This was mainly because of the unstable components in the liquefied product at high temperatures [35], and a series of complex reactions resulting in the formation of a large number of liquefied residues.

The comprehensive effects of the P/S ratio and liquefaction time on the residual content at a constant temperature are depicted in Figure 4a,b. It can be observed that the critical transition point was established when the P/S value was 4.21 and the liquefaction time was 2.24 h. With the increase in liquefaction time from 1 h to 2.24 h, the residual content gradually decreased to less than 5.6%. However, the residual content increased if the liquefaction time continued to extend. In the initial stage of the reaction, the degradation reaction played a dominant role, which decreased the residual content [32]; the residual content decreased. With the further extension of the reaction time, the polycondensation reaction and the degradation reaction among the liquefaction products gradually reach the balance until they took the dominant role, which led to the increase in the residue content [36].

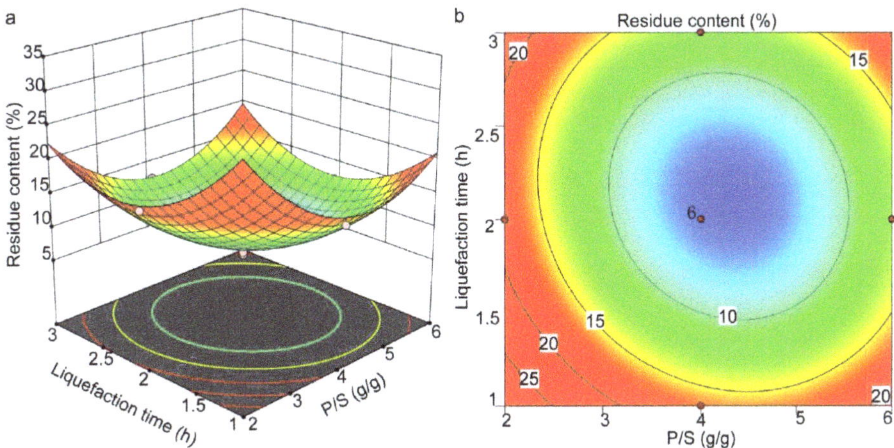

Figure 4. The response surface (**a**) and contour plots (**b**) showing the effects of P/S and liquefaction time on the residual content.

Figure 5a,b show the effects of liquefaction time and liquefaction temperature on the residual content at a constant P/S value. It can be seen that the effect of liquefaction time was closely related to the liquefaction temperature. Specifically, the variation in residual content over time was more pronounced at higher temperatures than at lower temperatures. At a constant liquefaction temperature (greater than 173.58 °C), when the liquefaction time increased from 1 h to 2.24 h, the residual content decreased significantly. Further extension of time to 2.6 h or longer resulted in an almost constant residual content or a slight increase. Similarly, compared with the shorter reaction times, the change in residual content with temperature was more significant under the longer reaction times. At a constant reaction time (greater than 2.24 h), the residual content decreased significantly with the temperature increase from 100 °C to 180 °C, followed by equilibrium and then a slight increase. This also confirmed that heating at 173.58 °C for 2.24 h facilitated waste sawdust liquefaction.

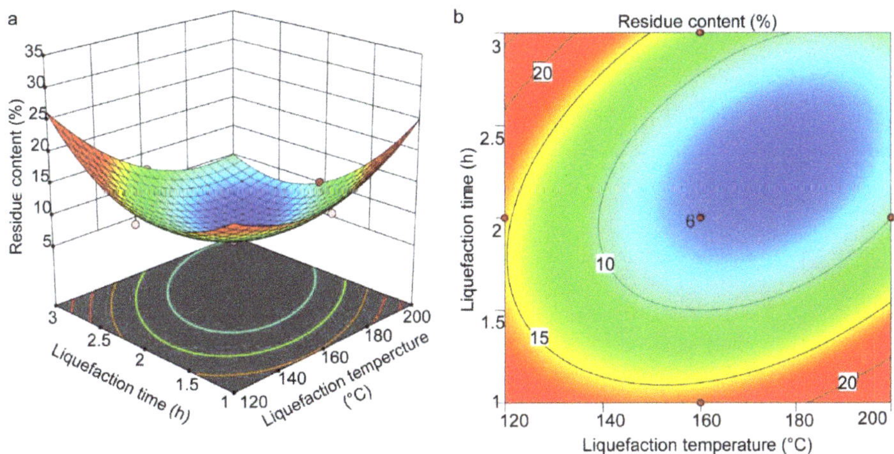

Figure 5. The response surface (**a**) and contour plots (**b**) showing the effects of liquefaction temperature and liquefaction time on the residual content.

To demonstrate the accuracy and reliability of the optimal liquefaction conditions determined from the fitted model, the experiment was repeated three times using the same method based on the predicted optimal point and the average value was calculated. The actual residual content obtained experimentally was 6.35%, while the residual content predicted based on Equation (1) was 6.15%. The results show that the experimental value agrees well with the model prediction, with an error of only 3.25%. It is thus clear that RSM can be used to optimize the process of waste sawdust phenol liquefaction. The optimal liquefaction occurred at a P/S ratio of 4.21, liquefaction temperature of 173.58 °C, and liquefaction time of 2.24 h.

2.3. SEM Analysis

SEM was used to assess the structural and morphological changes of sawdust after treatment (Figure 6). The surface of untreated sawdust was continuous and smooth, and the fibrous structure was relatively complete (Figure 6a), in which lignin was the encrusting material connecting the fibers and hemicellulose was the filling material distributed in the microfibers of the cell wall [37]. The pretreated sawdust (Figure 6b) revealed many microfibril aggregates with a rougher and more wrinkled surface due to the activation of lignocellulose with alkali, resulting in partial structural degradation [38]. Further, the rough surface of pretreated sawdust increased the specific surface area, which increased the accessibility of phenol during the subsequent liquefaction. Following phenol liquefaction, the structure of the pretreated sawdust was severely damaged, with a lower degree of polymerization and loose and irregular texture (Figure 6c,d), due to the surface coke generated by the condensed lignin and cellulose [39].

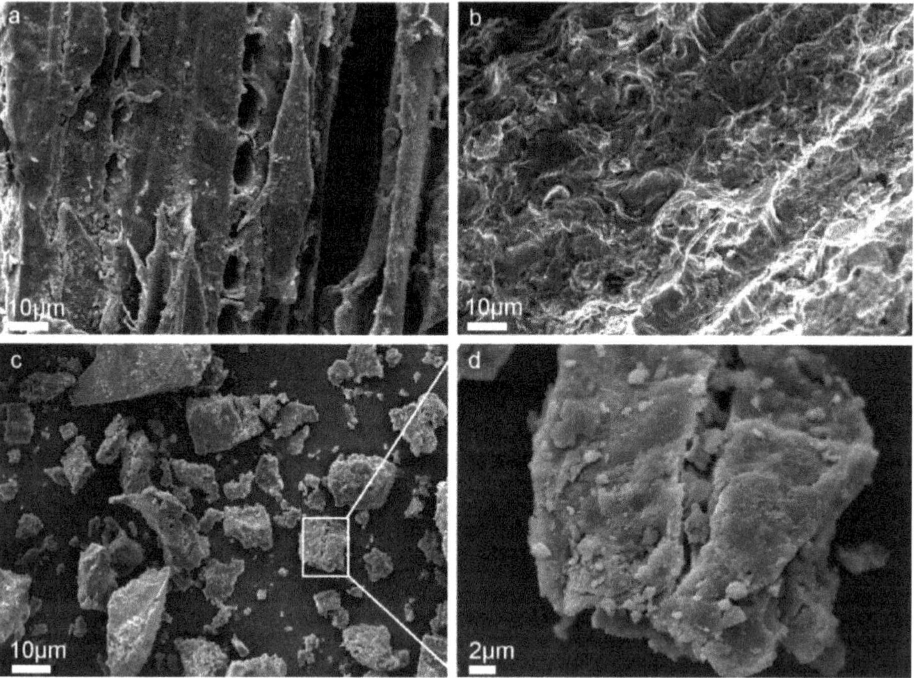

Figure 6. SEM micrographs of untreated sawdust (**a**), alkali–heat pretreated sawdust (**b**) liquefied residues (**c,d**).

2.4. FT-IR Analysis

The functional groups present in the untreated sawdust, pretreated sawdust, liquefied residues, and liquefied liquid were investigated by FT-IR. For untreated sawdust (Figure 7a), the broad peak at around 3400 cm^{-1} indicated O$-$H stretching vibrations [40]. The peak at 2925 cm^{-1} corresponded to the stretching vibration of C$-$H. The peaks between 1400 cm^{-1} and 1600 cm^{-1} are attributed to the aromatic skeleton in lignin [41]. Several peaks from 600 cm^{-1} to 900 cm^{-1} represent the characteristic peaks of aromatic monomers. Compared with untreated sawdust, the peak intensity of pretreated sawdust at 2925 cm^{-1} was significantly weakened, which suggested that pretreatment peels off most of the methylene groups in the aliphatic acid methylene group [42]. In addition, the weakening of the peaks at 1750 cm^{-1}, 1240 cm^{-1}, 1152 cm^{-1} and 1030 cm^{-1} indicates the breakage of cellulose glycosidic bonds and hemicellulose chains, suggesting that large amounts of hemicellulose and part of cellulose were degraded by hydrolysis and peeling reactions during the pretreatment [43]. Compared with untreated sawdust, the peaks between 1400 cm^{-1} and 1600 cm^{-1} and from 600 cm^{-1} to 900 cm^{-1} were enhanced, indicating that pretreatment promoted the decomposition of lignin into aromatic monomers and increased the formation of low-molecular-weight oligomers. It was noteworthy that the new peak of pretreated sawdust around 1332 cm^{-1} might correspond to the characteristic peak of NaOH.

After the liquefaction reaction, the disappearance of the peak at 1750 cm^{-1} for the liquefied residues (Figure 7b) indicated that the hemicellulose was further decomposed during the liquefaction. The strong peak of C$-$O at 1030 cm^{-1} associated with liquefied residues demonstrated the presence of a large amount of lignin and its derivatives remaining in the liquefied residues [39]. A strong peak at 1080 cm^{-1} attributed to the C–O peak in cellulose, indicated the presence of unliquefied cellulose in the liquefied residues. The peak enhancement at 1600 cm^{-1} and 1480 cm^{-1} indicated the presence of large amounts of aromatic compounds and their derivatives from lignin in liquefied liquid (Figure 7c) Similar phenomena were reported in previous studies [44,45]. The peak of the liquefied liquid around 1240 cm^{-1} may be due to the C–O stretching of the phenol or ester [46]. In addition, other peaks in the range of 600 cm^{-1} to 1000 cm^{-1} in the spectra of the liquefied liquid were attributed to C–H bending vibrations of aromatic hydrocarbons. The enhancement of these peaks can also be explained by the presence of a large number of phenolic compounds in the liquefied liquid.

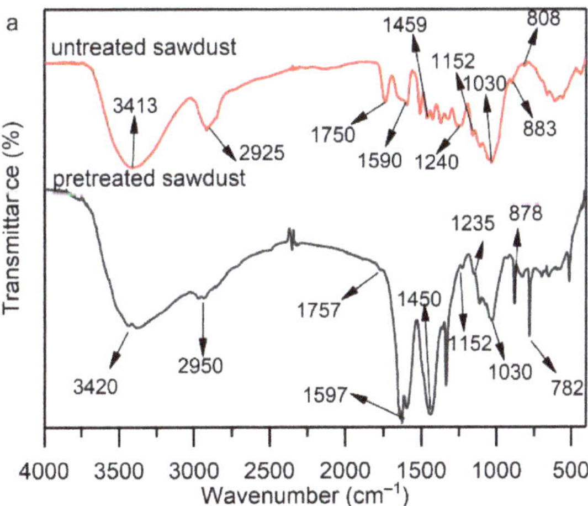

Figure 7. *Cont.*

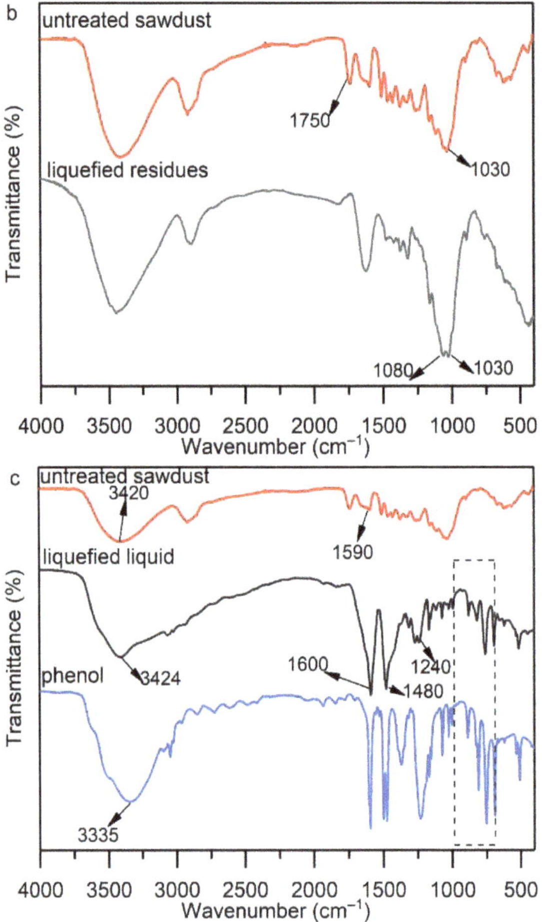

Figure 7. FT-IR spectra of untreated sawdust and pretreated sawdust (**a**), untreated sawdust and liquefied residues (**b**), untreated sawdust, liquefied liquid and phenol (**c**).

2.5. XRD Analysis

Waste sawdust showed two characteristic peaks of type I lignocellulose at 16.1° and 22.5°, which correspond to the lattice planes of (101) and (002), respectively [47]. The two peaks were both shifted after alkali–heat pretreatment due to the transformation of some cellulose crystals [48]. The Crystallinity Index (I_{Cr}) of the pretreated samples (18.21%) was reduced by 59.11% compared with untreated natural waste sawdust (44.53%) because a large number of hydrogen bonds between and within the cellulose were destroyed, resulting in the disruption of the crystal structure [49]. Besides, several small peaks between 28° and 50° resulted from the NaOH crystals rearranged in the hierarchical sawdust structures [50].

The XRD spectra of the liquefied residues exhibit many diffraction peaks (Figure 8), indicating that the liquefied residues contained several crystalline substances, which were difficult to liquefy. This was consistent with the results of previous studies [51]. The liquefied residues showed two peaks of type II cellulose at positions 20° and 22°, suggesting the synergistic effect of alkali–heat pretreatment and phenol liquefaction, resulting in altered cellulose crystals. Type II cellulose is stabler and less susceptible to liquefaction, which also explains its presence in the liquefied residues. A sharp peak at 26.5° was due to the formation of carbonaceous structures, such as graphite, during liquefaction. The peak at about 32.2° was attributed to oxidized lignin [52], which corresponded to the results of the FT-IR analysis.

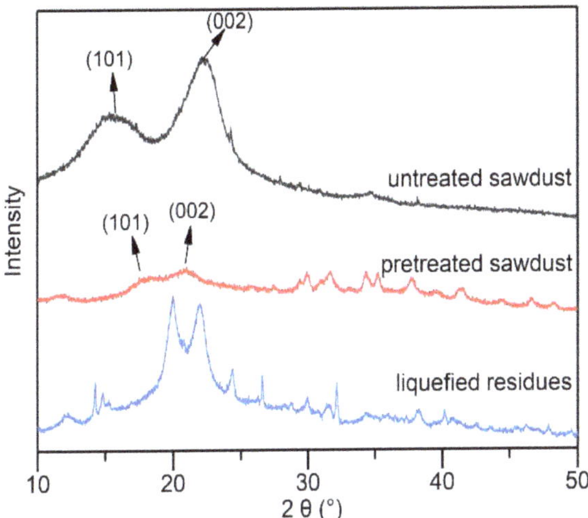

Figure 8. X-ray diffraction of untreated sawdust, alkali–heat pretreated sawdust and liquefied residues.

3. Materials and Methods

3.1. Materials

The waste sawdust with a size of 20 mesh was generously provided by the carpentry factory located at the South China Agricultural University. The chemical and elemental composition of the sawdust is shown in Table 3. Phenol (99%) and NaOH were purchased from Aladdin Reagent Company, Shanghai, China.

Table 3. Element and chemical composition of sawdust and the values were given on a dry basis.

Polymer Mass Fraction (%)			Element Mass Fraction (%)					Ash (%)
Cellulose	Hemicellulose	Klason Lignin	C	H	O	S	N	
42.11	24.32	28.06	46.77	5.97	46.52	0.00	1.01	1.6

3.2. Pretreatment

The sawdust was dried to absolute dryness in a drying oven at 103 °C ± 2 °C. The pretreatment was performed in a glass beaker (150 mL) with tinfoil, and heated in an oven. Briefly, 5 g dry sawdust, 25 g distilled water and some NaOH were loaded into the reaction vessel and mixed thoroughly. The mass ratios of dry sawdust to NaOH were 3:1, 2.5:1, 2:1, 1.5:1, 1:1, 1:1.5, and 1:2. The beaker was then transferred to a drying oven and heated to a fixed temperature (140 °C) until the distilled water was dried (about 280 min). At the end of the pretreatment, the beaker was stored in a glass desiccator for subsequent liquefaction experiments. The overall flow chart of the two step-liquefaction is presented in Figure 9.

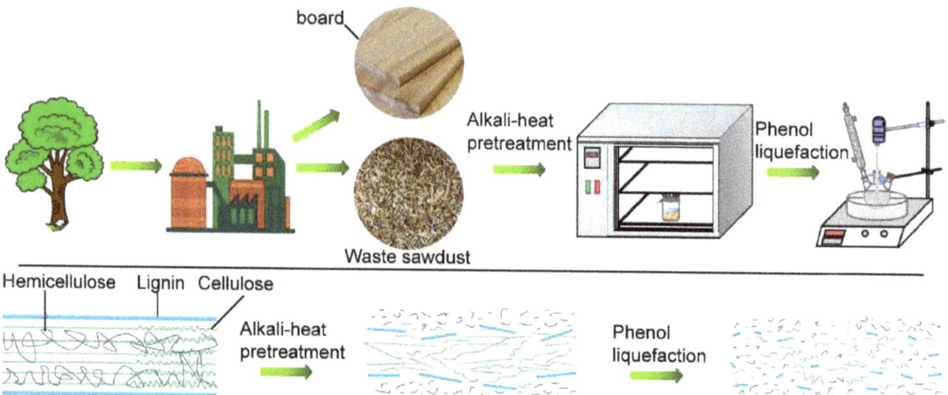

Figure 9. Overall flow chart of two step liquefaction of waste sawdust.

3.3. Phenol Liquefaction

The pretreated sawdust and phenol were loaded into a reactor equipped with a stirrer and a condenser. The whole assembly was immersed into an oil bath preheated to a given liquefaction temperature for a certain reaction time under continuous stirring. Then, the resulting reaction mixture was diluted and dissolved in hot distilled water (50 mL). The insoluble residues were separated by filtration with a G2 glass filter (30–50 μm) under vacuum (0.095 MPa). The filtrate was subjected to rotary evaporation under a vacuum at 70 °C to remove water. The residues were dried in an oven at 103 °C ± 2 °C to constant weight. All experiments were performed in triplicates and the average value was taken. All product yields were calculated using the following equations.

$$\text{Liquefied residue yield (wt \%)} = \frac{\text{Weight of Liquefaction Residue}}{\text{Weight of Sawdust}} \times 100 \qquad (2)$$

$$\text{Liquefaction yield (wt \%)} = (1 - \text{Liquefied residue yield}) \times 100 \qquad (3)$$

3.4. Experimental Design and Process Optimization Using Response Surface Methodology (RSM)

To investigate the effect of independent variables on the response value (residual content) and optimize the liquefaction conditions, experiments with three key variables and five levels were designed using the Box–Behnken design. The Design-Expert 12.0.3.0 software was used to design experiments, perform the statistical analysis, and create the regression model. Based on extensive pre-experiments, the independent variables that significantly influenced the residual content and the right levels were selected. The three independent variables were the mass ratio of phenol to pretreated sawdust (A), liquefaction temperature (B), and liquefaction time (C). The range of each value was chosen in the range of 2 to 6, 120 to 200 °C, and 1 to 3 h, as shown in Table 4. The experimental design included a total of 20 experiments, corresponding to eight factor points, six axial points, and six center point replications to ensure the accuracy of the experiment.

Table 4. Liquefaction variables and levels.

Variables	Level					
	Code	−2	−1	0	1	2
Mass ratio of phenol to pretreated sawdust, P/S	A	2	3	4	5	6
Liquefaction temperature, T (°C)	B	120	140	160	180	200
Liquefaction time, t (h)	C	1	1.5	2	2.5	3

What is more, the complete design matrix and actual residue are shown in Table 1, and the experimental data were analyzed using Design Expert 12.0.3.0. The test data were fitted to the following second order polynomial equation, as shown in Equation (4). The analysis of variance (ANOVA) and significance test was carried out for the residual content under different conditions to evaluate the quality of the model fitting. All these experiments were carried out in random order.

$$Y = \beta_0 + \sum_{i=1}^{n} \beta_i X_i + \sum_{i=1}^{n} \beta_{ii} X_i^2 + \sum_{i=1}^{n} \sum_{j=i+1}^{n} \beta_{ij} X_i X_j \tag{4}$$

where Y is the response function (residue content), β_0 is the model intercept, β_i, β_{ii} and β_{ij} represent coefficients of linear, quadratic, and interaction terms, respectively.

3.5. Characterization

3.5.1. Chemical and Elemental Composition Analysis

The chemical composition of sawdust was performed according to the Van Soest method [53]. Briefly, the neutral detergent fiber (NDF), acid detergent fiber (ADF), and acid detergent lignin (ADL) were prepared in turn by deterging waste sawdust sequentially with neutral detergent reagent, acid detergent reagent and 72% H_2SO_4. Another amount of dry sawdust was put in a muffle furnace at 600 °C for 6 h, and the ash content was calculated by the weight difference. The difference values between ADF and NDF, ADF and ADL, ADL and ash were considered as contents of hemicellulose, cellulose and Klason lignin, respectively [49]. The elemental analysis of sawdust was performed with an Elemental Analyzer (Vario EL cube, Elementar, Hanau, Germany). The chemical and elemental composition of the sawdust was analyzed three times and the average values were taken separately.

3.5.2. Scanning Electron Microscope (SEM) Analysis

The morphology of untreated sawdust, alkali–heat pretreated sawdust and liquefied residues were analyzed via SEM (EVO MA 15, ZEISS, Oberkochen, Germany). The working voltage was 10 kV.

3.5.3. Fourier Transform Infrared Spectroscopy (FT-IR) Analysis

FT-IR instrument (Vertex 70, Bruker, MA, USA) was used to analyze the functional groups in the samples (untreated sawdust, alkali–heat pretreated sawdust, liquefied residues and liquefied liquid). The sample was diluted nicely in KBr. The sample was scanned 32 times in the range of 400 cm^{-1} to 4000 cm^{-1} at a resolution of 4 cm^{-1}. Background spectra were recorded before every sampling.

3.5.4. X-ray Diffraction (XRD) Analysis

Samples ground to powder (100 mesh) were analyzed by X-ray diffraction (xrd-6000, Shimadzu, Kyoto, Japan) with an AlKα radiation source at 40 kV. The scanning range was 5 to 50° with a step of 0.02° at a scanning rate of 10°/min.

4. Conclusions

In this study, a facile method was used to liquefy waste sawdust via a two-step method at a significant liquefaction rate. The alkali–heat pretreatment was optimized by a temperature of 140 °C and a 1:1 mass ratio of sawdust to NaOH, resulting in a 4.2-fold higher liquefaction rate than that of untreated sawdust. Based on the response model established by RSM, it was found that the P/S ratio was the most important variable affecting the liquefaction yield. A P/S ratio of 4.21, a liquefaction temperature of 173.58 °C, and a liquefaction time of 2.24 h were the optimal conditions for phenol liquefaction of pretreated sawdust, resulting in corresponding liquefied residue yield of 6.35%. Thus, the liquefaction rate reached 93.65%. Based on SEM, FTIR, and XRD analyses, alkali–heat pretreatment is essential for subsequent phenol liquefaction.

Author Contributions: Conceptualization, J.S. and H.Z.; methodology, J.S. and S.L.; software, J.S. and S.L.; data curation, J.S., S.L., H.Z. and X.L.; validation, J.S. and S.L.; investigation, J.S., Z.G., S.L. and X.H.; formal Analysis, J.S. and S.L.; writing-original draft, J.S., S.L. and X.L.; writing—review and editing, J.S., S.L. and H.Z.; resources, J.S. and H.Z.; supervision, J.S., H.Z. and Z.G.; funding acquisition, J.S. and Z.G.; All authors have read and agreed to the published version of the manuscript.

Funding: This research was supported by the Science and Technology Program of Guangzhou (Project No. 202103000011).

Institutional Review Board Statement: Not applicable.

Informed Consent Statement: Not applicable.

Data Availability Statement: Not applicable.

Acknowledgments: The authors would like to acknowledge the Science and Technology Program of Guangzhou (Project No. 202103000011) for supporting this work. The authors also thank the support of the College of Materials and Energy, South China Agricultural University.

Conflicts of Interest: The authors declare no competing interest.

References

1. Guo, J.Y.; Yu, B. When fossil fuels run out, what then? *World Environ.* **2019**, *5*, 46–48.
2. Moriarty, P.; Honnery, D. Energy efficiency or conservation for mitigating climate change? *Energies* **2019**, *12*, 3543.
3. Holechek, J.L.; Geli, H.; Sawalhah, M.N.; Valdez, R. A global assessment: Can renewable energy replace fossil fuels by 2050? *Sustainability* **2022**, *14*, 4792.
4. Heal, G. Economic aspects of the energy transition. *Environ. Resour Econ.* **2022**, *83*, 5–21. [CrossRef]
5. Zhang, Z.C.; Song, Y.D. Research Progress in Biotechnology for Comprehensive Utilization of Lignocellulosic Materials. *Guangdong Agric. Sci.* **2021**, *48*, 150–159.
6. Kosmela, P.; Kazimierski, P.; Formela, K.; Haponiuk, J.; Piszczyk, U. Liquefaction of macroalgae enteromorpha biomass for the preparation of biopolyols by using crude glycerol. *J. Ind. Eng. Chem.* **2017**, *56*, 399–406.
7. Goncalves, D.; Orišková, S.; Matos, S.; Machado, H.; Vieira, S.; Bastos, D.; Gaspar, D.; Paiva, R.; Bordado, J.C.; Rodrigues, A.; et al. Thermochemical liquefaction as a cleaner and efficient route for valuing pinewood residues from forest fires. *Molecules* **2021**, *26*, 7156.
8. Athanasios, D.; Stella, B. Hydrothermal liquefaction of various biomass and waste feedstocks for biocrude production: A state of the art review. *Renew. Sust. Energy Rev.* **2017**, *68*, 113–125.
9. Gin, A.W.; Hassan, H.; Ahmad, M.A.; Hameed, B.H.; Din, A. Recent progress on catalytic co-pyrolysis of plastic waste and lignocellulosic biomass to liquid fuel: The influence of technical and reaction kinetic parameters. *Arab. J. Chem.* **2021**, *14*, 103035. [CrossRef]
10. Feng, S.; Yuan, Z.; Leitch, M.; Shui, H.; Xu, C.C. Effects of bark extraction before liquefaction and liquid oil fractionation after liquefaction on bark-based phenol formaldehyde resoles. *Ind. Crop. Prod.* **2016**, *84*, 330–336.
11. Cheng, S.N.; Yuan, Z.S.; Mathew, L.; Mark, A.; Xu, C.C. Highly efficient de-polymerization of organosolv lignin using a catalytic hydrothermal process and production of phenolic resins/adhesives with the depolymerized lignin as a substitute for phenol at a high substitution ratio. *Ind. Crop. Prod.* **2013**, *44*, 315–322. [CrossRef]
12. Maree, C.; Görgens, J.F.; Tyhoda, L. Lignin phenol formaldehyde resins synthesised using South African spent pulping liquor. *Waste Biomass Valori.* **2022**, *13*, 3489–3507.
13. Kumari, D.; Singh, R. Pretreatment of lignocellulosic wastes for biofuel production: A critical review. *Renew. Sust. Energy Rev.* **2018**, *90*, 877–891.
14. Mafe, O.; Davies, S.M.; Hancock, J.; Du, C. Development of an estimation model for the evaluation of the energy requirement of dilute acid pretreatments of biomass. *Biomass. Bioenerg.* **2015**, *72*, 28–38.
15. Kandasamy, S.; Zhang, B.; He, Z.; Chen, H.; Krishnamoorthi, M. Effect of low-temperature catalytic hydrothermal liquefaction of spirulina platensis. *Energy* **2020**, *190*, 116236. [CrossRef]
16. Hassan, S.N.A.M.; Ishak, M.A.M.; Ismail, K. Optimizing the physical parameters to achieve maximum products from co-liquefaction using response surface methodology. *Fuel* **2017**, *207*, 102–108.
17. Mazaheri, H.; Lee, K.T.; Bhatia, S.; Mohamed, A.R. Subcritical water liquefaction of oil palm fruit press fiber in the presence of sodium hydroxide: An optimisation study using response surface methodology. *Bioresour. Technol.* **2010**, *101*, 9335–9341. [CrossRef] [PubMed]
18. Knill, C.J.; Kennedy, J.F. Degradation of cellulose under alkaline conditions. *Carbohydr. Polym.* **2003**, *51*, 281–300. [CrossRef]

19. Sintamarean, I.M.; Grigoras, I.F.; Jensen, C.U.; Toor, S.S.; Pedersen, T.H.; Rosendahl, L.A. Two-stage alkaline hydrothermal liquefaction of wood to biocrude in a continuous bench-scale system. *Biomass Convers. Biorefinery* **2017**, *7*, 425–435. [CrossRef]
20. Kim, J.S.; Lee, Y.Y.; Kim, T.H. A review on alkaline pretreatment technology for bioconversion of lignocellulosic biomass. *Bioresour. Technol.* **2016**, *199*, 42–48.
21. Xin, L.; Yong, F. Research progress in pretreatment techniques for lignocellulosic materials. *Guangzhou Chem. Industry* **2014**, *42*, 16–18.
22. Grigoras, I.F.; Stroe, R.E.; Sintamarean, I.M.; Rosendahl, A.L. Effect of biomass pretreatment on the product distribution and composition resulting from the hydrothermal liquefaction of short rotation coppice willow. *Bioresour. Technol.* **2017**, *231*, 116–123. [PubMed]
23. Grace, T.M.; Malcolm, E.W. *Pulp and Paper Manufacture: Volume 5 Alkaline Pulping*, 3rd ed.; Joint Textbook Committee of the Paper Industry: Montreal, QC, Canada, 1989; p. 45.
24. Maldas, D.; Shiraishi, N. Liquefaction of biomass in the presence of phenol and H2O using alkalies and salts as the catalyst. *Biomass. Bioenerg.* **1997**, *12*, 273–279. [CrossRef]
25. Maldas, D.; Shiraishi, N.; Harada, Y. Phenolic resol resin adhesives prepared from alkali-catalyzed liquefied phenolated wood and used to bond hardwood. *J. Adhes. Sci. Technol.* **1997**, *11*, 305–316. [CrossRef]
26. Tran, M.H.; Lee, E. Development and optimization of solvothermal liquefaction of marine macroalgae Saccharina japonica biomass for biopolyol and biopolyurethane production. *J. Ind. Eng. Chem.* **2020**, *81*, 167–177. [CrossRef]
27. Bansode, A.; Barde, M.; Asafu-Adjaye, O.; Patil, V.; Hinkle, J.; Via, B.K.; Adhikari, S.; Adamczyk, A.J.; Farag, R.; Elder, T.; et al. Synthesis of Biobased Novolac Phenol–Formaldehyde Wood Adhesives from Biorefinery-Derived Lignocellulosic Biomass. *ACS. Sustain. Chem. Eng.* **2021**, *9*, 10990–11002. [CrossRef]
28. Wang, S.J.; Hou, X.F.; Huang, Y.S. Effect of Alkaline Ionic Liquid TBAH Pretreatment on Structure and Enzymatic Properties of Eucalypt. *Chem. Ind. For. Prod.* **2022**, *42*, 57–63.
29. Zhu, Z.; Rosendahl, L.; Toor, S.S.; Chen, G. Optimizing the conditions for hydrothermal liquefaction of barley straw for bio-crude oil production using response surface methodology. *Sci. Total. Environ.* **2018**, *630*, 560–569.
30. Wei, X.; Jie, D. Optimization to hydrothermal liquefaction of low lipid content microalgae *spirulina* sp. using response surface methodology. *J. Chem.* **2018**, *2018*, 2041812.
31. Li, G.; Hse, C.; Qin, T. Wood liquefaction with phenol by microwave heating and FTIR evaluation. *J. For. Res.* **2015**, *26*, 1043–1048.
32. Li, R.; Xu, W.; Wang, C.; Zhang, S.; Song, W. Optimization for the liquefaction of moso bamboo in phenol using response surface methodology. *Wood Fiber Sci.* **2018**, *50*, 220–227.
33. Hadhoum, L.; Loubar, K.; Paraschiv, M.; Burnens, G.; Awad, S.; Tazerout, M. Optimization of oleaginous seeds liquefaction using response surface methodology. *Biomass Convers. Bior.* **2021**, *11*, 2655–2667.
34. Zhou, R.; Zhou, R.; Wang, S.; Lan, Z.; Zhang, X.; Yin, Y.; Ye, L. Fast liquefaction of bamboo shoot shell with liquid-phase microplasma assisted technology. *Bioresour. Technol.* **2016**, *218*, 1275–1278. [CrossRef] [PubMed]
35. Singh, R.; Balagurumurthy, B.; Prakash, A.; Bhaskar, T. Catalytic hydrothermal liquefaction of water hyacinth. *Bioresour. Technol.* **2015**, *178*, 157–165. [PubMed]
36. Hadhoum, L.; Burnens, G.; Loubar, K.; Balistrou, M.; Tazerout, M. Bio-oil recovery from olive mill wastewater in sub-/supercritical alcohol-water system. *Fuel* **2019**, *252*, 360–370.
37. Zhao, L.; Zhou, J.H.; Sui, Z.J.; Zhou, X.G. Hydrogenolysis of sorbitol to glycols over carbon nanofiber supported ruthenium catalyst. *Chem. Eng. Sci.* **2010**, *65*, 30–35.
38. Mafu, L.D.; Neomagus, H.W.J.P.; Everson, R.C.; Carrier, M.; Strydom, C.A.; Bunt, J.R. Structural and chemical modifications of typical South African biomasses during torrefaction. *Bioresour. Technol.* **2016**, *202*, 192–197. [CrossRef]
39. Ma, Y.; Tan, W.H.; Feng, G.D.; Xu, J.M.; Wang, K.; Ying, H.; Jiang, J.C. Liquefaction of Bamboo Powder in Ethanol-phenol under High Pressure and Product Characterization. *Chem. Ind. For. Prod.* **2016**, *36*, 105–112.
40. Wang, W.X.; Zhang, L.; Li, A. Response relationship of hydrothermal humification products of waste biomass with acid base property of medium. *J. Dalian Univ. Technol.* **2022**, *62*, 9–17.
41. Zhao, J.; Xiuwen, W.; Hu, J.; Liu, Q.; Shen, D.; Xiao, R. Thermal degradation of softwood lignin and hardwood lignin by TG-FTIR and Py-GC/MS. *Polym. Degrad. Stabil.* **2014**, *108*, 133–138.
42. Biswas, B.; Singh, R.; Kumar, J.; Khan, A.A.; Krishna, B.B.; Bhaskar, T. Slow pyrolysis of prot, alkali and dealkaline lignins for production of chemicals. *Bioresour. Technol.* **2016**, *213*, 319–326. [CrossRef] [PubMed]
43. Biswas, B.; Kumar, A.; Krishna, B.B.; Bhaskar, T. Effects of solid base catalysts on depolymerization of alkali lignin for the production of phenolic monomer compounds. *Renew. Energ.* **2021**, *175*, 270–280. [CrossRef]
44. Sun, P.; Heng, M.; Sun, S.; Chen, J. Direct liquefaction of paulownia in hot compressed water: Influence of catalysts. *Energy* **2010**, *35*, 5421–5429.
45. Xu, E.; Sun, J.; Zhu, X.; Wang, X.; Gao, Z. Liquefaction of coconut fibers in alkaline hot-compressed water. *Energ. Source. Part A* **2016**, *38*, 1750–1755. [CrossRef]
46. Zuluaga, R.; Putaux, J.L.; Cruz, J.; Velez, J.; Mondragon, I.; Ganan, P. Cellulose microfibrils from banana rachis: Effect of alkaline treatments on structural and morphological features. *Carbohyd. Polym.* **2009**, *76*, 51–59. [CrossRef]
47. Liu, M.; Wang, H.; Han, J.; Niu, Y. Enhanced hydrogenolysis conversion of cellulose to C2–C3 polyols via alkaline pretreatment. *Carbohyd. Polym.* **2012**, *89*, 607–612. [CrossRef]

48. Eronen, P.; Österberg, M.; Jskelinen, A.S. Effect of alkaline treatment on cellulose supramolecular structure studied with combined confocal Raman spectroscopy and atomic force microscopy. *Cellulose* **2009**, *16*, 167–178. [CrossRef]
49. Wang, Z.; Hou, X.; Sun, J.; Li, M.; Chen, Z.; Gao, Z. Comparison of ultrasound-assisted ionic liquid and alkaline pretreatment of Eucalyptus for enhancing enzymatic saccharification. *Bioresour. Technol.* **2018**, *254*, 145–150. [CrossRef]
50. Warwicker, J.O.; Wright, A.C. Function of sheets of cellulose chains in swelling reactions on cellulose. *J. Appl. Polym. Sci.* **1967**, *11*, 659–671.
51. Ahmadzadeh, A.; Zakaria, S. Preparation of novolak resin by liquefaction of oil palm empty fruit bunches (EFB) and characterization of EFB residue. *Polym. Plast. Technol. Eng.* **2009**, *48*, 10–16.
52. Bakarudin, S.B.; Zakaria, S.; Chia, C.H.; Jani, S.M. Liquefied residue of kenaf core wood produced at different phenol-kenaf ratio. *Sains Malays.* **2012**, *41*, 225–231.
53. Van Soest, P.J. Use of detergents in the analysis of fibrous feeds. II. A rapid method for the determination of fiber and lignin. *J. Assoc. Off. Agric. Chem.* **1963**, *46*, 829–835.

Article

Use of Pineapple Waste as Fuel in Microbial Fuel Cell for the Generation of Bioelectricity

Segundo Rojas-Flores [1,*], Renny Nazario-Naveda [2], Santiago M. Benites [2], Moisés Gallozzo-Cardenas [3], Daniel Delfín-Narciso [4] and Félix Díaz [5]

1 Escuela de Ingeniería Mecánica Eléctrica, Universidad Señor de Sipán, Chiclayo 14000, Peru
2 Vicerrectorado de Investigación, Universidad Autónoma del Perú, Lima 15842, Peru
3 Universidad Tecnológica del Perú, Trujillo 13011, Peru
4 Grupo de Investigación en Ciencias Aplicadas y Nuevas Tecnologías, Universidad Privada del Norte, Trujillo 13007, Peru
5 Escuela Académica Profesional de Medicina Humana, Universidad Norbert Wiener, Lima 15046, Peru
* Correspondence: segundo.rojas.89@gmail.com

Abstract: The excessive use of fossil sources for the generation of electrical energy and the increase in different organic wastes have caused great damage to the environment; these problems have promoted new ways of generating electricity in an eco-friendly manner using organic waste. In this sense, this research uses single-chamber microbial fuel cells with zinc and copper as electrodes and pineapple waste as fuel (substrate). Current and voltage peaks of 4.95667 ± 0.54775 mA and 0.99 ± 0.03 V were generated on days 16 and 20, respectively, with the substrate operating at an acid pH of 5.21 ± 0.18 and an electrical conductivity of 145.16 ± 9.86 mS/cm at two degrees Brix. Thus, it was also found that the internal resistance of the cells was $865.845 \pm 4.726 \ \Omega$, and a maximum power density of 513.99 ± 6.54 mW/m^2 was generated at a current density of 6.123 A/m^2, and the final FTIR spectrum showed a clear decrease in the initial transmittance peaks. Finally, from the biofilm formed on the anodic electrode, it was possible to molecularly identify the yeast *Wickerhamomyces anomalus* with 99.82% accuracy. In this way, this research provides a method that companies exporting and importing this fruit may use to generate electrical energy from its waste.

Keywords: microbial fuel cell; waste; pineapple; bioelectricity; *Wickerhamomyces anomalus*

Citation: Rojas-Flores, S.; Nazario-Naveda, R.; Benites, S.M.; Gallozzo-Cardenas, M.; Delfín-Narciso, D.; Díaz, F. Use of Pineapple Waste as Fuel in Microbial Fuel Cell for the Generation of Bioelectricity. *Molecules* **2022**, *27*, 7389. https://doi.org/10.3390/molecules27217389

Academic Editors: Mohamad Nasir Mohamad Ibrahim, Patricia Graciela Vázquez and Mohd. Hazwan Hussin

Received: 21 September 2022
Accepted: 24 October 2022
Published: 31 October 2022

Publisher's Note: MDPI stays neutral with regard to jurisdictional claims in published maps and institutional affiliations.

1. Introduction

The enormous technological advances of humanity have made electrical energy a necessity in carrying out our daily tasks [1]. Currently, the main sources of fuel for electricity generation are fossil sources (natural gas, oil, and coal), which are the cause of many health and environmental problems and affect the quality of life of many people. Even so, in 2019, fossil fuels comprised 85.5% of the energy produced globally, with India, Japan, China, and the US using 92.5, 90.8, 87, and 85.3%, respectively, being the countries with the highest dependence [2–4]. In this sense, the International Energy Agency (IEA) reported that in the year 2020, the consumption of electrical energy increased by 3% because of the pandemic (COVID-19), and that this percentage would increase by approximately 22% by the year 2025 [5]. Due to the increase in demand and the problems that it generates, several research groups have worked on different ways of generating electrical energy in an environmentally friendly way, including reusing different forms of waste and applying a circular economy in the process [6,7].

Microbial fuel cells (MFCs) are electronic devices that have become more relevant in recent decades, mainly because they use different types of waste to generate electricity, thus reusing a wide variety of biomass produced from the different activities of humans [8–10]. This type of cell consists of anodic and cathodic chambers, which are almost always connected inside by a proton exchange membrane and outside by an external circuit [11].

This system converts chemical energy into electrical energy through the oxidation and reduction processes in the anodic and cathodic chambers, respectively [12–14]. A great variety of substrates have been used as fuel in MFCs for the generation of electricity, while, on the other hand, due to the problems generated in the whole process of harvesting and consuming agricultural products, these are becoming a potential source for use as fuel in MFCs as there are a large number of microorganisms present in the generated waste that can produce bioelectricity [15–17]. It is estimated that 140 Gtons of agricultural waste is produced each year. For example, from just cereals, in 2019, it was estimated that 2.8 Gtons of waste was generated worldwide; for this reason, it is necessary to use technologies to reuse this type of waste [18,19].

One of the most consumed agricultural products worldwide is the pineapple (*ananas comosus*), which generated nine billion dollars worldwide in 2019 [20]. This fruit is generated in large quantities in South America, making this continent the main exporter worldwide; for example, in Brazil alone, 28,179,348 tons were harvested in 2019 according to the Food and Agriculture Organization (FAO) [21,22]. The increase in pineapple consumption is due to the high nutritional content of water, carbohydrates, organic acids, dietary fibers, antioxidants, vitamins, and minerals, as well as minerals such as calcium (Ca), magnesium (Mg), phosphorus (P), sodium (Na), manganese (Mn), iron (Fe), copper (Cu), zinc (Zn), and selenium (Se), and vitamins such as B1, B2, B3, B5, B6, B9, and C [23–25]. Research has been reported in which bioelectricity is generated through different types of organic waste; for example, banana and orange peel have been used as fuel to generate bioelectricity in MFCs with zinc and graphite electrodes, managing to generate peaks of 0.586 and 0.492 V in the cells with banana and orange substrates, respectively [26]. Similarly, Manjrekar et al. (2018), in their research, used kitchen waste in their double-chamber MFCs with aluminum electrodes, managing to generate voltage peaks of 365 mV. It was also observed that these values decreased over time, mainly due to the sedimentation of the organic components [27]. Likewise, tomato waste has been used as a substrate for the generation of electricity in single-chamber MFCs with zinc and copper electrodes, with voltage peaks close to 10.8 V and an internal resistance of 0.148541 ± 0.012361 KΩ being observed at the optimal acid pH for the operation of MFCs; this study is one of the most promising works to date [28]. It was observed that, in the investigations carried out, metallic electrodes are of great help in generating higher current and voltage values, while single-chamber MFCs obtain higher current and power density values [29,30]. In the reviewed literature, research that uses different types of fruit waste as fuel was found; however, the information on pineapple waste is very limited, although this material is very promising for the generation of bioelectricity in microbial fuel cells due to its physical, chemical, and biological characteristics. Therefore, it is important to expand the knowledge in this field through research using this type of waste.

The main objective of this research was to generate bioelectricity through single-chamber microbial fuel cells at the laboratory scale, using pineapple waste as fuel, and zinc and copper electrodes as anode and cathode electrodes, respectively. The microbial fuel cells were monitored for voltage, current, pH, electrical conductivity, and Brix degrees for a period of 32 days. We also found the values for power density, current density, and internal resistance of microbial fuel cells. Additionally, the FTIR (Fourier transform infrared spectroscopy) patterns of the initial and final substrate were studied, as well as the molecular biology of the biofilm formed on the anode electrode.

2. Materials and Methods

2.1. Manufacture of Microbial Fuel Cells

Three (03) MFCs were manufactured with acrylic tubes (Poly/methyl 2-methypropenoate)/PMMA 10 × 30 cm in diameter and length, respectively. For the anodic and cathodic electrodes, copper (Cu) and zinc (Zn) 10 and 0.2 cm in diameter and thickness, respectively, were used. The Zn electrode was placed at one end of the tube with one side exposed to the environment and the other was exposed to the substrate used; while the anode electrode

(Cu) was placed inside the MFCs, both electrodes were joined by an external circuit of Cu wire (0.25 cm in diameter) and a 100 Ω resistor, whose configuration was similar to that used in the work of Segundo et al. (2022) [28].

2.2. Pineapple Waste Collection

Pineapple waste was selected by the people who sell it at Mercado La Hermelinda, Trujillo, Peru, who managed to collect six kilograms in total. The collected waste was taken to the laboratories of the university in airtight bags and washed 3 times with distilled water to remove any type of impurities (e.g., dirt obtained from the market) acquired from the environment, and then left to dry in an oven at 30 $\pm$ 2 °C for 12 h. The pineapple waste was passed through an extractor (Labtron, LDO-B10- Camberley, UK) to obtain juice from the waste. It was possible to obtain 2 L of juice, which was placed in a beaker and stored until use.

2.3. Characterization of Microbial Fuel Cells

The voltage and current values were monitored using a multimeter (Prasek Premium PR-85) for 32 days, which has an external circuit with 100 Ω resistance. Power density (PD) and current density (DC) values were obtained using external resistors of 0.3 ($\pm$0.1), 0.6 ($\pm$0.18), 1($\pm$0.3), 1.5($\pm$0.31), 3($\pm$0.6), 10 ($\pm$1.3), 20 ($\pm$6.5), 50($\pm$8.7), 60($\pm$8.2), 100($\pm$9.3), 120 ($\pm$9.8), 220 ($\pm$13), 240 ($\pm$15.6), 330 ($\pm$20.3), 390 ($\pm$24.5), 460 ($\pm$23.1), 531 ($\pm$26.8), 700 ($\pm$40.5), and 1000 ($\pm$50.6) Ω, using the method of Segundo et al. (2022) [31]. The monitored values of electrical conductivity (conductivity meter CD-4301), pH (pH meter 110Series Oakton), and degrees Brix (RHB-32 brix refractometer) were also measured for 32 days. The transmittance values were measured by FTIR (Thermo Scientific IS50), and for the resistance values of the MFCs, an energy sensor (Vernier- $\pm$ 30 V and $\pm$1000 mA) was used.

2.4. Isolation of Microorganisms from the Anode Chamber

To isolate the microorganisms, a swab of the anode plates was taken and then planted in McConkey agar and nutrient agar medium to isolate bacteria; on the other hand, Sabouraud agar was used to isolate fungi and yeasts with 4% glucose. Media were incubated for 24 h at 35 °C for bacterial isolations and 24 h at 30 °C for fungal and yeast isolations, and the procedure was performed in duplicate [32]. The reading consisted of observing the macroscopic characteristics of the colonies grown in the culture media, while methylene blue was used to observe the microscopic characteristics. Finally, pure cultures were made.

2.5. Molecular Identification of Fungi

Molecular identification was carried out in the BIODES laboratory (Laboratorio de Soluciones Integrales Comercial de Sociedad Limitada) where molecular biology techniques were used. The first users were the ITS (Internal Transcribed Spacer, USA) sequences, which are specific to fungi [33].

2.6. Statistical Analysis

The data points obtained from Figures 1–3 represent the average values obtained from the three replicates, and the error bars represent the corresponding standard deviations. Meanwhile, the bioinformatic software MEGA X (Molecular Evolutionary Genetics Analysis, USA) analyzed the sequence obtained by comparing it with the sequences of reference yeast species. For this, the sequence alignment tool BLAST (Basic Local Alignment Search Tool) was used to identify the species based on the percentage of identity [32].

3. Results and Analysis

The monitored voltage values are shown in Figure 1a, where it can be seen that the voltage values increased from the first day (0.5685 $\pm$ 0.01 V) to day 16, when the maximum voltage peak was found (0.99 $\pm$ 0.03 V), and later decayed to 0.624 $\pm$ 0.03 V on the last day of monitoring. The high voltage values found are directly attributed to the microbiota found on the surface of the anode electrode, whose influence was described by Khan et al. (2017) [34]. Similarly, Waheed et al. (2016) showed that the size of the particles influences the rate of hydrolysis, limiting the generation of voltage [35]. This research found values that exceeded those obtained by Kalagbor et al. (2020), who, with the use of the pineapple substrate, generated 0.8 V during the first days; however, in said study, a tendency towards a decrease in voltage production was observed [36]. Likewise, Priya and Setty (2019) generated a peak voltage of 0.4 V on the seventh day from the use of apple juice as a substrate in the anode chamber. It is worth mentioning that the values obtained in this study are not higher than those found in this investigation [37]. Figure 1b shows the electrical current values monitored throughout the investigation, observing that the values increased from 0.09667 ± 0.00577 mA on day 1 to the maximum peak of 4.95667 ± 0.54775 mA on day 20 and then decreased until the last day (1.97333 ± 0.50213 mA). The values of electrical currents in the MFCs are governed mainly by the fermentative microorganisms that convert the substrate (fermented fuel), such as glucose, into small-chain organic acids, hydrogen, and carbon dioxide; electricity is generated at the same time as an interaction is formed between the reduced compounds that are produced under redox conditions during fermentation or possibly in some direct transfer of electrons between the microorganisms and the anode surface [38–40]. On the other hand, the values of the electric current in Figure 1b showed a decrease during the last days, which would be due to the diffusion of oxygen from the cathode to the anode due to the lack of a membrane between them [41].

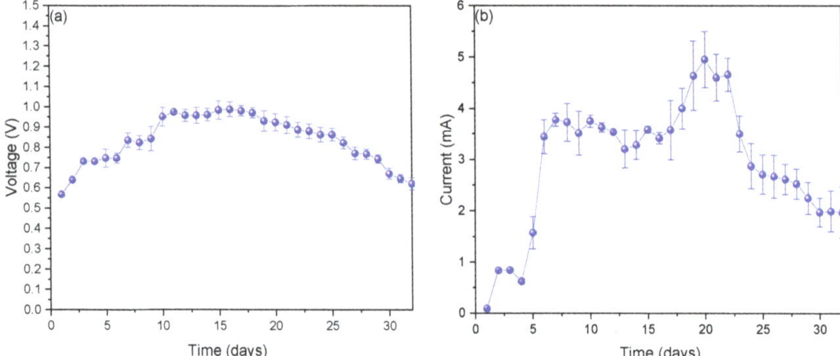

Figure 1. Monitoring of (**a**) voltage and (**b**) current values of microbial fuel cells (MFCs).

Figure 2a shows the values of the electrical conductivity of the substrates of the MFCs, observing that the values increased from the first day (189.3 ± 2.08 mS/cm) to the sixth day (199.33 ± 4.041 mS/cm) and then slowly decayed until the last day (135.51 ± 6.87 mS/cm). In previous studies, it was observed that the electrical conductivity values of different types of substrates (organic waste), when fermented, increase over time and decrease when observing substrate sedimentation (agglomeration of particles in the fermentation process) [42,43]. Figure 2b shows the pH values observed during the monitoring, where it is shown that throughout the monitoring, an acidic pH was maintained, although between days 16 and 19, maximum pH values were observed. For this work, it was observed that the optimal operating pH of the MFCs was 5.21 ± 0.18, which was achieved on day 16. In the literature, it has been found that the pH of the substrate affects the electrical efficiency of the MFCs because the transfer of electrons and protons occurs within them [44]. Thus, in the anode reactions, electrons are produced in the oxidation process, generating acidification, which consequently lowers the pH [45,46], while the reduction reactions that occur in the cathode produce alkalinization, increasing the pH, and the variation in the pH values over time is attributed to this [47,48]. In this sense, it is worth mentioning that high pH inhibits the growth of methanogens that indirectly improve the performance of MFCs [49]. The observed values of degrees Brix (°Brix) are shown in Figure 2c, which remained constant until the third day (6° Brix) and then gradually decreased until the twenty-fourth day, when they decreased to zero, and they remained constant until the last day. From monitoring, it has been observed in the literature that the °Brix values decrease mainly due to the decomposition of the nutrients of the substrates in the bioelectricity generation process of the MFCs [49,50].

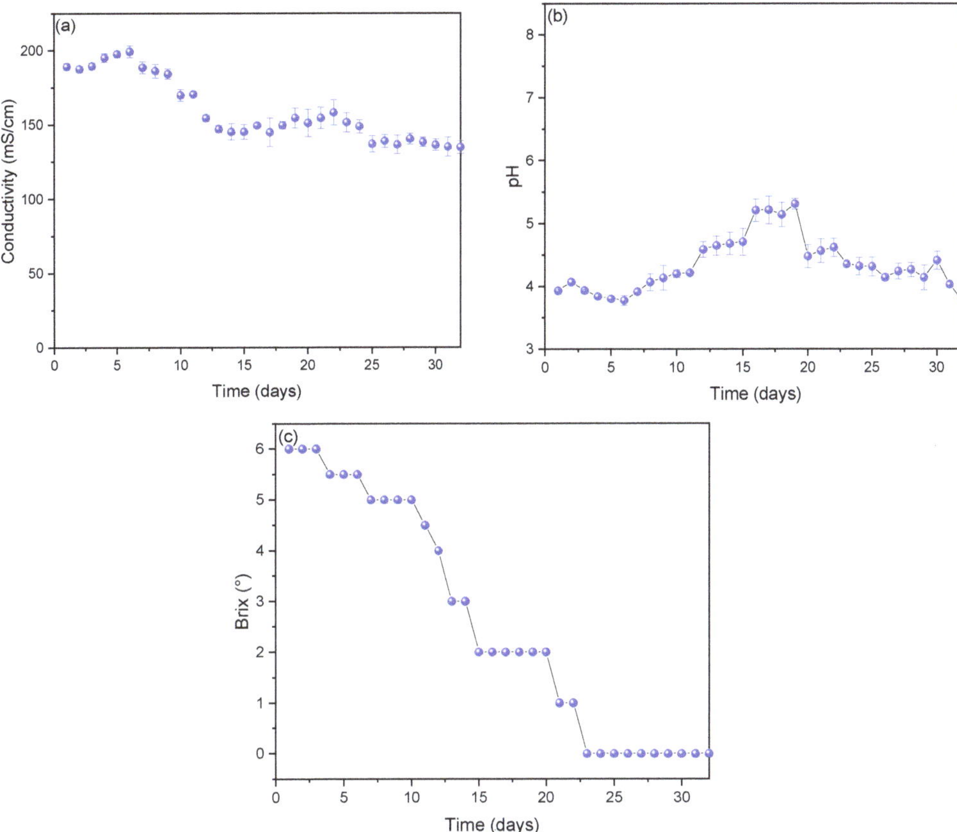

Figure 2. Monitoring of (**a**) conductivity, (**b**) pH, and (**c**) °Brix parameters of MFCs.

Figure 3a shows the value of the internal resistance of the microbial fuel cells, which are governed by Ohm's Law (V = RI). The X axis represents the electrical current values (I) and the Y axis the voltage values (V), whose the linear fit slope would represent the average internal resistance ($R_{int.}$) of the MFCs. The $R_{int.}$ found was 865,845 $\pm$ 47.23 Ω in the MFCs; these values were found on day 20 because it was the day that generated the most intense electrical current values. Compared with other investigations, the values found for the $R_{int.}$ are high, but the values of current and voltage are higher than those found by other investigations; for example, Rashid et al. (2021), in their research, managed to generate 330 mV and 2.75 mA using single-chamber MFCs with graphite electrodes and pharmaceutical effluents with 99 Ω internal resistance [51]. Likewise, Liu et al. (2020) investigated the use of electrogenic bacteria as substrates and carbon electrodes in MFCs, managing to generate peaks of 0.63 V in the internal resistance of 162.9 $\pm$ 3.5 Ω [52]. One of the most important and influential factors that can explain this phenomenon is the presence of organisms that form the biofilm, mainly in the anodic electrode, since it is in charge of receiving the electrons [53,54]. Figure 3b shows the values of the power density (PD) and voltage as a function of the current density (CD), where the DP_{MAX} can be observed to be 513.99 ± 6.54 mW/m^2 in a CD of 6.123 A/m^2, with a maximum voltage of 874.46 ± 19.64 mV. These obtained values are high compared to those obtained by Xin et al. (2018), who managed to obtain 0.20 W/m^2 in a CD of 0.27 A/cm^2 and a peak voltage of 0.58 V [55]. The values obtained according to Huang et al. (2021) would be due to the high content of glucose as a carbon source for the microorganisms present in the substrates [56]. Additionally, the DP value found by Yaqoob et al. (2022) was 0.30 mW/cm^2 at a DC of approximately 28 mA/cm^2 in their dual-chamber MFCs using mango (Mangifera indica) debris as the substrate. All these values are lower than those found in this research, which highlights the importance of research and the use of pineapple waste and the electrodes used [56].

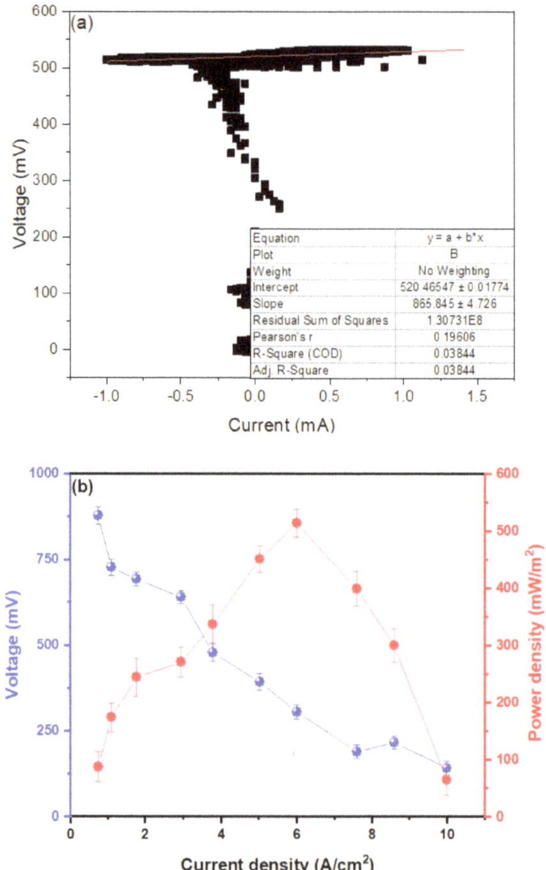

Figure 3. Characterization of (**a**) internal resistance and (**b**) power and voltage density in relation to current density of MFCs.

Figure 4 shows the transmittance spectra obtained by FTIR of the substrate used in the initial and final state of the monitoring performed. It was possible to observe that the most intense peak corresponds to the O-H bonds at 3360 cm^{-1}, while the peak at 2930 cm^{-1} belongs to the strong bonds of alkanes (C-H). Similarly, the 1610 cm^{-1} peak shows the presence of alkene compounds (C=C), and the 1425 and 1060 cm^{-1} peaks confirm the presence of NO$_2$ and C-H bonds [57,58]. As can be clearly seen, the intensity of the transmittance peaks decreases compared to the initial spectrum, which would be due to the degradation of microorganisms in the process of generating bioelectricity and sedimentation in the last days [59].

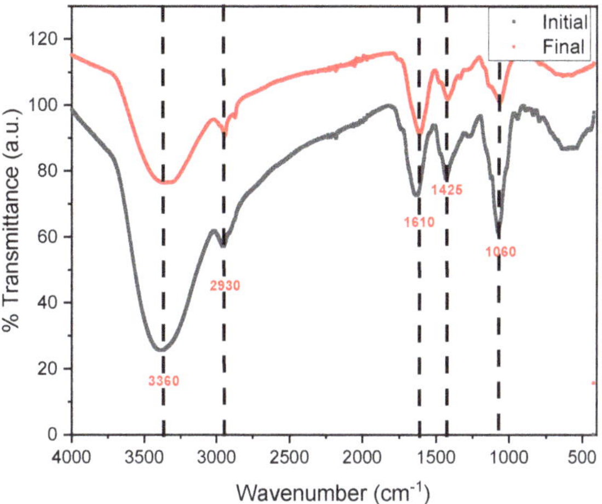

Figure 4. FTIR spectrophotometry of the initial and final pineapple waste.

The regions sequenced and analyzed in the BLAST program obtained an identity percentage of 99.82%, which corresponds to the species *Wickerhamomyces anomalus* (see Table 1). This is a widespread yeast in nature, present in habitats such as soil, plants, and fruits and as an opportunistic pathogen in humans and animals [60]. This yeast can grow under extreme conditions of environmental stress, due to which it can be a spoilage organism, especially in food products with a high sugar content [61].

Table 1. BLAST characterization of the rDNA sequence of yeast isolated from the MFC anode plate with pineapple debris substrate.

Blast Characterization	Length of Consensus Sequence (nt)	% Maximum Identidad	Accession Number	Phylogeny
Wickerhamomyces anomalus	545	99.82	KJ527063.1	Cellular organisms; Eukaryota; Opisthokonta; Fungi; Dikarya; Ascomycota; Saccharomyceta; Saccharomycetales; Phaffomycetaceae; Wickerhamomyces

The dendrogram was based on the ITS regions of the rDNA regions of a group of yeast strains isolated from the anode plate of pineapple microbial fuel cells (Figure 5), which were constructed using the MEGA program, which relates and groups sequences of species; from this, the species *Wickerhamomyces anomalus* was identified, which is among the "film-forming" yeasts. In this context, it is worth mentioning that in microbial fuel cells, the transfer of electrons from microorganisms to the electrode is produced by various mechanisms, including through pili or nanowires after the formation of a biofilm on the electrode [62,63]. It has been reported that this species exhibits a direct transfer of electrons without the help of mediators, using glucose as a carbon source, and that this yeast also has redox enzymes present in the cell membrane of the cell, which contribute to the production of current in an MFC [64,65]. Finally, Figure 6 shows the bioelectricity generation mechanism of using pineapple waste as fuel in single-chamber microbial fuel cells, where it is observed that the three fuel cells connected in series were capable of generating 2.85 V, which was enough to turn on an LED bulb (white) on the fifteenth day.

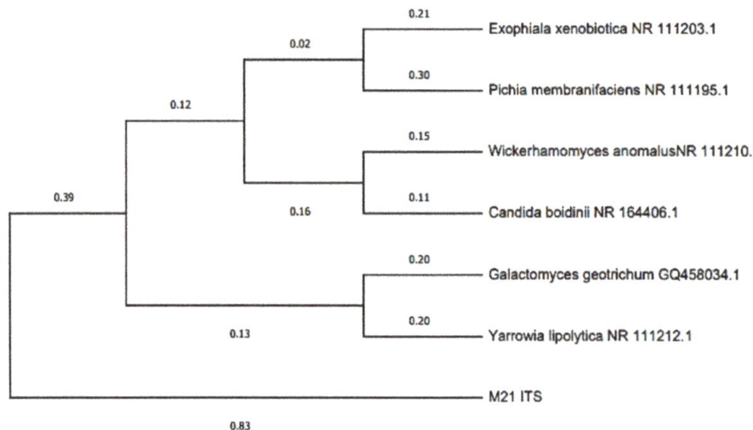

Figure 5. Dendrogram based on the ITS regions of the rDNA regions of a culture of *Wickerhamomyces anomalus* isolated from the MCC anode plate with pineapple substrate.

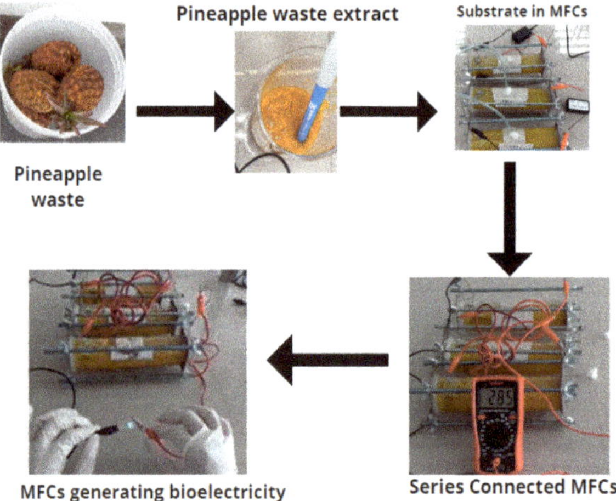

Figure 6. Mechanisms of the bioelectricity generation process from pineapple waste.

4. Conclusive Remarks and Future Perspectives

Bioelectricity was successfully generated through single-chamber microbial fuel cells on a laboratory scale using pineapple waste and zinc and copper electrodes as a substrate. It was possible to observe that the maximum peaks of voltage and electric current were 0.99 ± 0.03 V and 4.95667 ± 0.54775 mA on the sixteenth and twentieth days, and these values decreased slowly until the end of the monitoring. These peak values of voltage and electrical current were obtained at an optimal pH of 5.21 ± 0.18, a substrate electrical conductivity of 145.16 ± 9.86 mS/cm, and two ° Brix. Likewise, the maximum power density found was 513.99 ± 6.54 mW/m^2 with a current density of 6.123 A/m^2, and the internal resistance of the microbial fuel cells was 865.845 ± 4.726 Ω, while the initial and final FTIR spectra of the substrate used were obtained, achieving a decrease in the transmittance peaks, the most notable being the peak belonging to the O-H bond at 3360 cm^{-1}. Finally, the yeast *Wickerhamomyces anomalus* was molecularly identified as being present in the anode electrode with an identity percentage of 99.82%. This research highlights the importance of the use of pineapple waste in the generation of bioelectricity; with this substrate, electrical values higher than those found in the literature have been obtained. The molecular identification of the microorganisms present in the microbial fuel cells contributes to enriching the knowledge about the operation of microbial fuel cells. On the other hand, companies, society, and farmers can see an opportunity in which their waste can be reused, increasing their profits and benefits.

For future work, it is recommended that an optimal pH is used for the operation of the cells, since in this investigation, we worked with the pH of the pineapple waste itself; we further recommend using metal electrodes (due to their excellent electron-conducting properties) but coated with some type of non-toxic chemical compound so that microorganisms are not affected and thus increase the efficiency of microbial fuel cells. Carrying out these new investigations will clarify whether the metallic electrodes have a higher performance when they are coated or pure; this can be measured through the electrical values found in future investigations. On the other hand, the use of a cell system with air flows is recommended to increase power density, as it has been shown that the supply of O$_2$ increases the electrical values, mainly the PD values [66], which, combined with sucrose-enriched substrates, can increase the voltage and current values [67].

Author Contributions: Conceptualization, S.R.-F.; methodology S.M.B.; software, R.N.-N.; validation, R.N.-N. and M.G.-C.; formal analysis, S.R.-F.; investigation S.R.-F. and F.D.; data curation, M.G.-C.; writing—original draft preparation, D.D.-N.; writing—review and editing, S.R.-F. and M.G.-C.; project administration, S.R.-F. and R.N.-N. All authors have read and agreed to the published version of the manuscript.

Funding: This research received no external funding.

Institutional Review Board Statement: Not applicable.

Informed Consent Statement: Not applicable.

Data Availability Statement: Not applicable.

Conflicts of Interest: The authors declare no conflict of interest.

References

1. Bos, K.; Gupta, J. Climate change: The risks of stranded fossil fuel assets and resources to the developing world. *Third World Q.* **2018**, *39*, 436–453. [CrossRef]
2. Pourghasemi, A.; Akhbari, M. The role of fossil fuel (oil and gas) in the world geopolitics of energy (case study of Iran 2000–2015). *J. Ecophysiol. Occup. Health* **2018**, *18*, 18–23. [CrossRef]
3. Jackson, R.B.; Friedlingstein, P.; Andrew, R.M.; Canadell, J.G.; Le Quéré, C.; Peters, G.P. Persistent fossil fuel growth threatens the Paris Agreement and planetary health. *Environ. Res. Lett.* **2019**, *14*, 121001. [CrossRef]
4. Martins, F.; Felgueiras, C.; Smitkova, M.; Caetano, N. Analysis of fossil fuel energy consumption and environmental impacts in European countries. *Energies* **2019**, *12*, 964. [CrossRef]
5. Wan, D.; Xue, R.; Linnenluecke, M.; Tian, J.; Shan, Y. The impact of investor attention during COVID-19 on investment in clean energy versus fossil fuel firms. *Financ. Res. Lett.* **2021**, *43*, 101955. [CrossRef]
6. Mulvaney, D.; Richards, R.M.; Bazilian, M.D.; Hensley, E.; Clough, G.; Sridhar, S. Progress towards a circular economy in materials to decarbonize electricity and mobility. *Renew. Sustain. Energy Rev.* **2021**, *137*, 110604. [CrossRef]
7. Yaqoob, H.; Teoh, Y.H.; Din, Z.U.; Sabah, N.U.; Jamil, M.A.; Mujtaba, M.A.; Abid, A. The potential of sustainable biogas production from biomass waste for power generation in Pakistan. *J. Clean. Prod.* **2021**, *307*, 127250. [CrossRef]
8. Naseer, M.N.; Zaidi, A.A.; Khan, H.; Kumar, S.; bin Owais, M.T.; Jaafar, J.; Suhaimin, N.S.; Wahab, Y.A.; Dutta, K.; Asif, M.; et al. Mapping the field of microbial fuel cell: A quantitative literature review (1970–2020). *Energy Rep.* **2021**, *7*, 4126–4138. [CrossRef]
9. Gul, H.; Raza, W.; Lee, J.; Azam, M.; Ashraf, M.; Kim, K.H. Progress in microbial fuel cell technology for wastewater treatment and energy harvesting. *Chemosphere* **2021**, *281*, 130828. [CrossRef]
10. Munoz-Cupa, C.; Hu, Y.; Xu, C.; Bassi, A. An overview of microbial fuel cell usage in wastewater treatment, resource recovery and energy production. *Sci. Total Environ.* **2021**, *754*, 142429. [CrossRef]

11. Zhang, Y.; Liu, M.; Zhou, M.; Yang, H.; Liang, L.; Gu, T. Microbial fuel cell hybrid systems for wastewater treatment and bioenergy production: Synergistic effects, mechanisms and challenges. *Renew. Sustain. Energy Rev.* **2019**, *103*, 13–29. [CrossRef]

12. Cui, Y.; Lai, B.; Tang, X. Microbial fuel cell-based biosensors. *Biosensors* **2019**, *9*, 92. [CrossRef] [PubMed]

13. Al Lawati, M.J.; Jafary, T.; Baawain, M.S.; Al-Mamun, A. A mini review on biofouling on air cathode of single chamber microbial fuel cell; Prevention and mitigation strategies. *Biocatal. Agric. Biotechnol.* **2019**, *22*, 101370. [CrossRef]

14. Greenman, J.; Gajda, I.; Ieropoulos, I. Microbial fuel cells (MFC) and microalgae; photo microbial fuel cell (PMFC) as complete recycling machines. *Sustain. Energy Fuels* **2019**, *3*, 2546–2560. [CrossRef]

15. Anastopoulos, I.; Pashalidis, I.; Hosseini-Bandegharaei, A.; Giannakoudakis, D.A.; Robalds, A.; Usman, M.; Escudero, L.B.; Zhou, Y.; Colmenares, J.C.; Núñez-Delgado, A.; et al. Agricultural biomass/waste as adsorbents for toxic metal decontamination of aqueous solutions. *J. Mol. Liq.* **2019**, *295*, 111684. [CrossRef]

16. Bilandzija, N.; Voca, N.; Jelcic, B.; Jurisic, V.; Matin, A.; Grubor, M.; Kricka, T. Evaluation of Croatian agricultural solid biomass energy potential. *Renew. Sustain. Energy Rev.* **2018**, *93*, 225–230. [CrossRef]

17. Tripathi, N.; Hills, C.D.; Singh, R.S.; Atkinson, C.J. Biomass waste utilisation in low-carbon products: Harnessing a major potential resource. *Npj Clim. Atmos. Sci.* **2019**, *2*, 35. [CrossRef]

18. Hills, C.D.; Tripathi, N.; Singh, R.S.; Carey, P.J.; Lowry, F. Valorisation of agricultural biomass-ash with CO_2. *Sci. Rep.* **2020**, *10*, 13801. [CrossRef]

19. Siwal, S.S.; Zhang, Q.; Devi, N.; Saini, A.K.; Saini, V.; Pareek, B.; Gaidukovs, S.; Thakur, V.K. Recovery processes of sustainable energy using different biomass and wastes. *Renew. Sustain. Energy Rev.* **2021**, *150*, 111483. [CrossRef]

20. Roda, A.; Lambri, M. Food uses of pineapple waste and by-products: A review. *Int. J. Food Sci. Technol.* **2019**, *54*, 1009–1017. [CrossRef]

21. Vieira, I.M.M.; Santos, B.L.P.; Santos, C.V.M.; Ruzene, D.S.; Silva, D.P. Valorization of pineapple waste: A review on how the fruit's potential can reduce residue generation. *BioEnergy Res.* **2022**, *15*, 924–934. [CrossRef]

22. Aili Hamzah, A.F.; Hamzah, M.H.; Che Man, H.; Jamali, N.S.; Siajam, S.I.; Ismail, M.H. Recent updates on the conversion of pineapple waste (Ananas comosus) to value-added products, future perspectives and challenges. *Agronomy* **2021**, *11*, 2221. [CrossRef]

23. Ali, M.M.; Hashim, N.; Abd Aziz, S.; Lasekan, O. Pineapple (*Ananas comosus*): A comprehensive review of nutritional values, volatile compounds, health benefits, and potential food products. *Food Res. Int.* **2020**, *137*, 109675.

24. Banerjee, S.; Ranganathan, V.; Patti, A.; Arora, A. Valorisation of pineapple wastes for food and therapeutic applications. *Trends Food Sci. Technol.* **2018**, *82*, 60–70. [CrossRef]

25. Hikal, W.M.; Mahmoud, A.A.; Said-Al Ahl, H.A.; Bratovcic, A.; Tkachenko, K.G.; Kačániová, M.; Rodriguez, R.M. Pineapple (*Ananas comosus* L. Merr.), waste streams, characterisation and valorisation: An Overview. *Open J. Ecol.* **2021**, *11*, 610–634. [CrossRef]

26. Toding, O.S.L.; Virginia, C.; Suhartini, S. Conversion banana and orange peel waste into electricity using microbial fuel cell. *IOP Conf. Ser. Earth Environ. Sci.* **2018**, *209*, 012049.

27. Manjrekar, Y.; Kakkar, S.; Durve-Gupta, A. Bio-electricity generation using kitchen waste and molasses powered MFC. *IJSRSET* **2018**, *5*, 181–187.

28. Segundo, R.F.; Magaly, D.L.C.N.; Benites, S.M.; Daniel, D.N.; Angelats-Silva, L.; Díaz, F.; Luis, C.-C.; Fernanda, S.P. Increase in Electrical Parameters Using Sucrose in Tomato Waste. *Fermentation* **2022**, *8*, 335. [CrossRef]

29. Rojas Flores, S.J.; Benites, S.M.; Agüero Quiñones, R.; Enríquez-León, R.; Angelats Silva, L. Bioelectricity through microbial fuel cells from decomposed fruits using lead and copper electrodes [Bioelectricidad mediante Celdas de Combustible Microbiana a partir de frutas descompuestas usando electrodos de plomo y cobre]. In Proceedings of the 18th LACCEI International Multi-Conference for Engineering, Education, and Technology: "Engineering, Integration, and Alliances for a Sustainable Development" "Hemispheric Cooperation for Competitiveness and Prosperity on a Knowledge-Based Economy", Virtual Edition, Online, 27–31 July 2020.

30. Flores, S.J.R.; Benites, S.M.; Rosa, A.L.R.; Zoilita, A.L.Z.; Luis, A.S. The Using Lime (*Citrus × aurantiifolia*), Orange (*Citrus × sinensis*), and Tangerine (*Citrus reticulata*) Waste as a Substrate for Generating Bioelectricity: Using lime (*Citrus × aurantiifolia*), orange (*Citrus × sinensis*), and tangerine (*Citrus reticulata*) waste as a substrate for generating bioelectricity. *Environ. Res. Eng. Manag.* **2020**, *76*, 24–34.

31. Rojas-Flores, S.; De La Cruz-Noriega, M.; Milly Otiniano, N.; Benites, S.M.; Esparza, M.; Nazario-Naveda, R. Use of Onion Waste as Fuel for the Generation of Bioelectricity. *Molecules* **2022**, *27*, 625.

32. Rojas-Flores, S.; Benites, S.M.; De La Cruz-Noriega, M.; Cabanillas-Chirinos, L.; Valdiviezo-Dominguez, F.; Quezada Álvarez, M.A.; Vega-Ybañez, V.; Angelats-Silva, L. Bioelectricity production from blueberry waste. *Processes* **2021**, *9*, 1301. [CrossRef]

33. Raja, H.A.; Miller, A.N.; Pearce, C.J.; Oberlies, N.H. Fungal identification using molecular tools: A primer for the natural products research community. *J. Nat. Prod.* **2017**, *80*, 756–770. [CrossRef] [PubMed]

34. Khan, A.M.; Hussain, M.S. Conversion of Wastes to Bioelectricity, Bioethanol, and Fertilizer: Khan and Hussain. *Water Environ. Res.* **2017**, *89*, 676–686. [CrossRef]

35. Miran, W.; Nawaz, M.; Jang, J.; Lee, D.S. Conversion of orange peel waste biomass to bioelectricity using a mediator-less microbial fuel cell. *Sci. Total Environ.* **2016**, *547*, 197–205. [CrossRef] [PubMed]

36. Kalagbor, I.A.; Azunda, B.I.; Igwe, B.C.; Akpan, B.J. Electricity generation from waste tomatoes, banana, pineapple fruits and peels using single chamber microbial fuel cells (SMFC). *J. Waste Manag. Xenobiot.* **2020**, *3*, 000142.
37. Priya, A.D.; Setty, Y.P. Cashew apple juice as substrate for microbial fuel cell. *Fuel* **2019**, *246*, 75–78. [CrossRef]
38. Takahashi, S.; Miyahara, M.; Kouzuma, A.; Watanabe, K. Electricity generation from rice bran in microbial fuel cells. *Bioresour. Bioprocess.* **2016**, *3*, 50. [CrossRef] [PubMed]
39. Rojas-Flores, S.; De La Cruz-Noriega, M.; Nazario-Naveda, R.; Benites, S.M.; Delfín-Narciso, D.; Rojas-Villacorta, W.; Romero, C.V. Bioelectricity through microbial fuel cells using avocado waste. *Energy Rep.* **2022**, *8*, 376–382. [CrossRef]
40. Flores, S.R.; Pérez-Delgado, O.; Naveda-Renny, N.; Benites, S.M.; De La Cruz–Noriega, M.; Narciso, D.A.D. Generation of Bioelectricity Using Molasses as Fuel in Microbial Fuel Cells. *Environ. Res. Eng. Manag.* **2022**, *78*, 19–27. [CrossRef]
41. De La Cruz–Noriega, M.; Rojas-Flores, S.; Nazario-Naveda, R.; Benites, S.M.; Delfín-Narciso, D.; Rojas-Villacorta, W.; Diaz, F. Potential Use of Mango Waste and Microalgae *Spirulina* sp. for Bioelectricity Generation. *Environ. Res. Eng. Manag.* **2022**, *78*, 129–136. [CrossRef]
42. Shao, Y.; Guizani, C.; Grosseau, P.; Chaussy, D.; Beneventi, D. Biocarbons from microfibrillated cellulose/lignosulfonate precursors: A study of electrical conductivity development during slow pyrolysis. *Carbon* **2018**, *129*, 357–366. [CrossRef]
43. Zhang, S.; Tong, W.; Wang, M. Graphene-modified biochar anode on the electrical performance of MFC. *Ferroelectrics* **2021**, *578*, 1–14. [CrossRef]
44. Bagchi, S.; Behera, M. Evaluation of the effect of anolyte recirculation and anolyte pH on the performance of a microbial fuel cell employing ceramic separator. *Process Biochem.* **2021**, *102*, 207–212. [CrossRef]
45. Geng, Y.K.; Yuan, L.; Liu, T.; Li, Z.H.; Zheng, X.; Sheng, G.P. Thermal/alkaline pretreatment of waste activated sludge combined with a microbial fuel cell operated at alkaline pH for efficient energy recovery. *Appl. Energy* **2020**, *275*, 115291. [CrossRef]
46. Li, L.; Dai, Q.; Zhang, S.; Liu, H. Degradation efficiency and mechanism of sulfur-containing azo dye wastewater by microbial fuel cell under different pH conditions. *Chin. J. Environ. Eng.* **2021**, *15*, 115–125.
47. Zhang, Y.; Xu, Q.; Huang, G.; Zhang, L.; Liu, Y. Effect of dissolved oxygen concentration on nitrogen removal and electricity generation in self pH-buffer microbial fuel cell. *Int. J. Hydrogen Energy* **2020**, *45*, 34099–34109. [CrossRef]
48. Dewanckele, L.; Jing, L.; Stefańska, B.; Vlaeminck, B.; Jeyanathan, J.; Van Straalen, W.M.; Koopmans, A.; Fievez, V. Distinct blood and milk 18-carbon fatty acid proportions and buccal bacterial populations in dairy cows differing in reticulorumen pH response to dietary supplementation of rapidly fermentable carbohydrates. *J. Dairy Sci.* **2019**, *102*, 4025–4040. [CrossRef]
49. Yi, Y.; Xie, B.; Zhao, T.; Li, Z.; Stom, D.; Liu, H. Effect of external resistance on the sensitivity of microbial fuel cell biosensor for detection of different types of pollutants. *Bioelectrochemistry* **2019**, *125*, 71–78.
50. Wang, H.; Peng, Z.; Sun, H. Antifungal activities and mechanisms of trans-cinnamaldehyde and thymol against food-spoilage yeast *Zygosaccharomyces rouxii*. *J. Food Sci.* **2022**, *87*, 1197–1210. [CrossRef]
51. Rashid, T.; Sher, F.; Hazafa, A.; Hashmi, R.Q.; Zafar, A.; Rasheed, T.; Hussain, S. Design and feasibility study of novel paraboloid graphite based microbial fuel cell for bioelectrogenesis and pharmaceutical wastewater treatment. *J. Environ. Chem. Eng.* **2021**, *9*, 104502. [CrossRef]
52. Liu, Y.; Sun, X.; Yin, D.; Cai, L.; Zhang, L. Suspended anode-type microbial fuel cells for enhanced electricity generation. *RSC Adv.* **2020**, *10*, 9868–9877. [CrossRef]
53. Abbas, S.Z.; Yong, Y.C.; Chang, F.X. Anode materials for soil microbial fuel cells: Recent advances and future perspectives. *Int. J. Energy Res.* **2022**, *46*, 712–725. [CrossRef]
54. Yaqoob, A.A.; Ibrahim, M.N.M.; Yaakop, A.S.; Rafatullah, M. Utilization of biomass-derived electrodes: A journey toward the high performance of microbial fuel cells. *Appl. Water Sci.* **2022**, *12*, 99. [CrossRef]
55. Xin, X.; Ma, Y.; Liu, Y. Electric energy production from food waste: Microbial fuel cells versus anaerobic digestion. *Bioresour. Technol.* **2018**, *255*, 281–287. [CrossRef] [PubMed]
56. Huang, X.; Duan, C.; Duan, W.; Sun, F.; Cui, H.; Zhang, S.; Chen, X. Role of electrode materials on performance and microbial characteristics in the constructed wetland coupled microbial fuel cell (CW-MFC). A review. *J. Clean. Prod.* **2021**, *301*, 126951. [CrossRef]
57. Saragih, B.; Saragih, N.A.D. FTIR (Fourier Transform Infra Red) profile of banana corn flour, nutritional value and sensory properties of resulting brownies. *J. Phys. Conf. Ser.* **2021**, *1882*, 012112. [CrossRef]
58. Suryanto, H.; Wijaya, H.W.; Yanuhar, U. FTIR analysis of alkali treatment on bacterial cellulose films obtained from pineapple peel juice. *IOP Conf. Ser. Mater. Sci. Eng.* **2021**, *1034*, 012145. [CrossRef]
59. Nora, A.; Wilapangga, A.; Kahfi, A.; Then, M.; Fairus, A. Antioxidant Activity, Total Phenolic Content, and FTIR Analysis of Fermented Baduy Honey with Pineapple. *Bioedukasi* **2022**, *20*, 21–25. [CrossRef]
60. Satora, P.; Tarko, T.; Sroka, P.; Blaszczyk, U. The influence of *Wickerhamomyces anomaluskiller* yeast on the fermentation and chemical composition of apple wines. *FEMS Yeast Res.* **2014**, *14*, 729–740. [CrossRef] [PubMed]
61. Kurtzman, C.P. Chapter 80—*Wickerhamomyces* Kurtzman, Robnett & Basehoar-Powers (2008). In *The Yeasts*, 5th ed.; Elsevier Science: Amsterdam, The Netherlands, 2011; pp. 899–917. [CrossRef]
62. Roy, S.; Marzorati, S.; Schievano, A.; Pant, D. Microbial Fuel Cells. In *Encyclopedia of Sustainable Technologies*; Elsevier: Amsterdam, The Netherlands, 2017; pp. 245–259. [CrossRef]
63. Li, M.; Zhou, M.; Tian, X.; Tan, C.; McDaniel, C.T.; Hassett, D.J.; Gu, T. Microbial fuel cell (MFC) power performance improvement through enhanced microbial electrogenicity. *Biotechnol. Adv.* **2018**, *36*, 1316–1327. [CrossRef]

64. Hua, S.S.T.; Sarreal, S.B.L.; Chang, P.K.; Yu, J. Transcriptional regulation of aflatoxin biosynthesis and conidiation in Aspergillus flavus by *Wickerhamomyces anomalus* WRL-076 for reduction of aflatoxin contamination. *Toxins* **2019**, *11*, 81. [CrossRef]
65. Yaqoob, A.A.; Guerrero-Barajas, C.; Ibrahim, M.N.M.; Umar, K.; Yaakop, A.S. Local fruit wastes driven benthic microbial fuel cell: A sustainable approach to toxic metal removal and bioelectricity generation. *Environ. Sci. Pollut. Res.* **2022**, *29*, 32913–32928. [CrossRef] [PubMed]
66. Anjum, A.; Mazari, S.A.; Hashmi, Z.; Jatoi, A.S.; Abro, R. A review of role of cathodes in the performance of microbial fuel cells. *J. Electroanal. Chem.* **2021**, *899*, 115673. [CrossRef]
67. Anappara, S.; Senthilkumar, K.; Krishnan, H. Chapter 11—Nanomaterial and nanocatalysts in microbial fuel cells. In *Nanotechnology in Fuel Cells*; Elsevier: Amsterdam, The Netherlands, 2022; pp. 261–284.

Article

Introducing a Linear Empirical Correlation for Predicting the Mass Heat Capacity of Biomaterials

Reza Iranmanesh [1], Afham Pourahmad [2], Fardad Faress [3], Sevil Tutunchian [4], Mohammad Amin Ariana [5], Hamed Sadeqi [6], Saleh Hosseini [7], Falah Alobaid [8] and Babak Aghel [8,9,*]

[1] Faculty of Civil Engineering, K.N. Toosi University of Technology, Tehran 158754416, Iran
[2] Department of Polymer Engineering, Amirkabir University of Technology, Tehran 1591634311, Iran
[3] Department of Business, Data Analysis, The University of Texas Rio Grande Valley (UTRGV), Edinburg, TX 78539, USA
[4] Energy Institute, Energy Science and Technology Department, Istanbul Technical University, Istanbul 34469, Turkey
[5] Department of Petroleum Engineering, Gachsaran Branch, Islamic Azad University, Gachsaran 6387675818, Iran
[6] Department of Internet and Wide Network, Iran Industrial Training Center Branch, University of Applied Science and Technology, Tehran 1599665111, Iran
[7] Department of Chemical Engineering, University of Larestan, Larestan 7431813115, Iran
[8] Institut Energiesysteme und Energietechnik, Technische Universität Darmstadt, Otto-Berndt-Straße 2, 64287 Darmstadt, Germany
[9] Department of Chemical Engineering, Faculty of Energy, Kermanshah University of Technology, Kermanshah 6715685420, Iran
* Correspondence: babak.aghel@est.tu-darmstadt.de; Tel.: +49-6151-16-22673; Fax: +49-6151-16-22690

Citation: Iranmanesh, R.; Pourahmad, A.; Faress, F.; Tutunchian, S.; Ariana, M.A.; Sadeqi, H.; Hosseini, S.; Alobaid, F.; Aghel, B. Introducing a Linear Empirical Correlation for Predicting the Mass Heat Capacity of Biomaterials. *Molecules* **2022**, *27*, 6540. https://doi.org/10.3390/molecules27196540

Academic Editors: Mohamad Nasir Mohamad Ibrahim, Patricia Graciela Vázquez and Mohd Hazwan Hussin

Received: 5 September 2022
Accepted: 27 September 2022
Published: 3 October 2022

Publisher's Note: MDPI stays neutral with regard to jurisdictional claims in published maps and institutional affiliations.

Abstract: This study correlated biomass heat capacity (Cp) with the chemistry (sulfur and ash content), crystallinity index, and temperature of various samples. A five-parameter linear correlation predicted 576 biomass Cp samples from four different origins with the absolute average relative deviation (AARD%) of ~1.1%. The proportional reduction in error (REE) approved that ash and sulfur contents only enlarge the correlation and have little effect on the accuracy. Furthermore, the REE showed that the temperature effect on biomass heat capacity was stronger than on the crystallinity index. Consequently, a new three-parameter correlation utilizing crystallinity index and temperature was developed. This model was more straightforward than the five-parameter correlation and provided better predictions (AARD = 0.98%). The proposed three-parameter correlation predicted the heat capacity of four different biomass classes with residual errors between −0.02 to 0.02 J/g·K. The literature related biomass Cp to temperature using quadratic and linear correlations, and ignored the effect of the chemistry of the samples. These quadratic and linear correlations predicted the biomass Cp of the available database with an AARD of 39.19% and 1.29%, respectively. Our proposed model was the first work incorporating sample chemistry in biomass Cp estimation.

Keywords: biomass sample; heat capacity; empirical correlation; biomass crystallinity; feature reduction

1. Introduction

Global warming [1,2] and limitations of fossil fuel sources [3] have been two main problematic issues in recent decades. According to reports, the maximum allowable carbon dioxide (CO_2) concentration has exceeded 70 ppm in the atmosphere from the preindustrial period [4]. The combustion of coal and petroleum [5], natural gas industries [1], and petrochemical complexes are responsible for 80% of CO_2 emissions to the atmosphere [6,7]. Furthermore, cement, steel, and iron manufacturers are the subsequent sources of CO_2 emissions [4]. In this way, significant attention has been paid to carbon capture [8] and sequestration strategies [9] to reduce, control, and utilize greenhouse gases, including CO_2, methane, nitrogen, sulfur, chlorofluorocarbons, and so on [10,11]. To this end, according to the BLUE map scenario of the international energy agency [12], sustainable energy sources, including biomass [13], biogas [14], and solar energy [15], have been introduced as promising candidates to replace traditional fossil fuels.

Recently, biomass-to-energy processes have received growing interest because of the energy and global warming crises [16]. According to the United Nations Environment Program (UNEP) [17], 140 billion tons of biomass (mainly agricultural and wooden wastes) are produced throughout the world annually [18]. Wide ranges of added-value chemicals and biofuels may be synthesized from this low-cost, sustainable, and plentiful renewable feedstock [19]. A schematic illustration of synthesizing various products from lignocellulosic biomass is presented in Figure 1. Based on UNEP [12], around 20 times the available environmental yield of agricultural production technologies is required to achieve sustainable development in 2040 [20,21]. Accordingly, thermal processes of biomass, such as gasification and pyrolysis, have emerged as practical technologies to convert these types of biomass samples into valuable products [22,23]. It is worth noting that the mentioned processes include during the first pyrolysis step, in which biomass is converted to gas and a solid carbon in the presence of heat [24].

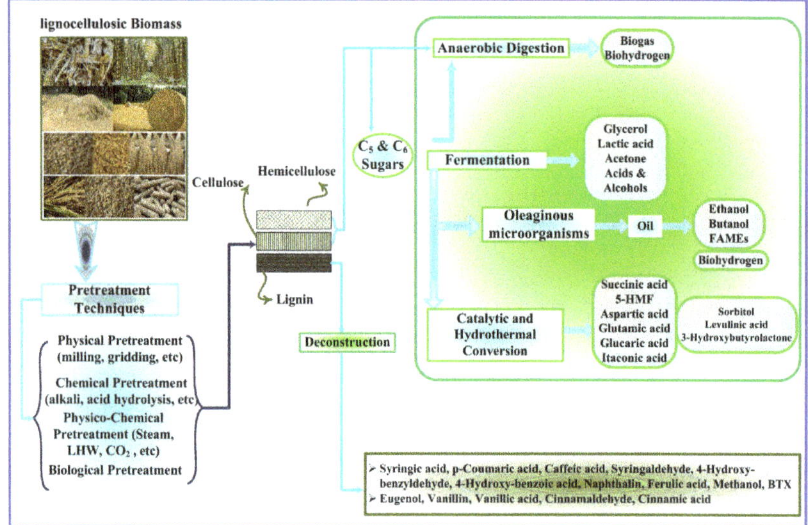

Figure 1. Schematic illustration of various valuable products from lignocellulosic biomass.

Reliable knowledge of the thermal characteristics of biomass is a crucial issue for biomass-to-energy process design [22]. Indeed, molecular kinetics govern the thermal behavior of biomass valorization processes [22,25]. Numerous experimental/modeling studies have been devoted to the determination of thermal properties of biomass, including elemental composition [26], higher heating value [27], thermal conductivity [28] and specific heat capacity [13].

Biomass heat capacity is often experimentally measured by differential scanning calorimetry (DSC) [29]. Although the DSC is an accurate method, utilizing a few milligrams of sample results in a shallow heat throws doubt on the measurement accuracy [22,24,25]. It is worth noting that measuring biomass heat capacity at a higher temperature than 423 K has some limitations due to sample decomposition [24]. Bitra et al. determined the heat capacity and thermal conductivity of kernels, peanut pods, and shells using a purpose-built vacuum flask calorimeter [30]. Furthermore, Mothée and De Miranda focused on the thermal analysis of sugarcane bagasse and coconut fiber as typical agricultural byproducts [31]. In addition, the heat capacity of cellulose regarding its applications in the pulp industry and tissue engineering has been investigated in many studies [32–34]. Ur'yash et al. employed the adiabatic calorimeter method to explain the water effect on biomass heat capacity in temperatures ranging from 80 K to 330 K [34]. Blokhin et al. experimentally measured the heat capacity of biomass samples obtained from four different sources [32].

Generally, laboratory-scale measurements are complicated and time-consuming, require economic expense, and often contain different uncertainty levels associated with human error and the wrong calibrations of apparatus. Furthermore, it is hard to directly incorporate experimentally measured data for computer-aided simulation purposes. Therefore, it is necessary to develop a correlation to estimate biomass heat capacity from some available features. This type of correlation reduces experimental cost, saves time, and can be easily coupled with computer-aided simulators. The heat capacity (Cp) of a wide range of pyrolysis chars obtained under conditions representative of industrial reactors was correlated in a temperature range of 40–80 °C [22]. Kollman and Cote suggested an empirical correlation for estimating the specific heat capacity of solid wood as a function of the temperature valid for 0–100 °C [35]. In another attempt, Gupta et al. proposed a model for softwood barks and their derived softwood chars using differential scanning calorimetry, showing acceptable Cp predictions in the temperature range of 40–140 °C [36]. The mathematical formulas of the literature suggesting correlations for estimating biomass heat capacity have been widely investigated in Section 2.2.

All these correlations only estimate the biomass heat capacity as a function of temperature and ignore the effect of biomass chemistry [22,35,36]. Since biomass chemistry influences heat capacity, it is necessary to include such information in the model's entry. Accordingly, this study developed a simple correlation to estimate biomass heat capacity by considering the sample's chemistry and operating conditions. To our knowledge, it is the first usage of the bio-sample composition (sulfur and ash contents), crystallinity index, and temperature to estimate the biomass heat capacity. This study also compared the accuracy of the developed correlation with those suggested in the literature. This type of correlation helped to enhance the understanding of biomass composition on heat capacity.

2. Results and Discussion

Multiple linear regression was employed to estimate the biomass heat capacity from the sample chemistry, crystallinity index, and temperature. Then, the proportional reduction in error was applied to simultaneously decrease the model size and increase its accuracy. Then, the accuracy of the constructed correlation was compared with those suggested in the literature. Finally, several graphical and numerical investigations were performed to monitor the prediction accuracy of the proposed model in real-field situations.

2.1. Developed Correlations in This Study

Equation (1) shows a simple linear correlation developed to estimate the heat capacity of biomass from four different origins in a temperature range of 81 to 368 K. The differential evolution algorithm [37] was used to adjust the coefficients of this correlation using 576 experimental datasets.

$$Cp^{cal} = -0.0158 \times CI - 0.0011 \times Ash + 0.022 \times S + 0.00407 \times T + 0.0165 \quad (1)$$

This correlation estimated the biomass heat capacity with the absolute average relative deviation (AARD%) of 1.1%. Equation (2) was used to calculate the AARD% from the experimental (Cp^{exp}) and calculated (Cp^{cal}) biomass heat capacities [38,39]. Here, N shows the number of available datasets, i.e., 576.

$$AARD\% = \sum_{i=1}^{N} 100 \times \left(\left| Cp_i^{exp} - Cp_i^{cal} \right| / Cp_i^{exp} \right) / N \quad (2)$$

The proportional reduction in error (PRE) is a practical method for conducting the sensitivity analysis on independent variables and ranking them based on their contribution to the model's accuracy [40]. The PRE uses the sum of squared errors (SSE) or mean squared errors (MSE) to conduct its duty. Equations (3) and (4) express the mathematical formulation of the SSE and MSE, respectively [41].

$$SSE = \sum_{i=1}^{N} \left(Cp_i^{\exp} - Cp_i^{cal} \right)^2 \tag{3}$$

$$MSE = \sqrt{\sum_{i=1}^{N} \left(Cp_i^{\exp} - Cp_i^{cal} \right)^2 / N} \tag{4}$$

The developed correlation (i.e., Equation (1)) has four parts, with each part including one independent variable. The PRE associated with each independent variable was obtained using Equation (4).

$$PRE = (SSE_{part} - SSE)/SSE_{part} \quad part = 1, 2, 3, 4 \tag{5}$$

Here, SSE and SSE_{part} indicate the sum of squared errors obtained by considering all terms (i.e., Equation (1)) and excluding the ith part of the original correlation, respectively. A high PRE shows that a considered independent variable has a more substantial role in model accuracy and vice versa.

Figure 2 represents the PRE related to the crystallinity index, ash and sulfur contents of the bio-samples, and temperature.

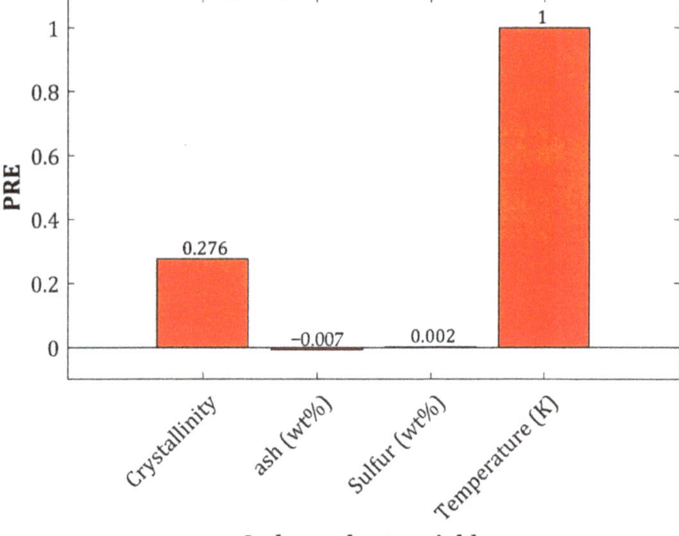

Figure 2. The relative importance of each feature on the prediction accuracy of the biomass heat capacity.

This figure justifies that the temperature and crystallinity index are the main parameters that improve the model's accuracy. On the other hand, the ash and sulfur contents only enlarged the model size and had a low contribution to its accuracy.

The previous analysis approved that it was better not to consider the ash and sulfur contents of the bio-samples in the model development, and re-design the correlation solely based on the temperature and crystallinity index. Equation (5) presents the mathematical shape of the developed correlation utilizing the most important independent variables.

$$Cp^{cal} = -0.0156 \times CI + 0.00407 \times T + 0.0162 \tag{6}$$

This three-parameter correlation was more straightforward than the previous five-parameter model, needing low entry information, and providing higher accuracy, i.e., AARD = 0.98%.

2.2. Comparison with the Literature Suggested Models

It was previously explained that the available correlations in the literature estimated the biomass heat capacity by either quadratic or linear relationships with the temperature. Therefore, it was necessary to adjust the coefficients of Equation (6) utilizing the available experimental database. It is worth noting that A equaled zero for the linear correlation.

$$Cp^{cal} = A \times T^2 + B \times T + C \tag{7}$$

Table 1 presents the numerical values of the adjustable coefficients of the linear and quadratic correlations based on the temperature. The last column of Table 1 indicates that the linear and quadratic correlations predict the heat capacity of the 576 bio-samples with the AARD of 1.29% and 39.19%, respectively. Not considering the effect of bio-sample chemistry on the heat capacity may have been responsible for this non-logical uncertainty [22,35,36].

Table 1. Adjusted coefficients and AARD% of the linear and quadratic correlations developed based on the temperature.

Correlation	A	B	C	AARD%
Linear	0	0.00406	0.0061	1.29
Quadratic	6.11×10^{-5}	-0.0210	2.232	39.19

2.3. Visually Inspecting the Performance of the Three-Parameter Correlation

Up to now, it has been approved that the three-parameter linear correlation based on the crystallinity index and temperature has the highest accurate prediction for biomass heat capacity. This section employs several graphical analyses to inspect the model performance further. The compatibility between experimental and calculated heat capacities for different biomass samples is depicted in Figure 3. It can be concluded that an outstanding level of agreement existed between the laboratory-measured and calculated heat capacities. The developed correlation encountered problems in predicting the fourth class's biomass heat capacity with excellent accuracy (wood amorphous cellulose).

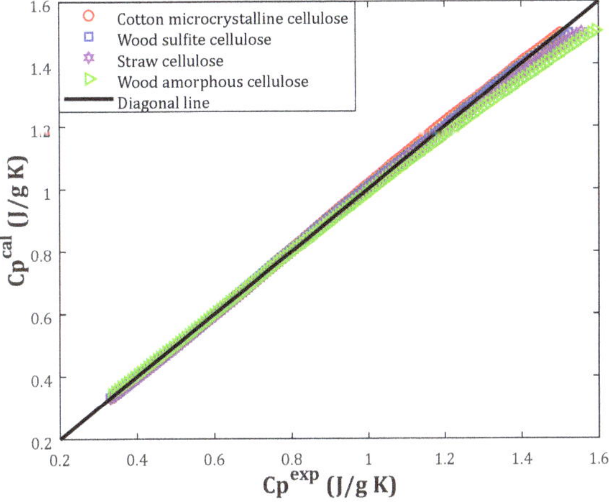

Figure 3. Correlation between actual and estimated heat capacities.

Since the crystallinity index of this biomass type was zero, the first part of the developed correlation diminished (i.e., Equation (5)). Indeed, the heat capacity of the biomass obtained from wood amorphous cellulose was estimated using a linear correlation solely based on the temperature. This explanation may justify the uncertainty observed in predicting the biomass heat capacity of the fourth class, especially at higher values.

The histogram of the arithmetic deviation between experimental and calculated biomass heat capacities (i.e., residual error) is illustrated in Figure 4. This figure states that the heat capacity of four different biomass classes was estimated with infinitesimal residual errors, mainly between -0.02 to 0.02 J/g·K. It was interesting to see that 200 biomass heat capacities were calculated with the residual error equaling zero.

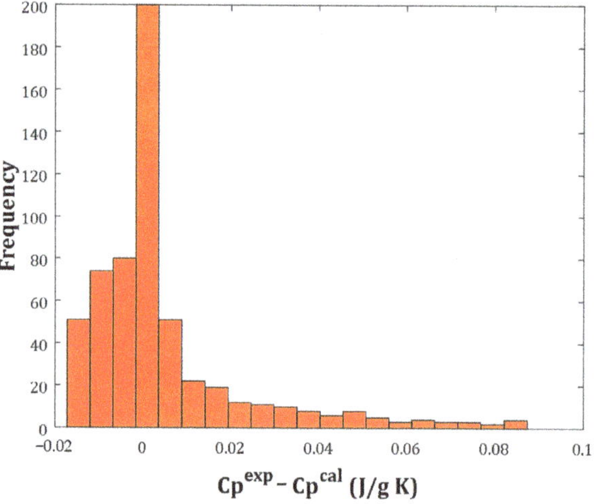

Figure 4. Histogram of the residual error between actual and predicted heat capacities.

Figure 5 shows the AARD observed for predicting the heat capacity of each biomass class. The heat capacity of the first to the fourth biomass classes was estimated with the AARD of 0.86%, 0.53%, 0.83% and 1.71%, respectively. The developed correlation presented the overall AARD = 0.98% for estimating the heat capacity of all 576 bio-samples.

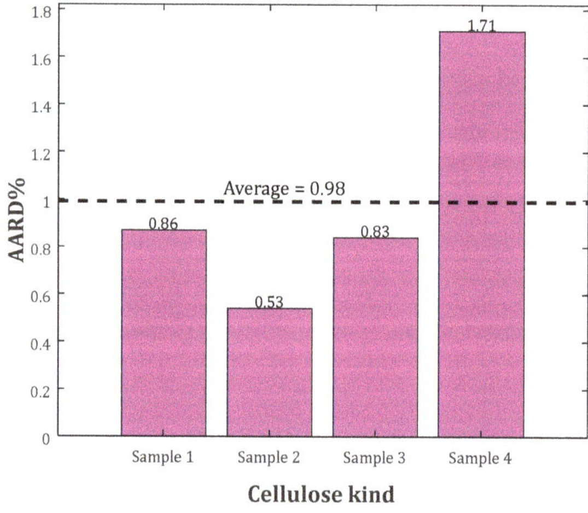

Figure 5. The AARD% associated with the prediction of the heat capacity of each biomass kind.

2.4. Analyzing the Cp of Bio-Samples with Different Origins

The experimental and modeling profiles of heat capacity versus temperature for the considered biomass classes are plotted in Figure 6a,d. Excellent compatibility between the laboratory measurements and the correlation predictions can be concluded from these figures. Furthermore, it can be seen that biomass heat capacities have a substantial direct relationship with temperature.

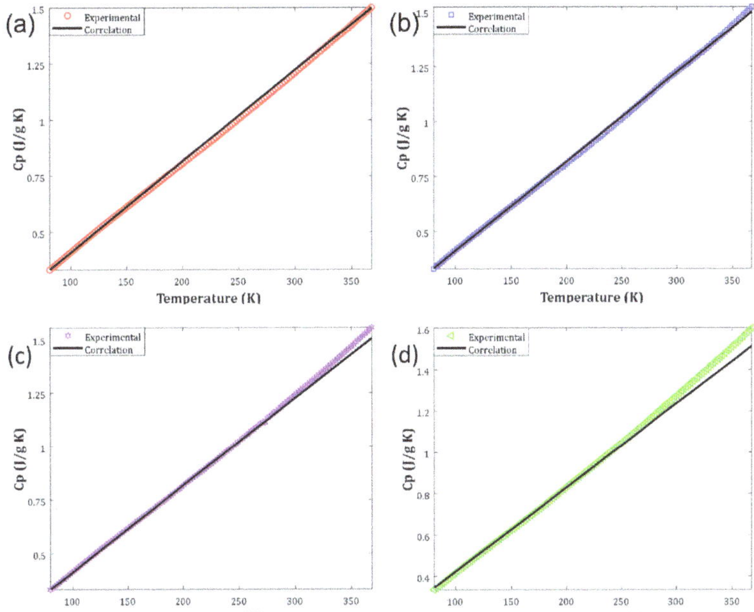

Figure 6. (**a**) Performance of the developed correlation for calculating the Cp versus temperature profile of the cotton microcrystalline cellulose. (**b**) The calculated and experimental profiles of the Cp versus temperature of the wood sulfite cellulose. (**c**)Investigating the temperature effect on the straw cellulose heat capacity from the experimental and modeling observations. (**d**) Performance of the developed correlation for approximating the Cp dependency of the wood amorphous cellulose on the temperature.

It should be highlighted that by reducing the crystallinity index from the first to the last biomass classes, the first term of the developed correlation (i.e., Equation (5)) gradually became weaker. As mentioned previously, the developed correlation finally appeared as a linear model based on the temperature only for the fourth biomass class (i.e., wood amorphous cellulose). The deviation between experimental and calculated heat capacities increased by decreasing the crystallinity index (compare Figure 6a–d). Although the observed deviation level was too small to distort the generalization ability of the developed correlation, it approved the importance of incorporating the sample's chemistry in the heat capacity estimation phase.

2.5. Pair Effect of Temperature and Crystallinity Index on the Biomass Cp

Furthermore, the variation in heat capacity of biomass based on temperature and crystallinity index is plotted in Figure 7. As can be observed, there is a linear relation between employed variables and related heat capacity. Additionally, temperature enhancement has a significant effect on the biomass heat capacity than crystallinity index, which was already proved by PRE analysis (Figure 2).

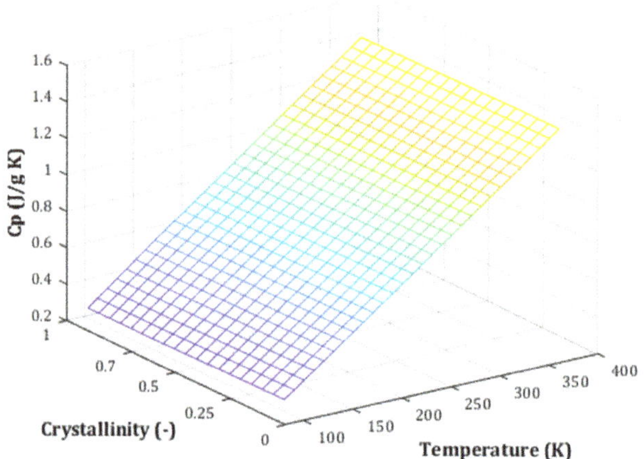

Figure 7. Color map plot of heat capacity of biomass based on temperature and crystallinity index variations.

2.6. Summary of the Study in the Flowchart Form

A summary of the steps followed in this study to develop and validate a linear correlation to estimate the biomass heat capacity is graphically introduced in Figure 8.

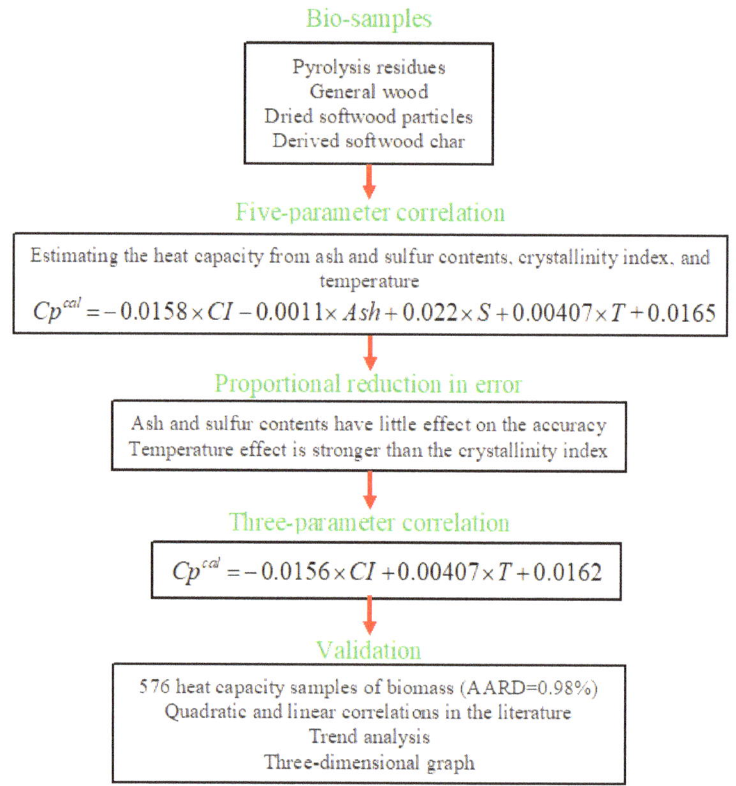

Figure 8. A flowchart of suggested methodology for estimating the biomass heat capacity.

3. Materials and Methods

This section presents experimental biomass heat capacities collected from the literature. Furthermore, the literature suggesting correlations for estimating biomass heat capacity has been reviewed.

3.1. Collected Database

This study focused on constructing a correlation to estimate the heat capacity of biomass from four origins, including cotton microcrystalline cellulose (sample 1), wood sulfite cellulose (sample 2), straw cellulose (sample 3), and wood amorphous cellulose (sample 4), were considered [32]. The chemical composition of bio-samples, crystallinity index, and temperature were considered in this estimation. Table 2 introduces numerical ranges of sulfur and ash mass fractions, crystallinity index, temperature, and the associated heat capacity of the considered bio-samples [32]. A multiple regression method was applied to linearly relate bio-sample heat capacity (Cp^{cal}) to the ash (Ash) and sulfur (S) contents, crystallinity index (CI), and temperature (T) based on Equation (8).

$$Cp^{cal} = f\left(CI,\ Ash,\ S,\ T\right) \tag{8}$$

Table 2. The heat capacity versus biomass chemical composition, crystallinity index and temperature [32].

Biomass Type	CI (-)	Temperature (K)	Ash (wt%)	S (wt%)	Cp (J/g·K)	Number Data
Sample 1	0.90	81.50–367.50	0.10	0.02	0.3335–1.500	143
Sample 2	0.80	80.73–367.40	0.10	0.43	0.3304–1.521	144
Sample 3	0.74	80.53–368.25	0.49	0.11	0.3314–1.554	145
Sample 4	0	80.61–368.09	0.07	0.02	0.3342–1.602	144

It should be mentioned that all constructed correlations in this study were only valid for estimating the heat capacity of those bio-samples with the composition listed in Table 2, in the temperature range of 80.53 to 368.25 K.

3.2. Literature Suggested Correlations

As previously mentioned, the literature suggested several correlations for estimating the bio-sample heat capacity. Table 3 summarizes the mathematical formulations of these correlations and the range of applications. This table shows that all the developed correlations only use temperature to estimate heat capacity. Generally, these correlations use quadratic or linear forms to correlate the bio-sample heat capacity to the temperature.

Table 3. Developed correlations in the literature to estimate biomass heat capacity.

Material	Correlation Shape	Temperature	Ref.
Various biomass	$Cp = 0.00534 \times T - 299$	40–80 °C	[22]
Pyrolysis residues	$Cp = 0.0014 \times T + 688$	40–80 °C	[22]
General wood	$Cp = 0.0046 \times T - 0.113$	0–100 °C	[35]
Dried softwood particles	$Cp = 0.00546 \times T - 0.524$	40–140 °C	[36]
Derived softwood char	$Cp = -3.8 \times 10^{-6} \times T^2 + 0.00598 \times T - 795$	40–140 °C	[36]

Since these correlations were only valid to estimate the heat capacity of the considered biomass in the specific operating ranges, their coefficients were needed to readjust to cover the utilized database in the current study (see Section 3.2).

4. Conclusions

This study aimed to develop an accurate and straightforward correlation for estimating biomass heat capacity, considering bio-sample chemistry and operating condition. The proportional reduction in error justified that temperature and crystallinity index of bio-samples were the most critical factors affecting biomass heat capacity. Multiple linear regression was applied to develop a three-parameter correlation based on the selected features. The proposed model was more accurate than the quadratic and linear models available in the literature. This model predicted 576 biomass heat capacities with the AARD = 0.98%, while the previously suggested models presented higher uncertainty for the same database. The constructed correlation in this study predicted the heat capacity of biomass samples from four different origins with an excellent AARD of lower than 1.71%. The sensitivity analysis approved that the deviation between experimental and calculated heat capacities increased by reducing the crystallinity index of biomass samples. This observation indicated the importance of incorporating biomass chemistry in the estimating phase of the heat capacity. Since this study estimated the heat capacity of biomass samples with low ash content, it is a good idea to consider straw/digestate as the independent variable in future research, where the influence of ash on the specific heat capacity would not be neglectable.

Author Contributions: R.I.: Writing—original draft, Writing—review & editing, Methodology; A.P.: Writing—original draft, Writing—review & editing, Conceptualization; F.F.: Writing—original draft, Writing—review & editing, Resources, Investigation; S.T.: Writing – original draft, Writing—review & editing, Validation; M.A.A.: Writing—original draft, Writing – review & editing, Methodology; H.S.: Writing—original draft, Writing—review & editing, Investigation; S.H.: Writing—original draft, Writing—review & editing; F.A.: Resources, Supervision; B.A.; Writing—Review. All authors have read and agreed to the published version of the manuscript.

Funding: This research received no external funding.

Institutional Review Board Statement: Not applicable.

Informed Consent Statement: Not applicable.

Data Availability Statement: The study data analyzed in this article can be obtained by request from the corresponding author.

Acknowledgments: We acknowledge support by the Deutsche Forschungsgemeinschaft (DFG—German Research Foundation) and the Open Access Publishing Fund of Technical University of Darmstadt.

Conflicts of Interest: The authors declare no conflict of interest.

Sample Availability: The study data analyzed in this article can be obtained by request from the corresponding author.

References

1. Karimi, M.; Rahimpour, M.R.; Rafiei, R.; Jafari, M.; Iranshahi, D.; Shariati, A. Reducing environmental problems and increasing saving energy by proposing new configuration for moving bed thermally coupled reactors. *J. Nat. Gas Sci. Eng.* **2014**, *17*, 136–150. [CrossRef]
2. Karimi, M.; Diaz de Tuesta, J.L.; Carmem, C.N.; Gomes, H.T.; Rodrigues, A.E.; Silva, J.A.C. Compost from Municipal Solid Wastes as a Source of Biochar for CO_2 Capture. *Chem. Eng. Technol.* **2020**, *43*, 1336–1349. [CrossRef]
3. Vaferi, B.; Eslamloueyan, R.; Ghaffarian, N. Hydrocarbon reservoir model detection from pressure transient data using coupled artificial neural network—Wavelet transform approach. *Appl. Soft Comput. J.* **2016**, *47*, 63–75. [CrossRef]
4. Karimi, M.; Silva, J.A.C.; Gonçalves, C.N.D.P.; Diaz De Tuesta, J.L.; Rodrigues, A.E.; Gomes, H.T. CO_2 Capture in Chemically and Thermally Modified Activated Carbons Using Breakthrough Measurements: Experimental and Modeling Study. *Ind. Eng. Chem. Res.* **2018**, *57*, 11154–11166. [CrossRef]
5. Esmaeili-Faraj, S.H.; Vaferi, B.; Bolhasani, A.; Karamian, S.; Hosseini, S.; Rashedi, R. Design of a Neuro-Based Computing Paradigm for Simulation of Industrial Olefin Plants. *Chem. Eng. Technol.* **2021**, *44*, 1382–1389. [CrossRef]
6. Karimi, M.; Zafanelli, L.F.A.S.; Almeida, J.P.P.; Ströher, G.R.; Rodrigues, A.E.; Silva, J.A.C. Novel Insights into Activated Carbon Derived from Municipal Solid Waste for CO_2 Uptake: Synthesis, Adsorption Isotherms and Scale-up. *J. Environ. Chem. Eng.* **2020**, *8*, 104069. [CrossRef]

7. Henrique, A.; Karimi, M.; Silva, J.A.C.; Rodrigues, A.E. Analyses of Adsorption Behavior of CO_2, CH_4, and N_2 on Different Types of BETA Zeolites. *Chem. Eng. Technol.* **2019**, *42*, 327–342. [CrossRef]
8. Aghel, B.; Behaein, S.; Alobaid, F. CO_2 capture from biogas by biomass-based adsorbents: A review. *Fuel* **2022**, *328*, 125276. [CrossRef]
9. Karimi, M.; Shirzad, M.; Silva, J.A.C.; Rodrigues, A.E. Biomass/Biochar carbon materials for CO_2 capture and sequestration by cyclic adsorption processes: A review and prospects for future directions. *J. CO_2 Util.* **2022**, *57*, 101890. [CrossRef]
10. Karimi, M.; Rodrigues, A.E.; Silva, J.A.C. Designing a simple volumetric apparatus for measuring gas adsorption equilibria and kinetics of sorption. Application and validation for CO_2, CH_4 and N_2 adsorption in binder-free beads of 4A zeolite. *Chem. Eng. J.* **2021**, *425*, 130538. [CrossRef]
11. Esmaeili-Faraj, S.H.; Hassanzadeh, A.; Shakeriankhoo, F.; Hosseini, S.; Vaferi, B. Diesel fuel desulfurization by alumina/polymer nanocomposite membrane: Experimental analysis and modeling by the response surface methodology. *Chem. Eng. Process.-Process Intensif.* **2021**, *164*, 108396. [CrossRef]
12. Figueroa, J.D.; Fout, T.; Plasynski, S.; McIlvried, H.; Srivastava, R.D. Advances in CO_2 capture technology-The U.S. Department of Energy's Carbon Sequestration Program. *Int. J. Greenh. Gas Control* **2008**, *2*, 9–20. [CrossRef]
13. Karimi, M.; Hosin Alibak, A.; Seyed Alizadeh, S.M.; Sharif, M.; Vaferi, B. Intelligent modeling for considering the effect of bio-source type and appearance shape on the biomass heat capacity. *Meas. J. Int. Meas. Confed.* **2022**, *189*, 110529. [CrossRef]
14. Rafiee, A.; Khalilpour, K.R.; Prest, J.; Skryabin, I. Biogas as an energy vector. *Biomass Bioenergy* **2021**, *144*, 105935. [CrossRef]
15. Cao, Y.; Kamrani, E.; Mirzaei, S.; Khandakar, A.; Vaferi, B. Electrical efficiency of the photovoltaic/thermal collectors cooled by nanofluids: Machine learning simulation and optimization by evolutionary algorithm. *Energy Rep.* **2022**, *8*, 24–36. [CrossRef]
16. da Costa, T.P.; Quinteiro, P.; da Cruz Tarelho, L.A.; Arroja, L.; Dias, A.C. Environmental impacts of forest biomass-to-energy conversion technologies: Grate furnace vs. fluidised bed furnace. *J. Clean. Prod.* **2018**, *171*, 153–162. [CrossRef]
17. UNEP. Converting Waste Agricultural Biomass into a Resource. In *Compendium of Technologies. Osaka, United Nations Environment Programme*; United Nations Environment Programme (UNEP): Nairobi, Kenya, 2009; pp. 1–437.
18. Dobele, G.; Dizhbite, T.; Gil, M.V.; Volperts, A.; Centeno, T.A. Production of nanoporous carbons from wood processing wastes and their use in supercapacitors and CO_2 capture. *Biomass Bioenergy* **2012**, *46*, 145–154. [CrossRef]
19. Yoo, C.G.; Meng, X.; Pu, Y.; Ragauskas, A.J. The critical role of lignin in lignocellulosic biomass conversion and recent pretreatment strategies: A comprehensive review. *Bioresour. Technol.* **2020**, *301*, 122784. [CrossRef]
20. Tan, J.; Li, Y.; Tan, X.; Wu, H.; Li, H.; Yang, S. Advances in Pretreatment of Straw Biomass for Sugar Production. *Front. Chem.* **2021**, *9*, 696030. [CrossRef]
21. Singh, S.; Sinha, R.; Kundu, S. Role of organosolv pretreatment on enzymatic hydrolysis of mustard biomass for increased saccharification. *Biomass Convers. Biorefinery* **2022**, *12*, 1657–1668. [CrossRef]
22. Dupont, C.; Chiriac, R.; Gauthier, G.; Toche, F. Heat capacity measurements of various biomass types and pyrolysis residues. *Fuel* **2014**, *115*, 644–651. [CrossRef]
23. Lewandowski, W.M.; Ryms, M.; Kosakowski, W. Thermal biomass conversion: A review. *Processes* **2020**, *8*, 516. [CrossRef]
24. Karimi, M.; Aminzadehsarikhanbeglou, E.; Vaferi, B. Robust intelligent topology for estimation of heat capacity of biochar pyrolysis residues. *Meas. J. Int. Meas. Confed.* **2021**, *183*, 109857. [CrossRef]
25. Shirzad, M.; Karimi, M.; Silva, J.A.C.; Rodrigues, A.E. Moving Bed Reactors: Challenges and Progress of Experimental and Theoretical Studies in a Century of Research. *Ind. Eng. Chem. Res.* **2019**, *58*, 9179–9198. [CrossRef]
26. Olatunji, O.O.; Akinlabi, S.; Madushele, N.; Adedeji, P.A. Estimation of the Elemental Composition of Biomass Using Hybrid Adaptive Neuro-Fuzzy Inference System. *Bioenergy Res.* **2019**, *12*, 642–652. [CrossRef]
27. Suleymani, M.; Bemani, A. Application of ANFIS-PSO algorithm as a novel method for estimation of higher heating value of biomass. *Energy Sources, Part A Recover. Util. Environ. Eff.* **2018**, *40*, 288–293. [CrossRef]
28. Zhu, L.T.; Gu, Y.F.; Wu, G.S. Biomass thermal conductivity measurement system design. *J. For. Eng.* **2020**, *5*, 97–102. [CrossRef]
29. Pecchi, M.; Patuzzi, F.; Benedetti, V.; Di Maggio, R.; Baratieri, M. Kinetic analysis of hydrothermal carbonization using high-pressure differential scanning calorimetry applied to biomass. *Appl. Energy* **2020**, *265*, 114810. [CrossRef]
30. Bitra, V.S.P.; Banu, S.; Ramakrishna, P.; Narender, G.; Womac, A.R. Moisture dependent thermal properties of peanut pods, kernels, and shells. *Biosyst. Eng.* **2010**, *106*, 503–512. [CrossRef]
31. Mothé, C.G.; De Miranda, I.C. Characterization of sugarcane and coconut fibers by thermal analysis and FTIR. *J. Therm. Anal. Calorim.* **2009**, *97*, 661–665. [CrossRef]
32. Voitkevich, O.V.; Kabo, G.J.; Blokhin, A.V.; Paulechka, Y.U.; Shishonok, M.V. Thermodynamic properties of plant biomass components. Heat capacity, combustion energy, and gasification equilibria of lignin. *J. Chem. Eng. Data* **2012**, *57*, 1903–1909. [CrossRef]
33. Kiiskinen, H.T.; Lyytikäinen, K.; Hämäläinen, J.P. Specific heats of dry scandinavian wood pulps. *J. Pulp Pap. Sci.* **1998**, *24*, 219–223.
34. Uryash, V.F.; Larina, V.N.; Kokurina, N.Y.; Novoselova, N.V. The thermochemical characteristics of cellulose and its mixtures with water. *Russ. J. Phys. Chem. A* **2010**, *84*, 915–921. [CrossRef]
35. Wilfred Jr, A.; Kollmann, F.F.P. *Principles of Wood Science and Technology*; Springer: Berlin/Heidelberg, Germany, 1968.
36. Gupta, M.; Yang, J.; Roy, C. Specific heat and thermal conductivity of softwood bark and softwood char particles. *Fuel* **2003**, *82*, 919–927. [CrossRef]

37. Iranshahi, D.; Jafari, M.; Rafiei, R.; Karimi, M.; Amiri, S.; Rahimpour, M.R. Optimal design of a radial-flow membrane reactor as a novel configuration for continuous catalytic regenerative naphtha reforming process considering a detailed kinetic model. *Int. J. Hydrog. Energy* **2013**, *38*, 8384–8399. [CrossRef]
38. Alibak, A.H.; Khodarahmi, M.; Fayyazsanavi, P.; Alizadeh, S.M.; Hadi, A.J.; Aminzadehsarikhanbeglou, E. Simulation the adsorption capacity of polyvinyl alcohol/carboxymethyl cellulose based hydrogels towards methylene blue in aqueous solutions using cascade correlation neural network (CCNN) technique. *J. Clean. Prod.* **2022**, *337*, 130509. [CrossRef]
39. Madani, M.; Lin, K.; Tarakanova, A. DSResSol: A sequence-based solubility predictor created with dilated squeeze excitation residual networks. *Int. J. Mol. Sci.* **2021**, *22*, 13555. [CrossRef]
40. Judd, C.M.; McClelland, G.H.; Ryan, C.S. *Data Analysis: A Model Comparison Approach*; Routledge: Oxfordshire, UK, 2011; ISBN 0203892054.
41. Hosseini, S.; Vaferi, B. Determination of methanol loss due to vaporization in gas hydrate inhibition process using intelligent connectionist paradigms. *Arab. J. Sci. Eng.* **2022**, *47*, 5811–5819. [CrossRef]

Article

Nypa fruticans Frond Waste for Pure Cellulose Utilizing Sulphur-Free and Totally Chlorine-Free Processes

Evelyn [1,*], Sunarno [1], David Andrio [1], Azka Aman [1,2] and Hiroshi Ohi [3]

[1]   Department of Chemical Engineering, University of Riau, Pekanbaru 28293, Indonesia
[2]   Kerinci Mill, PT. Riau Andalan Pulp and Paper, Pangkalan, Kerinci 28300, Indonesia
[3]   Faculty of Life and Environmental Science, University of Tsukuba, 1-1-1 Tennodai, Tsukuba 305-8572, Japan
*   Correspondence: evelyn@eng.unri.ac.id

Abstract: The search for alternative methods for the production of new materials or fuel from renewable and sustainable biomass feedstocks has gained increasing attention. In this study, *Nypa fruticans* (nipa palm) fronds from agricultural residues were evaluated to produce pure cellulose by combining prehydrolysis for 1–3 h at 150 °C, sulfur-free soda cooking for 1–1.5 h at 160 °C with 13–25% active alkali (AA), 0.1% soluble anthraquinone (SAQ) catalyst, and three-stage totally chlorine-free (TCF) bleaching, namely oxygen, peroxymonosulfuric acid, and alkaline hydrogen peroxide stages. The optimal conditions were 3 h prehydrolysis and 1.5 h cooking with 20% AA. Soda cooking with SAQ was better than the kraft and soda process without SAQ. The method decreased the kappa number as a residual lignin content index of pulp from 13.4 to 9.9–10.2 and improved the yields by approximately 6%. The TCF bleaching application produced pure cellulose with a brightness of 92.2% ISO, 94.8% α-cellulose, viscosity of 7.9 cP, and 0.2% ash content. These findings show that nipa palm fronds can be used to produce pure cellulose, serving as a dissolving pulp grade for viscose rayon and cellulose derivatives.

Keywords: *Nypa fruticans* fronds; cellulose; soda cooking; peroxymonosulfuric acid; alkaline hydrogen peroxide

Citation: Evelyn; Sunarno; Andrio, D.; Aman, A.; Ohi, H. *Nypa fruticans* Frond Waste for Pure Cellulose Utilizing Sulphur-Free and Totally Chlorine-Free Processes. *Molecules* **2022**, *27*, 5662. https://doi.org/10.3390/molecules27175662

Academic Editors: Alicia Prieto, Mohamad Nasir Mohamad Ibrahim, Patricia Graciela Vázquez and Mohd Hazwan Hussin

Received: 5 July 2022
Accepted: 30 August 2022
Published: 2 September 2022

Publisher's Note: MDPI stays neutral with regard to jurisdictional claims in published maps and institutional affiliations.

1. Introduction

Nypa fruticans (nipa palm) is a widely grown palm on peat soils and wetlands in several countries, including India, Myanmar, Thailand, Malaysia, Indonesia, Borneo, Philippines, the Ryukyu Islands, Papua New Guinea, countries in West Africa, the Solomon Islands, and northern Australia [1,2]. Previous studies also revealed that it has several advantages, such as the use of its leaves as traditional thatching materials [3]. Furthermore, this plant has been utilized for food crops in agriculture, such as the sap, which is often used for the production of toddy, vinegar, and sugar. It has also been used to produce fruit preserves, which serve as a local dessert due to their high sucrose, glucose, and fructose content [4,5]. The sap is considered a potential feedstock for ethanol or butanol production [3,5]. It can serve as a raw material for the production of medium-density fiberboards [6], heavy metal adsorbents [7], pulping materials [8], cellulose derivatives [9], and fuels [3]. The utilization of nipa palm for medicine has also been registered [10]. However, most of its parts are left as residues and can serve as a biomass source for fuels or producing new materials. It has also been considered a source of biomass for renewable energy, along with oil palm [11]. A previous study revealed that its total cultivation area in Indonesia is approximately 700,000 ha [12].

At present, non-wood plant fibers are often considered an important source of materials for the production of pulp and its derivatives [13]. They can be classified into agricultural residues including sugarcane bagasse, corn stalks, rice straw, wheat straw, and cereal straw, as well as natural growing plants, such as bamboo, reeds, and sabai grass.

These fibers can also be categorized as non-wood crops, such as jute, cotton fiber, ramie, and cotton linters [14]. A previous study revealed that 80% of China and India's total pulp was obtained from non-wood materials [15]. The utilization of their fibers was facilitated by the rapid growth of the pulp and paper industry, a shortage of wood fibers, the level of technical equipment, environmental contamination, and their abundant availability [14,15]. Therefore, it is important to find suitable resources, the most efficient methods, and the most environmentally friendly production techniques.

Lignocellulosic raw materials containing ≥34% α-cellulose are considered suitable for pulp and paper production [16]. Previous studies reported that different parts of nipa palm including the fronds, husk, shell, and leaf have an α-cellulose content of 28.9–48.2%; hence, they are potential raw materials for fuels and chemicals [5,17], as shown in Figure 1. The slenderness ratio, namely the fiber length/fiber diameter of its fronds and petiole, is very high compared to other non-wood fibers, such as bagasse and jute fiber, which indicates that it has high strength properties [1,18].

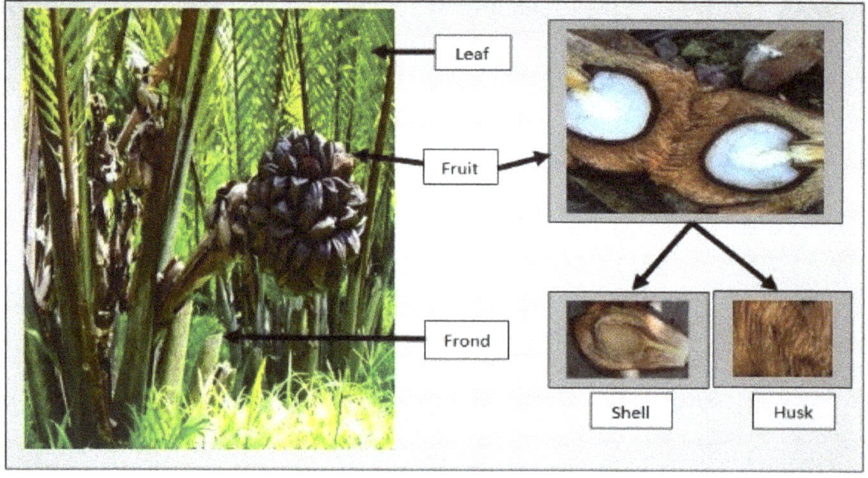

Figure 1. Potential parts of the nipa palm as raw materials for fuels and chemicals.

Dissolving/bright pulp is currently an important commodity in the global market, which contains >90% pure cellulose, <6–7% hemicellulose, and a low lignin content with a high brightness of >90% [19,20]. It is often used in the production of textile fibers, such as viscose rayon and lyocell as well as other derivatives, including cellulose nitrate, cellulose acetate, carboxymethylcellulose, and ethylcellulose. The high cellulose content can be obtained through the addition of a prehydrolysis step to extract most of the hemicelluloses before kraft, alkaline sulfite, or soda cooking. The process is then continued with bleaching or purification. The separated hemicellulose, namely hexose and pentose, can potentially be converted into valuable products, such as furfural, xylitol, and ethanol, which are often called potential biorefinery units [21].

Several studies explored the beneficial effect of anthraquinone (AQ) as a catalyst in the pulp and paper industries [22,23]. Furthermore, the mechanism of AQ catalysis is illustrated as a redox cycle. AQ is reduced to anthrahydroquinone (AHQ) by carbohydrates in the wood, and the product is oxidized back to the anthraquinone through a reaction, which involves reducing the active site (ß-O-4) in lignin during alkaline cooking [24]. Several studies also reported the benefits of AQ addition in pulping, namely a lower kappa number and improved pulp yield for kraft/soda cooking of hardwood and non-wood materials [25–28]. Utami et al. [29] revealed the advantages of prehydrolysis followed by soda with soluble anthraquinone (soda-SAQ) cooking over kraft cooking for *Acacia crassicarpa*, namely a higher pulp yield of 1.8% and a higher brightness level by 3.8–3.9 points. Jahan et al. [30]

also stated that the method led to additional decreases of 1.2–6.3 points in the kappa numbers of *Acacia auriculiformis* compared to the kraft and soda processes after bleaching. These studies reported higher yields of soda-SAQ compared to kraft cooking, namely by 2.8–4.3%. However, Salaghi et al. [31] did not observe the advantages of prehydroyzed soda-SAQ over prehydrolyzed kraft for *Eucalyptus globulus* wood. All of these dissolving pulp studies only used hardwood materials, and hence no study has compared kraft, soda, and soda-SAQ pulping to produce purified pulp or dissolving pulp using non-wood materials.

There are only a few studies on the potential of totally chlorine-free (TCF) bleaching with peroxymonosulfuric acid (P_{sa}) to produce high-quality dissolving pulp [23–25]. Soda-SAQ cooking and TCF bleaching are considered environmentally friendly due to the sulfur-free process in the cooking and reduction in the halogenated organic compounds emissions during bleaching [32–34]. Jahan et al. [8] and Dewi et al. [12] evaluated *Nypa fruticans* wastes, namely nipa palm fronds (NPFs) and petioles, as pulping materials without prehydrolysis and bleaching. A previous study obtained a low pulp yield of 36.2–37.2% based on the raw materials as well as a high residual lignin content of 27.2–29.1 using the soda-AQ and kraft processes because the conditions were not optimal. Other studies only performed soda processing on nipa palm petioles without the prehydrolysis and bleaching processes, and the results showed a 27.9% yield and 7.2 pulp lignin content. This implies that there are no reports on the production of pure cellulose as a dissolving pulp grade from nipa palm using a process which includes the prehydrolysis and purification stages. The final properties and performance of pure cellulose/derivatives were affected by the raw material, processes, and other previously reported factors. Therefore, this study aims to find efficient and safe methods to produce dissolving pulp from NPF fibers.

NPFs obtained as agricultural residues were processed using prehydrolysis followed by sulfur-free cooking and TCF bleaching to produce pure cellulose as dissolving pulp grades. The chemical characterization of the sample was carried out along with the determination of the optimum prehydrolysis conditions. The possibility of pure cellulose production using a soda process with an AQ-catalyst (SAQ) and modified P_{sa}-TCF bleaching was also performed. This is the first study to report the production of pure cellulose from NPF fiber using these processes. Therefore, this study aims to: (i) determine and compare the chemical composition of NPF used for pure cellulose; (ii) investigate the effect of prehydrolysis time of 1–3 h at 150 °C on kappa number, which comprises the residual lignin content and pulp yield after soda-SAQ cooking; (iii) compare the effect of cooking methods, namely kraft, soda, and soda-SAQ at 160 °C, 1–1.5 h cooking time, and 13–25% active alkali/AA dosage on pulp yields and kappa number; and (iv) apply three stages of modified P_{sa}-TCF bleaching, namely oxygen, P_{sa}, and alkaline hydrogen peroxide to the selected soda-SAQ pulp and compare the product to the dissolving pulp standard.

2. Results and Discussion

2.1. Potential Characteristics of Nypa fruticans Fronds for Pure Cellulose

Table 1 shows that NPFs contain 61.3 ± 2.4% holocellulose, 24.0 ± 1.9% pentosan, 17.5 ± 0.7% acid-insoluble lignin, 0.8 ± 0.5% acid-soluble lignin, 1.5 ± 0.6% extractives, and 16.5 ± 2.2% ash. Furthermore, a previous study on the characterization of NPFs for pulp showed similar results for hollocellulose, pentosan, lignin, and extractives [17]. High carbohydrate contents of lignocellulosic materials have the potential to be exploited as raw materials for pulp production, fuels, and chemicals. The obtained 60% holocellulose content is comparable to that of several non-wood plants, such as vine stem [35], grapevine stalks [36], rice, wheat straw [37–39], vine shoots [40], tobacco, cotton stalks [41], barley straw [41], and sugarcane whole bagasse [42].

The α-cellulose content of NPF in this study was 37.3 ± 2.1%, which is also comparable to other non-wood species. Figure 2 shows that a range of 31.6–38.2% was recorded for banana peduncle/kadhi and stem [43], okra sticks [44], and rice straw [14,45]. Furthermore, the α-cellulose content of bamboo [46], oil palm fronds [47], date palm rachis [48], softwoods [49], and most hardwoods [50] is 40–56%, which is higher than non-wood materials.

Table 1. Chemical composition of *Nypa fruticans* fronds used in this study and the literature.

Components (%)	This Study *	Tamunaidu and Saka (2011) [17]
Holocellulose	61.3 ± 2.4	61.5
α-Cellulose	37.3 ± 2.1	35.1
Pentosan	24.0 ± 1.9	26.4
Acid-insoluble lignin	17.5 ± 0.7	17.8
Acid-soluble lignin	0.8 ± 0.5	1.9
Ash	16.5 ± 2.2	11.4–11.7
Extractives	1.5 ± 0.6 (Dichloromethane)	1.9 (Acetone)

* Values presented for this study are averages ± standard deviations.

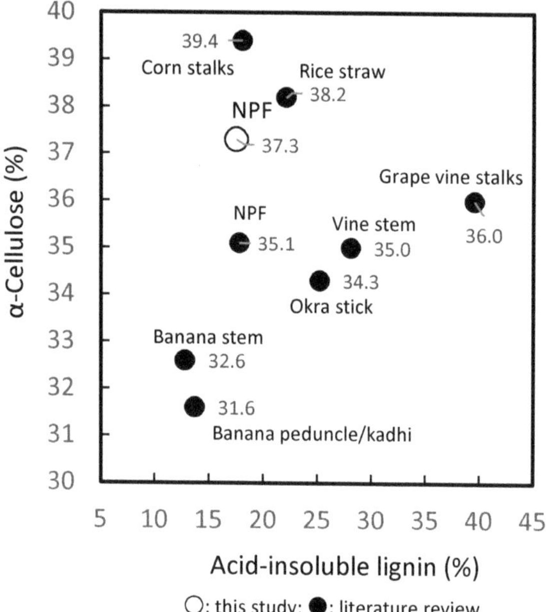

Figure 2. Comparison of α-cellulose and acid-insoluble lignin contents of *Nypa fruticans* fronds with those of other non-wood plants reported in the literature.

The acid-insoluble lignin content of NPF is similar to corn stalks [51], oil palm fronds (15.2%) [47], rice straw and banana stem/peduncle [43,45], but lower compared to vine stem [35], grapevine stalks [36], okra stick [44], and dhaincha (23.2%) [52], as shown in Figure 2. Tamunaidu and Saka [17] revealed that various parts of nipa palm contain a syringyl (S)-, guaiacyl (G)-, or *p*-hydroxyphenyl (P)-type lignin, and a large S to G ratio. The proportion of S units is greater than that of G units of lignin, which indicates that the wood material can easily be delignified [49]. These studies also showed that the chemical composition of all parts of nipa palm was very similar to that of oil palm. The acid-soluble lignin content obtained in this study using the TAPPI method at 205 nm and 110 L/g cm absorption coefficient is lower than the values recorded in previous studies for the same material, namely 0.8 ± 0.5% vs. 1.9% [17]. The differences in the measured values using the UV-spectrophotometric method are caused by the biomass feedstock, solubility in the solvent, differences in absorbance of the G-S-P type of lignin, and experimental factors [53]. However, the acid-insoluble lignin content is higher than the acid-soluble content in the NPFs, which contributed to the total lignin. Future studies can include a determination of the absorption coefficient, because it is important for biorefinery.

Table 1 shows that the ash content of NPF is higher than that of other non-wood plants, but it is relatively comparable to banana plant stem [54] and rice straw, namely 18.3% and 17.2%, respectively [37]. Its presence in silica can cause a variety of operational problems in dissolving pulp processes as well as limiting its uses, and hence it must be removed [16]. The silica content was not determined in this study, but previous investigations showed that it is comparatively low compared to the total ash content [17]. The extractives are also relatively low and comparable to Jahan et al., namely 1.7% [17].

In conclusion, NPF can potentially be used for the production of cellulose derivatives as well as biofuels and chemicals due to its α-cellulose and lignin content. The high α-cellulose content is expected to produce pulp with good mechanical properties, while relatively low lignin content often requires short reaction times or small amounts of reagent [39]. Therefore, this study examined the biomass's suitability for the production of pure cellulose as a dissolving pulp.

2.2. Effect of Prehydrolysis Time on Residual Lignin Content and Pulp Yield after Cooking

Figures 3 and 4 show that increasing the time of prehydrolysis from 1 h to 2 h when cooking for 1 h contributed to a decrease in kappa number and pulp yield. These conditions also caused an increase in the active alkali (AA) dosage from 13% to 20%. Similar results were obtained after increasing the prehydrolysis time from 2 h to 3 h and the AA dosages when cooking for 1.5 h. The combination of prehydrolysis and soda cooking causes the hemicellulose dissolution of non-wood lignocellulosic materials and delignification. Extension of the reaction time also contributed to the decrease in kappa number.

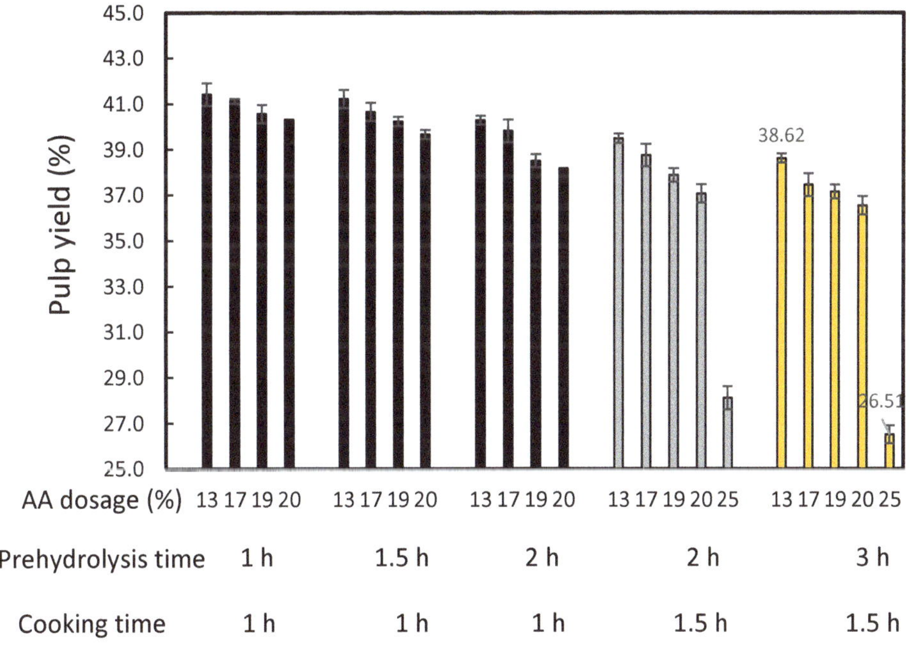

Figure 3. Effect of prehydrolysis time on the pulp yield of *Nypa fruticans* frond soda-SAQ pulp. The values presented are averages ± standard deviations.

There are no studies on the production of dissolving pulp using NPF as the raw material. Jahan et al. [6] studied soda-AQ pulp production with a 6 L/kg liquor to solid ratio, 18% AA, and 170 °C maximum cooking temperature. The findings showed that a yield of 36.2–37.2% was obtained with a kappa number of 27.2 and 29.1. However, these results cannot be compared because the processes were carried out without the prehydrolysis stage. The findings of Dewi et al. [12] with nipa palm petioles cannot also be compared. This implies that the effect of prehydrolysis time on the NPF pulp yield and kappa number after cooking was first reported in this study. No attempts were carried out to measure the pulp yield and kappa number after the prehydrolysis stage; nevertheless, Figure 3 shows quite good cellulose yields at long prehydrolysis times with low alkali dosages. Additionally, Figure 4 illustrates the positive effect on kappa number of increasing the prehydrolysis time. However, a decrease in these parameters has previously been reported with 1.0–7.0 h prehydrolysis time using Moso bamboo and oil palm empty fruit bunches [30,34,55]. The low pulp yield from the process was caused by the removal of hemicellulose [55]. Harsono et al. revealed that extension of the time did not affect the complete delignification and dissolution of hemicellulose [34]. The effect of AA on decreasing the kappa number and pulp yield in various AA dosages was also observed in the previous studies [30,32,34], which indicates that the increase in AA led to lignin removal.

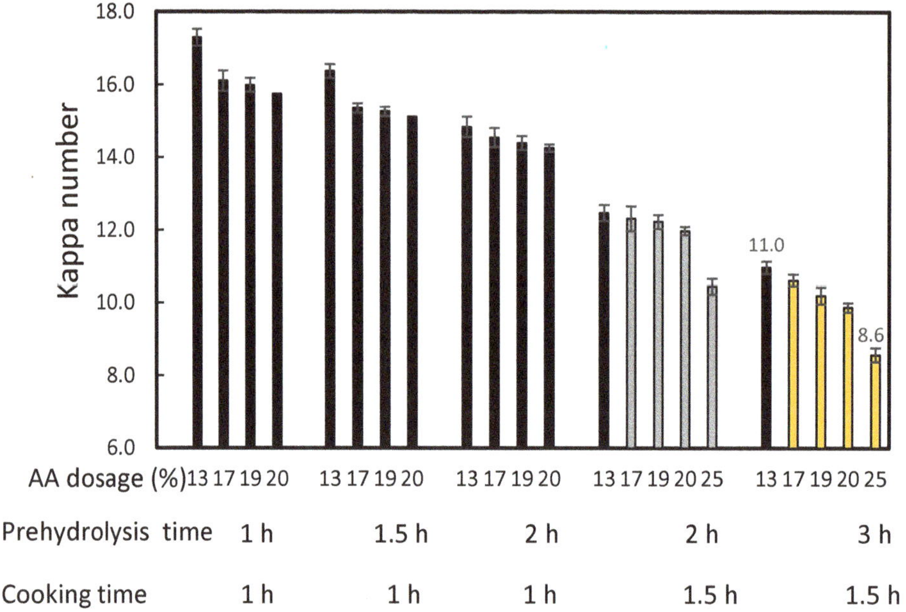

Figure 4. Effect of prehydrolysis time on the kappa number of *Nypa fruticans* frond soda-SAQ pulp. The values presented are averages ± standard deviations.

Prehydrolysis for 3 h and cooking for 1.5 h were better for lignin removal than the other conditions, namely 1–2 h prehydrolysis and 1–1.5 h cooking.

2.3. Effect of Cooking Methods on Pulp Yields and Kappa Number

A comparison of the prehydrolysis soda-SAQ cooking with the prehydrolysis kraft and soda cooking methods is presented in Figure 5. Soda-SAQ cooking (•) reduced the kappa number to 9.9–10.2 at an AA dosage of 19–20%. Compared to kraft cooking (•) with the same AA dosage, the kappa number was lower by approximately nine points, but high yields of 40.5–40.9% were obtained, which was approximately four points higher than the soda-SAQ cooking, namely 36.5–37.2% ($p < 0.05$). Utami et al. [29] also reported high kappa numbers for the prehydrolysis kraft compared to the prehydrolysis soda-SAQ cooking of *A. crassicarpa* at an AA dosage of 17–20%.

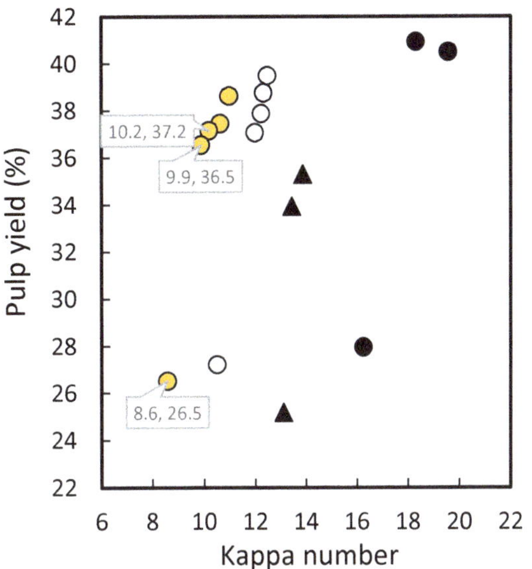

● Prehydrolysis 3 h/Kraft cooking (AA = 19, 20, and 25%)
▲ Prehydorolysis 3 h/Soda cooking (AA = 19, 20, and 25%)
○ Prehydrolysis 3 h/Soda-SAQ cooking (AA = 13, 17, 19, 20, and 25%)
○ Prehydrolysis 2 h/Soda-SAQ cooking (AA = 13, 17, 19, 20, and 25%)

Figure 5. Comparison of kappa number and pulp yield in prehydrolysis cooking of *Nypa fruticans* fronds between kraft, soda, and soda-SAQ processes.

The kappa number obtained from soda cooking with 0.1% SAQ was approximately 30% lower than soda cooking (▲) at the same AA dosage, namely 8.6–10.2 and 19–25%, respectively ($p < 0.05$). Furthermore, the yield recorded was approximately 6% higher than soda cooking at AA dosage of 19–20%, namely 26.5–36.5% and 33.9–35.3%, respectively. This finding indicates that the soda-SAQ was more effective at $p < 0.05$. Jahan and Mun [56] stated that the soda-anthraquinone (AQ) process is suitable for non-wood cooking due to relatively low-average molecular weight of lignin. Several studies revealed that soda-AQ provided better results than soda cooking for pulp production. For example, Salehi et al. [26] reported that the AQ use in the process for wheat and rye straw increased the screen pulp yields by 2–14% and simultaneously enhanced the delignification rate by approximately 30% compared to soda cooking without AQ. Similar results were also obtained in other studies where there was an increase in pulp yield and a decrease in kappa number for cotton stalks at 0.075–0.2% AQ dosage [27] as well as bast kenaf fibers at 0.2% AQ dosage [28]. The higher pulp yield in soda-AQ cooking was caused by the retention of hemicellulose and cellulose along with the enhancement of lignin degradation. Although Utami et al. [29] reported comparable yields of *A. crassicarpa* soda-SAQ pulp to soda pulp, lower kappa numbers were observed after adding 0.1% SAQ. Additional decreases after the prehydrolysis soda cooking with AQ were also observed by Jahan et al. [57] for *A. auriculiformis* soda-AQ pulp.

The results obtained for NPF were comparable to those of Harsono et al. [34] for oil palm empty fruit bunches with a kappa number of 9.4–10.7 and 29.0–33.5% pulp yield using prehydrolysis soda-SAQ cooking, namely 3 h prehydrolysis at 150 °C and 3 hours of cooking at 160 °C with a 19–21% AA dosage. Putra et al. [58] used the same conditions as Harsono et al. [34] and obtained a 31.1% yield of pulp with a kappa number of 9.6 at 20% AA dosage for *A. crassicarpa*. Furthermore, Maryana et al. [32] recorded 33.9–34.7% yields with low kappa numbers of 5.3–5.7 using prehydrolysis for 3 h at 150 °C and soda-SAQ cooking for 2 h at 150 °C with a 17–19% AA dosage for sugarcane bagasse. Chem et al. [30] reported a higher kappa number of 12.8–22.9 from Moso bamboo stem pulp than this study, but the yields of 32.6–34.0% were comparable when using 3 h prehydrolysis at 150 °C and 3 h cooking at 160 °C with 17–23% AA dosage. Banana plant stem gave high kappa numbers of 28.3–37.6 and 27–30% pulp yields using prehydrolysis for 3 h at 150 °C and cooking for 0.5–1.5 h at 150 °C with 25% AA dosage [54]. The high residual lignin content indicated by the kappa number in Moso bamboo and banana plant stem was caused by the initial contents of the two raw materials. The yields obtained from non-wood raw materials are comparable and depend on the selected/optimum condition of the process as well as the initial α-cellulose content of the materials. Meanwhile, the kappa number recorded seemed to be affected by the initial lignin content. The comparison results between prehydrolysis soda-SAQ and prehyrolysis kraft/soda of nipa palm fronds definitely provide new information to the available literature, because no previous studies were found.

2.4. Application to Totally Chlorine-Free Bleaching

The combination of prehydrolysis for 3 h and soda-SAQ cooking for 1.5 h produced 26.5–36.9% pulp yields with a low kappa number of 8.6–11.0 for totally chlorine-free bleaching. NPF pulp obtained through prehydrolysis and soda-SAQ cooking at a 20% AA dosage was selected for application in TCF bleaching because the conditions produced a kappa number of 9.9 ± 0.1, a yield of $36.5 \pm 0.6\%$, $34.8 \pm 0.5\%$ ISO brightness, and 22.1 ± 0.7 cP viscosity, as shown in Table 2.

Table 2. The brightness and viscosity of *Nypa fruticans* frond pulp obtained through prehydrolysis for 3 h at 150 °C followed by soda-SAQ cooking for 1.5 h at 160 °C *.

Active Alkali Dosage (%)	Screened Pulp Yield (%)	Brightness (% ISO)	Viscosity (cP)
13	38.6 ± 0.7	29.6 ± 0.4	21.8 ± 0.4
17	37.5 ± 0.3	32.1 ± 0.6	22.7 ± 0.4
19	37.2 ± 0.4	34.2 ± 0.4	23.0 ± 0.3
20	36.5 ± 0.6	34.8 ± 0.5	22.1 ± 0.7
25	26.5 ± 0.4	35.1 ± 0.6	20.6 ± 0.5

* Values presented are average $\pm$ standard deviations.

The optimum conditions for the soda-SAQ and TCF bleaching produced kappa numbers < 10 and a reasonable level of pulp yield, as suggested by previous studies [31,34]. Furthermore, there were significant differences between the screened pulp yields, brightness, and viscosity as the AA dosage was increased from 13 to 20% in the selected prehydrolysis–cooking process, namely $38.6 \pm 0.7\%$ vs. $26.5 \pm 0.4\%$, $29.6 \pm 0.4\%$ vs. $35.1 \pm 0.6\%$, and 35.1 ± 0.6 cP vs. 20.6 ± 0.5 cP, respectively ($p < 0.05$).

In this study, the bleaching procedure proposed by previous studies [31,32,34] with a five-stage sequence of O-P_{sa}-E_p-P_{sa}-E_p was modified to increase the dosage of NaOH in the O-stage to 2%. P_{sa} was selected instead of ozone because of its better selectivity in TCF bleaching [33]. Modifying the NaOH dosage can shorten the stages involved (O-P_{sa}-E_p) and thus reduce the number of processes, as shown in Figures 6 and 7. There was also a gradual increase in pulp brightness at each stage starting from $72.6 \pm 0.8\%$ ISO after the O stage, followed by $77.7 \pm 0.7\%$ ISO at the P_{sa} stage, and $92.2 \pm 1.0\%$ ISO at the third E_p stage. Meanwhile, the viscosity decreased from 11.0 ± 0.4 to 7.9 ± 0.6 cP after the whole process. There was clear evidence of differences in the brightness and viscosity values between the two consecutive bleaching stages ($p < 0.05$). The viscosity is related to the degree of polymerization (DP, Equation (1)) with values of 20.6–22.3 after pulping followed by a subsequent decrease to 7.2–10.3 after TCF bleaching, which indicates the degradation of NPF cellulose fibers after these processes. The normal DP values in wood are reported to be between 300 and 1700 [59]. Similar results were obtained in the five-stage TCF sequence for sugarcane bagasse pulp, namely 89.1% ISO with 6.4 cP [23] as well as *E. globulus* pulp with 88.4% ISO and 6.0 cP [31].

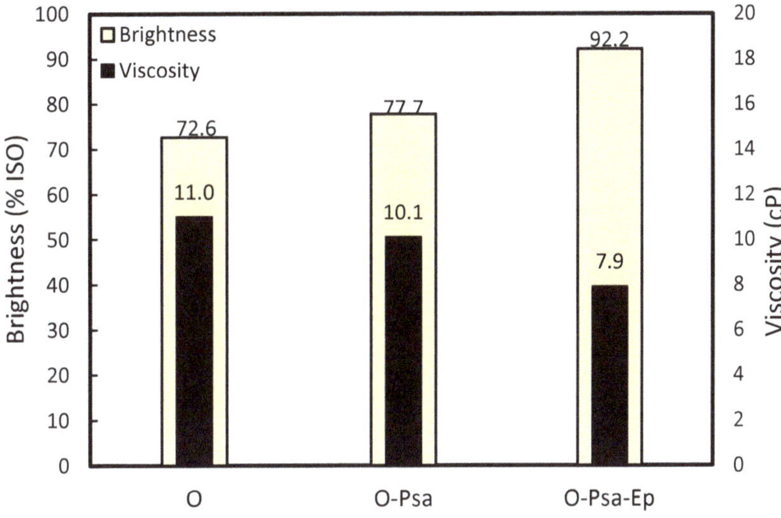

Figure 6. The brightness and viscosity profiles during three-stage totally chlorine-free bleaching of *Nypa fruticans* frond pulp at 20% AA dosage (O: oxygen; P_{sa}: peroxymonosulfuric acid; E_p: alkaline hydrogen peroxide).

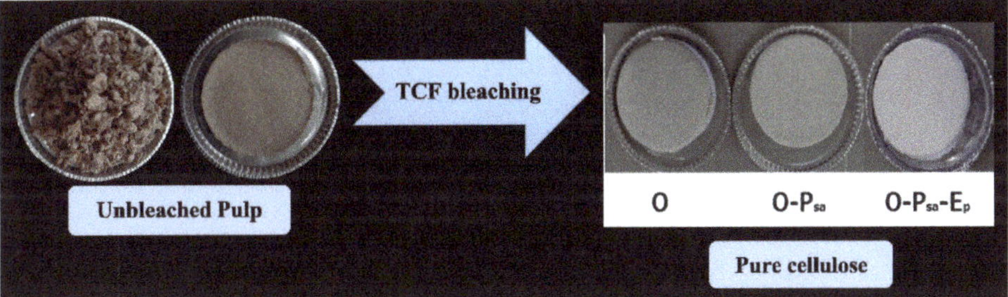

Figure 7. Transformation of the *Nypa fruticans* unbleached pulp into pure cellulose during three-stage totally chlorine-free bleaching at 20% AA dosage (O: oxygen; P_{sa}: peroxymonosulfuric acid; E_p: alkaline hydrogen peroxide).

The α-cellulose and ash contents of the bleached pulp as pure cellulose in this study were 94.8 ± 2.8% and 0.2 ± 0.07%, respectively, as shown in Table 3. The final yield of the purified pulp was 28.0 ± 1.3% based on the material weight. The properties of pure cellulose include 92.2 ± 1.0% ISO brightness, 94.8 ± 2.8% α-cellulose, 7.9 ± 0.6 cP viscosity, which are acceptable levels for viscose rayon and cellulose derivatives. Regarding the ash content obtained in this study, namely 0.2 ± 0.07%, it was reported that dissolving pulps from non-woody materials with an ash content of approximately 0.7% still showed potential for the production of viscose rayon, carboxymethylcellulose, and other derivatives [60,61]. Therefore, pure cellulose was successfully produced from nipa palm fronds using prehydrolysis, soda-SAQ and TCF bleaching. The processes can further be utilized for the production of other useful products or materials.

The results show that only a small amount of pentosan, namely 1.0 ± 0.4%, remained in the purified pulp. Most of the pentosan in the materials can be isolated from the prehydrolysate for furfural production. From the sulfur-free soda cooking liquor, dissolved lignin can easily be precipitated based on a proposed method, such as the lignoboost process for raw materials of biopolymers [62]. The sulfur-free and totally chlorine-free processes for cellulose purification from NPF can be developed in future into a promising biorefinery.

Table 3. Chemical composition of pure cellulose from *Nypa fruticans* fronds.

Components	α-Cellulose (%)	Pentosan (%)	Ash (%)
Materials	37.3 ± 2.1	24.0 ± 1.9	16.5 ± 2.2
Sulfur-free unbleached pulp	84.1 ± 2.7	3.3 ± 1.2	1.0 ± 0.5
Totally chlorine-free pulp	94.8 ± 2.8	1.0 ± 0.4	0.2 ± 0.07

3. Materials and Methods

3.1. Materials

Nypa fruticans fronds were collected from Muntai village, Bantan district of Bengkalis island, Indonesia. The fiber fragments with lengths of 0.5–1.0 cm were prepared after washing and sun-drying to approximately 90%. Figure 8a–c shows the nipa palm fronds and their chipped form. SAQ (1,4-dihydro-9,10-dihydroxyanthraxene sodium salt) was provided by Air Water Performance Chemical Inc., Kawasaki, Japan. Furthermore, P_{sa} (H_2SO_5) was synthesized based on the method proposed by Kuwabara et al. [63] using 98% sulfuric acid (Wako Pure Chemical Industries, Ltd., Osaka, Japan) and a 50% hydrogen peroxide aqueous solution (Mitsubishi Gas Chemical Company, Inc., Tokyo, Japan) at a molar ratio of 1:3 and 70 °C.

(a)   (b)   (c)

Figure 8. Nipa palm fronds received (**a**); peeled (**b**), and; chipped (**c**).

3.2. Methods

3.2.1. Prehydrolysis and Sulfur-Free Cooking

The prehydrolysis and cooking of the NPF were carried out in a 350 mL stainless-steel reactor (Taiatsu Techno Corporation, Tokyo, Japan), with a maximum temperature of 150 °C. The parameters used for the process include a distilled water to biomass ratio of 7 L/kg, a temperature of 150 °C, and a reaction time of 1–3 h. The wet solid residues without washing were subjected to soda-SAQ cooking using sodium hydroxide solution as fresh alkaline cooking liquor. The parameters used were a 0.1% SAQ dosage on the raw material weight, a 7 L/kg liquor to solid ratio, a temperature of 160 °C, 1–1.5 h cooking time, and active alkali (AA), namely Na_2O, with dosages of 13, 17, 19, 20, and 25%. The residue obtained was processed into pulp using a disintegrator (FRANK-PTI GmbH), followed by screening, washing, and drying at 105 °C. Subsequently, the yield (%) of pulp was determined based on the raw materials' weight. The optimal prehydrolysis and cooking times were selected based on the yield and kappa number. The selected parameters, procedures, and analyses were carried out based on the method from previous studies for the production of dissolving pulp from non-wood materials with similar characteristics, such as oil palm empty fruit bunch [34]. Cooking with soda without SAQ as well as the kraft method using a mixture of sodium hydroxide and sodium sulfide (sulfidity of 30% as Na_2O) were also carried out to compare the results from the selected soda-SAQ process.

3.2.2. Totally Chlorine-Free Bleaching

A modified and simple three-stage sequence of TCF bleaching, including oxygen (O), P_{sa}, and alkaline hydrogen peroxide (E_p) treatments and procedures, was carried out for the soda-SAQ pulp based on the multi-stage sequence proposed by a previous study [34]. This TCF sequence has been suggested as a solution for the pulp and paper industry to prevent pollution during bleaching [32,34]. The conditions of each stage are presented in Table 4. Depending on the TCF bleaching stage, a pulp consistency (PC) of 10 or 30% was prepared by carefully mixing the pulps with distilled water. The particles of wet pulp were mixed with NaOH in a polyethylene bag by hand, providing limited homogeneity of the treatment specifically to the samples with 30% PC; however, successful results for bleaching were obtained in this study. The PC value was then determined with the TAPPI method T 240-om 93 [64]. Regarding the oxygen delignification step, a portion of oxygen was added to the reaction mixture at the beginning of the process until the pressure was reached, followed by the reaction at the determined temperature and time (T-t). We did not use a high-share mixer for medium-consistency oxygen bleaching. Meanwhile, at the P_{sa} stage, a target amount of P_{sa} and a small amount of NaOH aqueous solution for pH adjustment were added to the pulp suspension in a polyethylene bag at the determined T-t. With respect to the E_p stage, target amounts of H_2O_2 and NaOH were also added to the pulp suspension in a polyethylene bag and at the determined T-t. These two later steps of bleaching at lower consistency (10%) were carried out for easier comparison with past results in the literature (Appendix A). The brightness and viscosity after each stage were obtained using methods explained in the next section.

Table 4. Conditions of three-stage totally chlorine-free bleaching for *Nypa fruticans* frond soda pulp.

Bleaching Stage	Conditions
O: Oxygen	O_2 pressure: 0.5 MPa, NaOH dosage: 2.0%, 60 min, 115 °C, pulp consistency (PC): 30%.
P_{sa}: Peroxymonosulfuric acid	H_2SO_5 dosage: 0.2%, 70 min, 70 °C, pH 3.0, PC: 10%.
E_p: Alkaline hydrogen peroxide	H_2O_2 dosage: 2.0%; NaOH dosage: 1.4%, 60 min, 70 °C, PC: 10%.

3.2.3. Chemical Analysis of Materials and Pulp

For raw materials, the contents of holocellulose, alpha (α)-cellulose, pentosan, acid-insoluble lignin (Klason lignin), extractives, and ash were determined using the TAPPI methods, namely T 249 om-00, T 203 cm-99, T 223 cm-10, T 222 om-06, T 204 cm-07, and T 211 om-07, respectively [65–70]. Meanwhile, the acid-soluble lignin was determined with UV-vis spectrophotometry at 205 nm using a gram extinction coefficient of 110 L/g cm [TAPPI method: T 250 um-62]. The methods T236 om-99, T452 om-08, and T230 om-04 were used to determine the kappa number, brightness, and viscosity, respectively [71–74]. The pulp viscosity was also converted into the degree of polymerization or DP based on the equation proposed by Shi et al. [75] (Equation (1)):

$$DP^{0.905} = 0.75V \tag{1}$$

where V represents the pulp viscosity (mPa·s or cP) after the process.

3.2.4. Statistical Analysis

In this study, at least two independent experiments and two measurements for each were carried out for each condition/method. The average values ± standard deviations were calculated and presented. Furthermore, a t-test with significance at $p < 0.05$ was used to compare any two pulp yield or kappa number values with different processing conditions/methods (Systat v13.2.01, Systat Statsoft Inc., San Jose, CA, USA), as described by Hart and Sharp [76]. It was also used to compare any two brightness or viscosity values in the TCF bleaching experiment.

4. Conclusions

Nypa fruticans frond waste is a promising raw material for producing pure cellulose and its derivatives, as well as biofuels and chemicals, due to a reasonable amount of α-cellulose and a relatively low lignin content, namely 37.3% and 18.3%, respectively. The selected prehydrolysis sulfur-free soda process with soluble anthraquinone (SAQ) catalyst has more advantages compared to the kraft and soda processes due to the decrease in the kappa number by $\geq$3.6 points at a 19–20% active alkali or AA dosage. Soda-SAQ also exhibited higher pulp yields of 6% at the same dosage. The optimal conditions for the soda-SAQ process include 3 hours of prehydrolysis at 150 °C and cooking for 1.5 h at 160 °C with a 20% AA concentration. Prehydrolysis sulfur-free soda cooking with the SAQ catalyst followed by three-stage totally chlorine-free bleaching with oxygen, peroxymonosulfuric acid, and alkaline hydrogen peroxide stages can produce pure cellulose as a dissolving pulp from NPF waste. The final properties of the product obtained include 92.2% ISO brightness, 94.8% α-cellulose, 7.9 cP viscosity, and 0.2% ash content. The results show the benefits and details of the prehydrolysis sulfur-free soda process with SAQ to produce pure cellulose that is acceptable for viscose rayon and cellulose derivatives. Further studies can be conducted to investigate and develop its potential for biorefinery because 99% of the pentosan was removed in the prehyrolysate and cooking liquor. The production of the viscose rayon and cellulose derivatives, such as cellulose nitrate/acetate from the obtained prehydrolysis soda-SAQ pulps, can also be explored.

Author Contributions: Conceptualization, E. and H.O.; methodology, E., S. and D.A.; software, E. and S.; validation, E. and S.; formal analysis, E., A.A. and H.O.; investigation and supervision, S. and D.A.; resources, E., A.A. and H.O.; data curation, E., S. and D.A.; writing—original draft preparation, E.; writing—review and editing, E., A.A. and H.O.; project administration, E. and D.A. All authors have read and agreed to the published version of the manuscript.

Funding: This study was funded by the Directorate General of Education, Culture, Research, and Technology, Indonesia, through the World Class Research (WCR) Scheme with Project Agreement number 1395/UN19.5.1.3/PT.01.03/2021.

Institutional Review Board Statement: Not applicable.

Informed Consent Statement: Not applicable.

Data Availability Statement: Not applicable.

Acknowledgments: The authors are grateful to Retno Dwi Astutik and Donny Tusary, students at the Chemical Engineering Department, Faculty of Engineering, for their valuable contributions in the preparation and investigation of this study within the framework of the WCR project.

Conflicts of Interest: The authors declare no conflict of interest.

Appendix A

High Consistency of Oxygen Bleaching

1. Measure the water content of the air-dried pulp sample.
2. Measure the pulp sample with 10.0 g of oven-dried weight, and then tear it, and put them in a polyethylene plastic bag.
3. Prepare 20 mL (or A: 10 mL for $MgSO_4 \cdot 7H_2O_2$ and B: 10 mL for NaOH) of distilled water for 30% of pulp consistency.
4. Prepare the targeted NaOH and Mg dosage. Example 1.0% and 0.1% respectively.
5. If the pulp sample was 10.0 g, then add 2.5 mL of 1 N (1 mol/L) NaOH (0.1 g of NaOH) into the distilled water.
6. If the NaOH is 2N (2 mol/L), then add 1.25 ml.
7. Drop A: 10 mL of Mg solution containing 100 mg of $MgSO_4 \cdot 7H_2O_2$ to the pulp sample, and mix the pulp sample in a polyethylene bag with hands (Figure A1). Drop B: 12.5 mL or 11.25 mL of the NaOH solution to the pulp sample, and mix the pulp sample in a polyethylene bag with hands (Figure A1).

8. Make pulp into small, bulky and fluffy particles (diameter 5–10 mm).
9. Put pulp in a polypropylene plastic bottle and put it in a reactor (Figure A2).
10. Charge oxygen gas into the reactor under 0.35–0.50 MPa (3.5–5.0 bars) at room temperature for several minutes (Figure A3). Release the oxygen gas from the reactor, and repeat the charging twice for purging air inside sufficiently.
11. Set the reactor in 123 °C oil bath.
12. Count the reaction time from 91–115°C.
13. Cool the reactor after 60 minutes in a water bath.
14. Transfer the pulp into a jar of a disintegrator and wash the pulp with 1.5 L of water. Dewater the pulp suspension using a Büchner funnel.
15. Wash the pulp twice more, and make 2 pieces of hand-sheets and air-dry them.

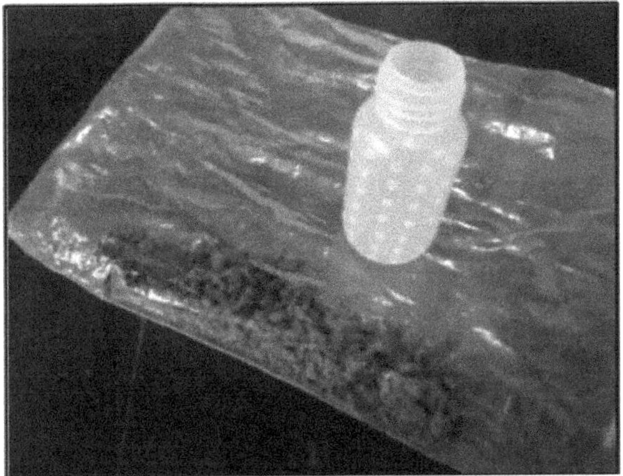

Figure A1. Put pulp in polyethylene plastic bag and make pulp into small particles (diameter 5–10 mm).

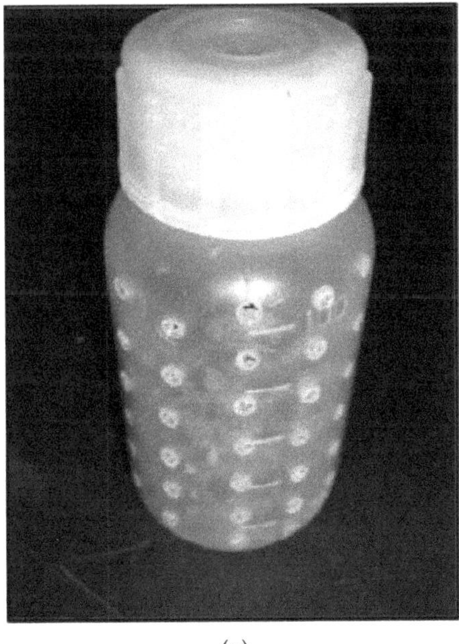

(**a**) (**b**)

Figure A2. (**a**) Put pulp in a polypropylene plastic bottle, and (**b**) then put the plastic bottle into the reactor.

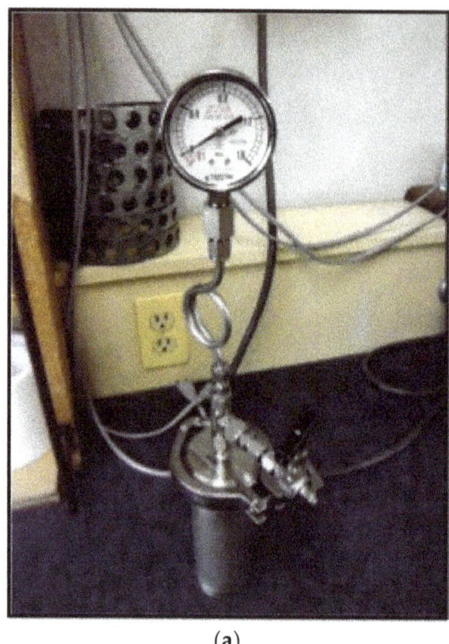 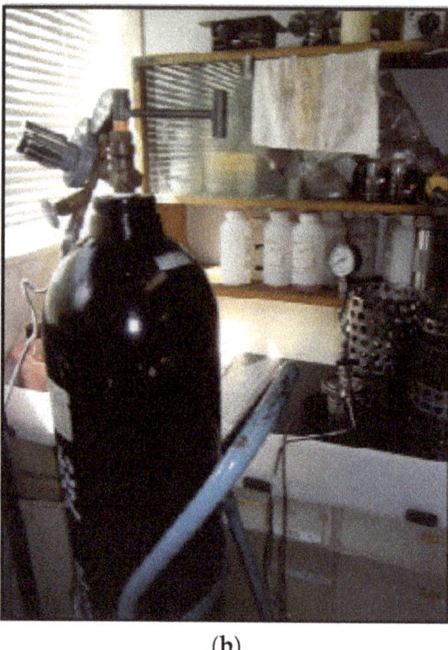

(a)   (b)

Figure A3. Charge oxygen gas under 0.35–0.50 MPa at room temperature.

References

1. Akpakpan, A.E.; Akpabio, U.D.; Obot, I.B. Evaluation of physicochemical properties and soda pulping of *Nypa fruticans* frond and petiole. *Elixir Appl. Chem.* **2012**, *45*, 7664–7668.
2. CABI. *Nypa fruticans* . Available online: https://www.cabi.org/isc/datasheet/36772 (accessed on 3 July 2022).
3. Harun, N.Y.; Saeed, A.A.H.; Vegnesh, A.; Ramachandran, L.A. Abundant nipa palm waste as bio-pellet fuel. *Mater. Today.* **2021**, *42*, 436–443.
4. Cheablam, O.; Chanklap, B. Sustainable nipa palm (*Nypa fruticans* Wurmb.) product utilization in Thailand. *Scientifica* **2020**, *2020*, 1–10. [CrossRef]
5. Tamunaidu, P.; Matsui, N.; Okimori, Y.; Saka, S. Nipa (*Nypa fruticans*) sap as a potential feedstock for ethanol production. *Biomass Bioenergy* **2013**, *52*, 96–102. [CrossRef]
6. Kruse, K.; Frühwald, A. *Properties of Nipa and Coconut Fibers and Production and Properties of Particle and MDF Boards Made from Nipa and Coconut*; Bundesforschungsanstalt für Forst-und Holzwirtschaft Nr 04: Hamburg, Germany, 2011.
7. Wankasi, D.; Horsfall, M.J.; Spiff, A.I. Desorption of Pb^{2+} and Cu^{2+} from nipa palm (*Nypa fruticans* wurmb) biomass. *Afr. J. Biotechnol.* **2005**, *4*, 923–927.
8. Jahan, M.S.; Nasima, D.A.; Chowdhury, M.; Islam, K. Characterization and evaluation of golpata fronds as pulping raw materials. *Bioresour. Technol.* **2006**, *97*, 401–406. [CrossRef]
9. Akpabio, U.D.; Effiong, I.E.; Akpakpan, A.E. Preparation of pulp and cellulose acetate from nypa palm leaves. *Int. J. Environ. Bioenerg.* **2012**, *1*, 179–194.
10. Yusoff, N.A.; Yam, M.F.; Beh, H.K.; Razak, K.N.A.; Widyawati, T.; Mahmud, R.; Ahmad, M.; Asmawi, N.Z. Antidiabetic and antioxidant activities of *Nypa fruticans* Wurmb. vinegar sample from Malaysia. *Asian Pac. J. Trop. Med.* **2015**, *8*, 595–605. [CrossRef]
11. Hakimien, A.K.; Alshareef, I.; Ho, W.K.; Soh, A.C. Potential of nypa palm as a complementary biomass crop to oil palm in Malaysia. *Acta Hortic.* **2016**, *1128*, 285–290. [CrossRef]
12. Dewi, I.A.; Ihwah, A.; Wijana, S. Optimization on pulp delignification from nypa palm (*Nypa fruticans*) petioles fibre of chemical and microbiological methods. *IOP Conf. Ser. Earth Environ. Sci.* **2019**, *187*, 1–10. [CrossRef]
13. Abd El-Sayed, E.S.; El-Sakhawy, M.; El-Sakhawy, M.A. Non-wood fbers as raw material for pulp and paper industry. *Nordic Pulp Paper Res. J.* **2020**, *35*, 215–230. [CrossRef]
14. Hammett, A.L.; Youngs, R.L.; Sun, X.; Chandra, M. Non wood fiber as an alternative in China's pulp and paper industry. *Holzforschung* **2001**, *55*, 219–224. [CrossRef]
15. Liu, Z.; Wang, H.; Hui, L. Pulping and Papermaking of non-wood fibers, pulp and paper processing. In *Pulp and Paper Processing*; Kazi, S.N., Ed.; IntechOpen: London, UK, 2018; Available online: https://www.intechopen.com/chapters/62223 (accessed on 15 July 2022).

16. Jahan, M.S.; Uddin, M.N.; Rahman, A.; Rahman, M.M.; Amin, M.N. Soda pulping of umbrella palm grass (*cyperus flabettiformic*). *J. Bioresour. Bioprod.* **2016**, *1*, 85–91.

17. Tamunaidu, P.; Saka, S. Chemical characterization of various parts of nipa palm (*Nypa fruticans*). *Ind. Crops Prod.* **2011**, *34*, 1423–1428. [CrossRef]

18. Sarkar, A.M.; Farzana, M.; Rahman, M.M.; Jin, Y.; Jahan, S. Future cellulose based industries in Bangladesh—A mini review. *Cellul. Chem. Technol.* **2021**, *55*, 443–459. [CrossRef]

19. Sixta, H. Pre-Hydrolysis. In *Handbook of Pulp*; Sixta, H., Ed.; Wiley-VchVerlag Gmbh and Co. Kgaa: Weinheim, Germany, 2006; pp. 325–345.

20. Shokri, S.; Hedjazi, S.; Lê, H.Q.; Abdulkhani, A.; Sixta, H. High-purity cellulose production from birch wood by γ-valerolactone/water fractionation and IONCELL-P process. *Carbohydr. Polym.* **2022**, *288*, 1–8. [CrossRef]

21. Kumar, H.; Christopher, L.P. Recent trends and developments in dissolving pulp production and application. *Cellulose* **2017**, *24*, 2347–2365. [CrossRef]

22. Rullifank, K.; Roefinal, M.E.; Konstanti, M.; Sartika, L.; Evelyn. Pulp and paper industry: An overview on pulping technologies, factors, and challenges. *IOP Conf. Ser. Mater. Sci. Eng.* **2020**, *845*, 1–9. [CrossRef]

23. Hart, P.W.; Rudie, A.W. Anthraquinone a review of the rise and fall of a pulping catalyst. *Tappi J.* **2014**, *13*, 23–31. [CrossRef]

24. Ohi, H. Function of anthraquinone as pulping additive and its possibilities. *Jpn. Tappi J.* **1994**, *48*, 531–544. [CrossRef]

25. Anita, Y.; Utami, S.P.; Ohi, H.; Evelyn, E.; Nakagawa-Izumi, A. Mutagenicity of *Tectona grandis* wood extracts and their ability to improve carbohydrate yield for kraft cooking *Eucalyptus* wood. *Molecules* **2017**, *26*, 7171. [CrossRef] [PubMed]

26. Salehi, K.; Kordsachia, O.; Patt, R. Comparison of MEA/AQ, soda and soda/AQ pulping of wheat and rye straw. *Ind. Crops Prod.* **2014**, *52*, 603–610. [CrossRef]

27. Akgül, M.; Tozluolu, A. A comparison of soda and soda-AQ pulps from cotton stalks. *Afr. J. Biotechnol.* **2009**, *8*, 6127–6133.

28. Shakhes, J.; Zeinaly, F.; Marandi, M.; Saghaf, T. The effect of processing variables on the soda and soda-AQ pulping of kenaf basf fiber. *Bioresour.* **2011**, *6*, 4626–4639.

29. Utami, S.P.; Tanifuji, K.; Putra, A.S.; Nakagawa-Izumi, A.; Ohi, H.; Evelyn, E. Effects of soluble anthraquinone application on prehydrolysis soda cooking of *Acacia crassicarpa* wood. *Jpn. TAPPI J.* **2021**, *75*, 373–379. [CrossRef]

30. Chem, M.; Tanifuji, K.; Utami, S.P.; Putra, A.S.; Ohi, H.; Nakagawa-Izumi, A. Development of dissolving pulp from *Phyllostachys pubescens* stem by prehydrolysis soda cooking with 2-methylanthraquinone. *Ind. Crops Prod.* **2022**, *178*, 1–9. [CrossRef]

31. Salaghi, A.; Putra, A.S.; Rizaluddin, A.T.; Kajiyama, M.; Ohi, H. Totally chlorine-free bleaching of prehydrolysis soda pulp from plantation hardwoods consisting of various lignin structures. *J. Wood Sci.* **2019**, *65*, 1–12. [CrossRef]

32. Maryana, R.; Nakagawa-Izumi, A.; Kajiyama, M.; Ohi, H. Environmentally friendly non-sulfur cooking and totally chlorine free bleaching for preparation of sugarcane bagasse cellulose. *J. Fiber Sci. Technol.* **2017**, *73*, 182–191. [CrossRef]

33. Rizaluddin, A.T.; Liu, Q.; Panggabean, P.R.; Ohi, H.; Nakamata, K. Application of peroxymonosulfuric acid as a modification of the totally chlorine-free bleaching of acacia wood prehydrolysis-kraft pulp. *J. Wood Sci.* **2015**, *61*, 292–298. [CrossRef]

34. Harsono, H.; Putra, A.S.; Maryana, R.; Rizaluddin, A.T.; H'ng, Y.Y.; Nakagawa-Izumi, A. Preparation of dissolving pulp from oil palm empty fruit bunch by prehydrolysis soda-anthraquinone cooking method. *J. Wood Sci.* **2015**, *62*, 65–73. [CrossRef]

35. Mansouri, S.; Khiari, R.; Bendouissa, N.; Saadallah, S.; Mhenni, F.; Mauret, E. Chemical composition and pulp characterization of Tunisian vine stems. *Ind. Crops Prod.* **2012**, *36*, 22–27. [CrossRef]

36. Ping, L.; Brosse, N.; Sannigrahi, P.; Ragauskas, A. Evaluation of grape stalks as a bioresource. *Ind. Crops Prod.* **2017**, *33*, 200–204. [CrossRef]

37. Jahan, M.S.; Haris, F.; Rahman, M.M.; Samaddar, P.R.; Sutradhar, S. Potassium hydroxide pulping of rice straw in biorefinery initiatives. *Bioresour. Technol.* **2016**, *219*, 445–450. [CrossRef]

38. Tutus, A.; Eroğlu, H. A practical solution to the silica problem in straw pulping. *Appita J.* **2003**, *56*, 111–115.

39. Rodríguez, A.; Moral, A.; Serrano, L.; Labidi, J.; Jiménez, L. Rice straw pulp obtained by using various methods. *Bioresour. Technol.* **2008**, *99*, 2881–2886. [CrossRef] [PubMed]

40. Jimenez, L.; Angulo, V.; Ramos, E.; De la Torre, M.J.; Ferrer, J.L. Comparison of various pulping processes for producing pulp from vine shoots. *Ind. Crops Prod.* **2006**, *23*, 122–130. [CrossRef]

41. Ates, S.; Deniz, I.; Kirci, H.; Atik, C.; Okan, O.T. Comparison of pulping and bleaching behaviors of some agricultural residues. *Turkish J. Agric. For.* **2015**, *39*, 144–153. [CrossRef]

42. Andrade, M.F.; Colodette, J.L. Dissolving pulp production from sugar cane bagasse. *Ind. Crops Prod.* **2014**, *52*, 7664–7668. [CrossRef]

43. Rahman, M.M.; Islam, T.; Nayeem, J.; Jahan, M. Variation of chemical and morphological properties of different parts of banana plant (Musa paradisica) and their effects on pulping. *Intl. J. Lignocell. Prod.* **2014**, *1*, 93–103.

44. Jahan, M.S.; Rahman, M.; Rahman, M.M. Characterization and evaluation of okra fibre (*Abelmoschus esculentus*) as a pulping material. *J. Sci. Technol. Forest Prod. Proc.* **2012**, *2*, 12–17.

45. Jahan, M.S.; Shamsuzzamana, M.; Rahman, M.M.; Moeiz, I.; Ni, Y. Effect of pre-extraction on soda-anthraquinone (AQ) pulping of rice straw. *Ind. Crops Prod.* **2012**, *37*, 164–169. [CrossRef]

46. Sugesty, S.; Kardiansyah, T.; Hardiani, H. Bamboo as raw materials for dissolving pulp with environmental friendly technology for rayon fiber. *Proc. Chem.* **2015**, *17*, 194–199. [CrossRef]

47. Wanrosli, W.D.; Zainuddin, Z.; Law, K.N.; Asro, R. Pulp from oil palm fronds by chemical processes. *Ind. Crops Prod.* **2007**, *25*, 89–94. [CrossRef]
48. Khiari, R.; Mhenni, M.F.; Belgacem, M.N.; Mauret, E. Chemical composition and pulping of date palm rachis and Posidonia oceanica—A comparison with other wood and non-wood fibre sources. *Bioresour. Technol.* **2010**, *101*, 775–780. [CrossRef]
49. Chen, C.; Duan, C.; Li, J.; Liu, Y.; Ma, X.; Zheng, L.; Stavik, J.; Ni, Y. Cellulose (dissolving pulp) manufacturing processes and properties: A mini-review. *Bioresour.* **2016**, *11*, 5553–5564. [CrossRef]
50. Sharma, A.K.; Dutt, D.; Upadhyaya, J.S.; Roy, T.K. Anatomical, morphological, and chemical characterization of bambusa tulda, dendrocalamus hamiltoni, bambusa balcoa, malocana baccifera, bambusa arundinacea and eucalyptus tereticornis. *Bioresour.* **2011**, *6*, 5062–5073.
51. Li, Z.; Zhai, H.; Zhang, Y.; Yu, L. Cell morphology and chemical characteristics of corn stover fractions. *Ind. Crops Prod.* **2012**, *37*, 130–136. [CrossRef]
52. Sarkar, M.; Sutradhar, S.; Sarwar, A.G.; Uddin, M.N.; Chanda, S.C.; Jahan, M.S. Variation of chemical characteristics and pulpability of dhaincha (*Sesbania bispinosa*) on location. *J. Bioresour. Bioprod.* **2017**, *2*, 24–29.
53. Kline, L.M.; Douglas, G.H.; Womac, A.R.; Labbé, N. Simple determination of lignin content in hard and softwood via UV-Spectrophotometric analysis of biomass dissolved in ionic liquids. *BioResources* **2010**, *5*, 1366–1383.
54. Das, A.; Nakagawa-Izumi, A.; Ohi, H. Quality evaluation of dissolving pulp fabricated from banana plant stem and its potential for biorefinery. *Carbohydr. Polym.* **2016**, *147*, 133–138. [CrossRef]
55. Jahan, M.S.; Mun, S.P. Effect of tree age on the soda anthraquinone pulping of Nalita wood (*Trema orientalis*). *Korean J. Ind. Eng. Chem.* **2004**, *10*, 766–771.
56. Wanrosli, W.D.; Peng, L.C.; Zainuddin, Z.; Tanaka, R. Effects of prehydrolysis on the production of dissolving pulp from empty fruit bunches. *J. Trop. For. Sci.* **2004**, *16*, 343–349.
57. Jahan, M.S.; Haque, M.; Arafat, K.M.Y.; Jin, Y.; Chen, H. Effect of prehydrolysis on pulping and bleaching of *Acacia auriculiformis* A. Cunn. ex Benth. *Biomass Conv. Bioref.* **2022**, *12*, 2369–2376. [CrossRef]
58. Putra, A.S.; Nakagawa-Izumi, A.; Ohi, H. Biorefinery of oil palm empty fruit bunch by nitric acid prehydrolysis soda cooking. Production of furfural and dissolving pulp. *Jpn. TAPPI J.* **2018**, *72*, 641–649. [CrossRef]
59. Blanco, A.; Monte, M.C.; Campano, C.; Balea, A.; Merayo, N.; Negro, C. Nanocellulose for industrial use. In *Handbook of Nanomaterials for Industrial Applications*; Hussain, C.M., Ed.; Elsevier: Amsterdam, The Netherlands, 2018.
60. Barba, C.; Montané, D.; Rinaudo, M.; Farriol, X. Synthesis and characterization of carboxymethylcelluloses (CMC) from non-wood fibers I. Accessibility of cellulose fibers and CMC synthesis. *Cellulose* **2002**, *9*, 319–326. [CrossRef]
61. Batalha, R.; Colodette, L.A.; Gomide, J.L.; Barbosa, J.L.; Maltha, L.C.A.; Borges, C.R.A.; Gomes, F.J. Dissolving pulp production from bamboo, Bioresour. 2012, 7, 640–651. *Bioresour* **2012**, *7*, 640–651.
62. Tomani, P. The lignoboost process. *Cellul. Chem. Technol.* **2010**, *44*, 53–58.
63. Kuwabara, E.; Koshitsuka, T.; Kajiyama, M.; Ohi, H. Impact of filtrate from bleached pulp treated with peroxymo-nosulfuric acid for effective removal of hexenuronic acid. *Japan TAPPI J.* **2011**, *65*, 97–101. [CrossRef]
64. Tappi, T. *Consistency (Concentration) of Pulp Suspensions*; TAPPI Press: Atlanta, GA, USA, 1993.
65. Tappi, T. *Holocellulose in Wood*; TAPPI Press: Atlanta, GA, USA, 2006.
66. Tappi, T. *Alpha-, Beta- and Gamma-cellulose in Pulp*; TAPPI Press: Atlanta, GA, USA, 1999.
67. Tappi, T. *Pentosans in Wood and Pulp*; TAPPI Press: Atlanta, GA, USA, 2010.
68. Tappi, T. *Acid-Insoluble Lignin in Wood and Pulp*; TAPPI Press: Atlanta, GA, USA, 2006.
69. Tappi, T. *Solvent Extractives of Wood and Pulp*; TAPPI Press: Atlanta, GA, USA, 2007.
70. Tappi, T. *Ash in Wood, Pulp, Paper, and Paperboard: Combustion at 525C*; TAPPI Press: Atlanta, GA, USA, 2007.
71. Tappi, T. *Kappa Number of Pulp*; TAPPI Press: Atlanta, GA, USA, 2006.
72. Tappi, T. *Acid Soluble Lignin in Wood and Pulp*; TAPPI Press: Atlanta, GA, USA, 1991.
73. Tappi, T. *Brightness of Pulp, Paper, and Paperboard (Directional Reflectance at 457 nm)*; TAPPI Press: Atlanta, GA, USA, 2008.
74. Tappi, T. *Viscosity of Pulp (Capillary Viscometer Method)*; TAPPI Press: Atlanta, GA, USA, 2004.
75. Shi, S.L.; He, F.W.; Zhang, Z.; Yang, R.N. *Pulp and Paper Analysis and Testing*; China Light Industry Press: Beijing, China, 2015; pp. 65–73.
76. Hart, P.W.; Sharp, H.F. Statistical determination of the effects of enzymes on bleached pulp yield. *TAPPI J.* **2005**, *4*, 3–6.

molecules

Article

Electric Current Generation by Increasing Sucrose in Papaya Waste in Microbial Fuel Cells

Segundo Rojas-Flores [1,*] , Magaly De La Cruz-Noriega [1], Santiago M. Benites [1], Daniel Delfín-Narciso [2], Angelats-Silva Luis [3], Felix Díaz [4], Cabanillas-Chirinos Luis [5] and Gallozzo Cardenas Moises [6]

[1] Vicerrectorado de Investigación, Universidad Autónoma del Perú, Lima 15842, Peru
[2] Grupo de Investigación en Ciencias Aplicadas y Nuevas Tecnologías, Universidad Privada del Norte, Trujillo 13007, Peru
[3] Laboratorio de Investigación Multidisciplinario, Universidad Privada Antenor Orrego (UPAO), Trujillo 13008, Peru
[4] Escuela Académica Profesional de Medicina Humana, Universidad Norbert Wiener, Lima 15046, Peru
[5] Instituto de Investigación en Ciencias y Tecnología de la Universidad Cesar Vallejo, Trujillo 13001, Peru
[6] Universidad Tecnológica del Perú, Trujillo 13011, Peru
* Correspondence: segundo.rojas.89@gmail.com

Citation: Rojas-Flores, S.; De La Cruz-Noriega, M.; Benites, S.M.; Delfín-Narciso, D.; Luis, A.-S.; Díaz, F.; Luis, C.-C.; Moises, G.C. Electric Current Generation by Increasing Sucrose in Papaya Waste in Microbial Fuel Cells. *Molecules* **2022**, *27*, 5198. https://doi.org/10.3390/molecules27165198

Academic Editors: Mohamad Nasir Mohamad Ibrahim, Patricia Graciela Vázquez and Mohd Hazwan Hussin

Received: 22 July 2022
Accepted: 12 August 2022
Published: 15 August 2022

Abstract: The accelerated increase in energy consumption by human activity has generated an increase in the search for new energies that do not pollute the environment, due to this, microbial fuel cells are shown as a promising technology. The objective of this research was to observe the influence on the generation of bioelectricity of sucrose, with different percentages (0%, 5%, 10% and 20%), in papaya waste using microbial fuel cells (MFCs). It was possible to generate voltage and current peaks of 0.955 V and 5.079 mA for the cell with 20% sucrose, which operated at an optimal pH of 4.98 on day fifteen. In the same way, the internal resistance values of all the cells were influenced by the increase in sucrose, showing that the cell without sucrose was 0.1952 ± 0.00214 KΩ and with 20% it was 0.044306 ± 0.0014 KΩ. The maximum power density was 583.09 mW/cm^2 at a current density of 407.13 A/cm^2 and with a peak voltage of 910.94 mV, while phenolic compounds are the ones with the greatest presence in the FTIR (Fourier transform infrared spectroscopy) absorbance spectrum. We were able to molecularly identify the species Achromobacter xylosoxidans (99.32%), Acinetobacter bereziniae (99.93%) and Stenotrophomonas maltophilia (100%) present in the anode electrode of the MFCs. This research gives a novel use for sucrose to increase the energy values in a microbial fuel cell, improving the existing ones and generating a novel way of generating electricity that is friendly to the environment.

Keywords: saccharose; microbial fuel cells; waste; papaya; bioelectricity

1. Introduction

Due to the exponential increase in society, it has generated two main problems, the need for new sources of electricity generation and a way to reuse the waste produced by human consumption. This last problem has generated a problem for the collection centers of the different municipalities of large and small cities [1,2]. In 2016, waste production exceeded 2.02 billion tons, and it was estimated that by 2050 it would be approximately 3.4 billion tons, which will make waste management increase from 1.61 million dollars in 2020 to 2.50 million by 2030 [3,4]. In this sense, part of the waste is that from agricultural production, which in its process of sowing, harvesting, sale and consumption, generates different types of waste that in recent years have begun to be reused in order to give them a second use and take advantage of them in other activities of society [5]. One of the waste products with the highest production in Latin America and consumed worldwide is papaya derivatives (*Carica papaya* L.). In its different presentations, this product represents approximately 15.36% (11.22 metric tons) of tropical fruits produced each year; approximately 24 countries produce this type of product [6]. The high consumption of this fruit is mainly due to the high amounts of vitamins A, C and E that it presents, besides being a natural diuretic. In the last decade, its production has increased by 85% for South American countries, mainly in their tropical zones [7]. On the other hand, new technologies have emerged to generate electricity in a sustainable way, for example, microbial fuel cells (MFCs). These types of devices generate electricity through the oxidation and reduction processes that occur inside them, converting chemical energy into electrical energy [8,9]. These systems are generally composed of two chambers (anodic and cathodic). In the chamber where the anodic electrode is placed, the microorganisms oxidize the organic matter, producing electrons, which travel through an external circuit to the cathodic chamber where they are oxidized. Due to this process, a flow of electrons occurs, producing electricity [10–12].

There is a wide variety of substrates used as fuels in MFCs, with biological factors being a fundamental characteristic because they are used as an environment for microbial growth and other metabolic activities, therefore, the composition and degradability of the substrate promote the rate of activity for the generation of electrons, which translates into a better performance of the MFCs [13–15]. Various investigations have reported the use of a wide range of substrates, from domestic, industrial and municipal wastewater to simpler substrates such as glucose, which serves as a carbon source in MFCs for electricity production [16,17]. In this sense, the concentration of glucose as a substrate establishes the maximum amount of chemical energy available to convert into electrical energy, so a substrate with a high content of sugars can improve the generation of electrons in MFCs [18]. In this sense, in their research, Kamau et al. (2020) used waste from avocados, tomatoes, bananas, watermelons and mangos as substrates, mainly monitoring the voltage and current values generated in MFCs, reporting that the tomato produced the highest voltage (0.702 V) and, in terms of current values, it increased linearly over time for all residues. On the other hand, that study indicated that moisture content and carbohydrate level were the main factors influencing electricity generation [19]. Likewise, Kalagbor and Akpotayire (2020) evaluated the generation of electricity from tropical fruit residues (watermelon and papaya) in single-chamber MFCs. The cells were monitored for a period of four weeks, and the maximum voltage generated was 139.5 mV in the use of the watermelon substrate and 222.9 mV produced by the papaya. In terms of the power density of the watermelon substrate, it was 0.2452 mW/cm^2. For the papaya substrate, on the other hand, the values of dissolved oxygen (DO) and biological oxygen demand (BOD) showed that the medium was conducive to the proliferation of microorganisms [20,21]. These results demonstrated that single-chamber MFCs are capable of generating electricity from tropical fruit residues, so the use of these systems was recommended as a sustainable alternative since they represent an option to increase electricity supply in urban and rural areas [22,23]. Likewise, Utami and Yenti (2018) studied the generation of electrical energy from papaya peel waste in MFCs. In the anaerobic compartment of the anode, the layer of this substrate was used as an electron donor, while in the cathode compartment, $KMnO_4$ was used as an electron acceptor. Regarding the results, the power density was 121 mW/m^2 and a current of 179 mA with a voltage of 1.095 V [24]. In this sense, it has been proven that carbohydrates rich in glucose and fructose are of essential importance for the generation of electricity in MFCs. In this sense, sucrose is a natural component present in all-natural juices.

In this sense, the main objective of this research was to evaluate the generation of electricity using papaya residues as a substrate by adding different concentrations of sucrose (0%, 5%, 10%, and 20%) in a single-chamber microbial fuel cell manufactured at low cost with Zn-Cu electrodes, monitoring their voltage, current, current density, power density and pH for 30 days. Thus, the values of the internal resistance and absorbance spectrum were also measured by FTIR (Fourier transform infrared spectroscopy).

2. Results and Discussion

Figure 1 shows the influence of sucrose concentrations on the generation of voltage, current and pH, with the concentration of sucrose at 20% being the one with the highest value when used as a substrate, reaching a maximum value (0.955 V) on day 15. From the first day, the measurements were remarkable, with the highest voltage values when the sucrose concentration was 20% generating 0.19 V more than the MFCs with 10% and 0.26 V with the MFCs used as blank (Figure 1a). While in Figure 1b it is observed that the MFCs with 20% sucrose generated a higher electric current with a peak value of 5.079 mA on the sixteenth day, all electrical current values have their maximum peaks between the eleventh and sixteenth day. The main reason for the increase in current parameters, according to Fujimura et al. (2022), is because sucrose, being a disaccharide (glucose and fructose), has been used for fermentation, releasing electrons in the process, generating electric current [24]. On the other hand, when glucose is coupled in respiratory chains, it is oxidized to gluconate by glucose dehydrogenase and is subsequently oxidized to 2-cetagluconate by gluconate dehydrogenase [25]. While microorganisms consume glucose as a source of carbon electrons and protons, previous research mentions that 24 mol of electrons and hydrogen ions are generated by the oxidation of one mole of glucose under anaerobic conditions [26]. On the other hand, the pH values increased from the first day of monitoring to the last, as shown in Figure 1c; the optimum operating pH of 4.45, 4.61, 4.77 and 4.98 for the MFCs with 0%, 5%, 10% and 20% sucrose on days 17, 15, 14 and 15, respectively. Leiva et al. (2018) mention that the low pH values in microbial fuel cells are due to the accumulation of protons in the anode electrode and hydroxide ions in the cathode electrode [27]. The pH values influence the generation of voltage and current in MFCs mainly because the microorganisms present in each cell need the ideal conditions for their growth and acclimatization [28].

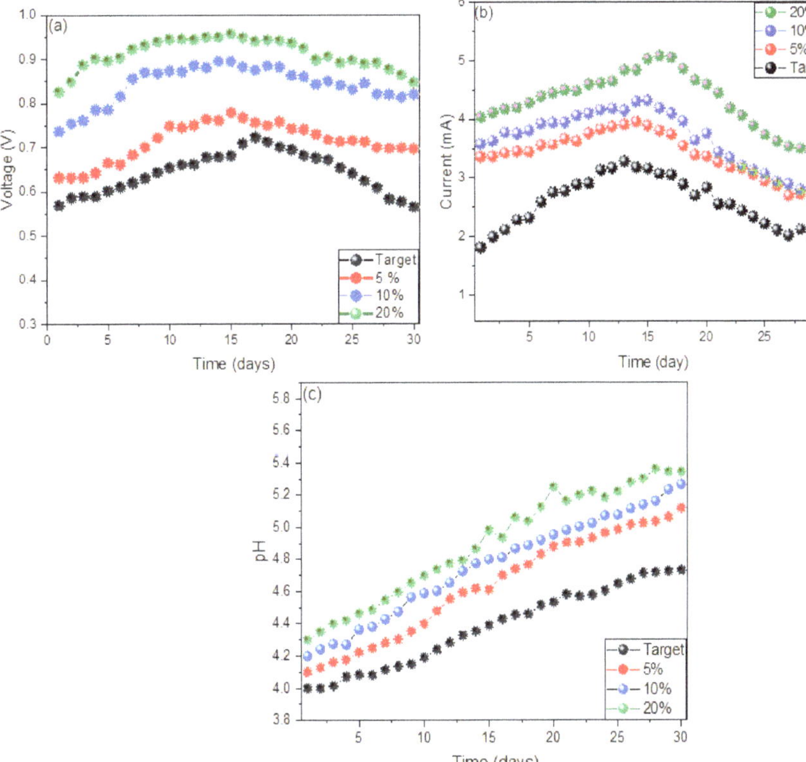

Figure 1. Values of (**a**) voltage, (**b**) electric current and (**c**) pH obtained from the monitoring of microbial fuel cells.

Figure 2 shows the values of the electrical resistance obtained from the microbial fuel cells at different percentages of sucrose, where the experimental data is adjusted to Ohm's law (V = RI), where the x-axis is the current (I) and the y axis is the voltage (V), for which the slope of the linear fit is the internal resistance ($R_{int.}$) of the cells. The $R_{int.}$ values found were 0.044306 ± 0.0014, 0.03572 ± 0.00716, 0.02269 ± 0.0015 and 0.1952 ± 0.00214 KΩ and stop the MFCs with 0%, 5%, 10% and 20% sucrose. As clearly observed in Figure 3, the values decrease with increasing sucrose concentration. According to Ueda et al. (2022), the time required for the decomposition of the substrates has a dependence on the resistance of the microbial fuel cells. It is known that when the resistance is low, the electrons flow more freely, generating a greater electric current, and it is probable that this affects the microbes in the electrode biofilm [29,30].

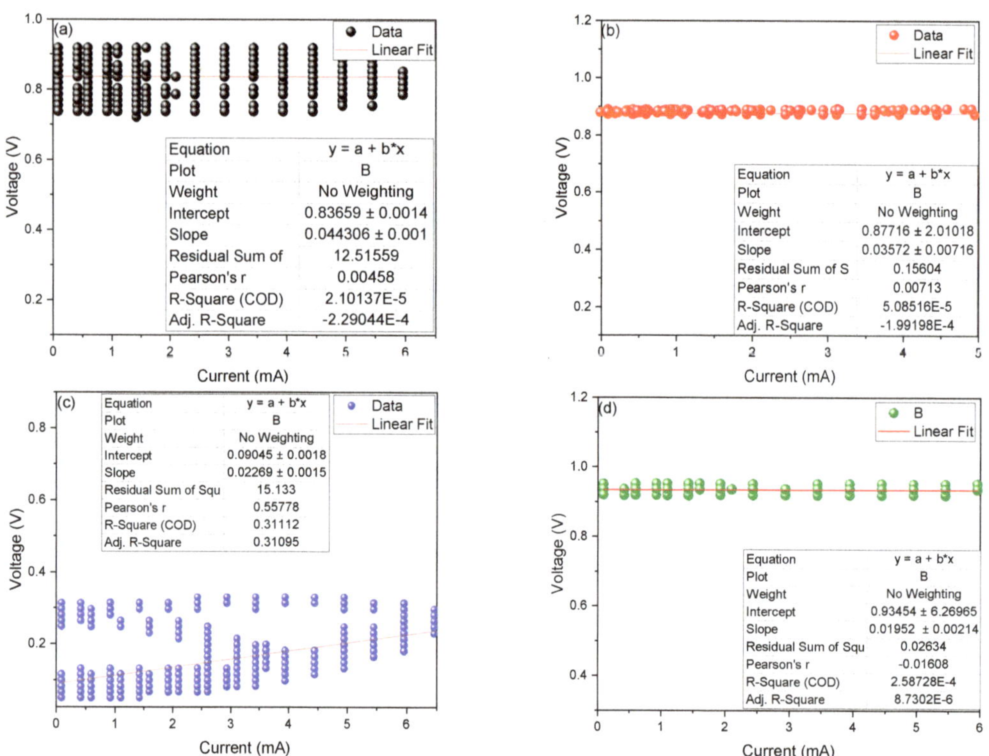

Figure 2. Internal resistance values of the microbial fuel cells at (**a**) 0, (**b**) 5, (**c**) 10 and (**d**) 20% sucrose.

Figure 3 shows the values of power density (PD) and maximum voltage as a function of current density (CD), being the MFCs with 20% sucrose the one that generated a higher value of PD with 583.09 mW/cm^2 at a CD of 407.13 A/cm^2 and a peak voltage of 910.94 mV; with a 36% higher than that generated by the MFCs used as target (0% sucrose) that generated a PD of 427.14 mW/cm^2 in a CD of 4.920 A/cm^2 with a peak voltage of 537.72 mV. These values are higher than those generated by Mohamed et al. (2020), where they used kitchen wastewater and photosynthetic algae as fuel in their dual-chamber MFCs, managing to generate maximum PD peaks of 31.6 ± 0.5 mW/cm^2 in a CD of 172 mA/cm^2 with a maximum voltage of 600 mV [31]. In the same way, Kondaveeti et al. (2019) used citrus peels as fuel in their single-chamber MFCs, managing to generate PD peaks of 63.4 mW/cm^2 at a CD of 280.56 mA/cm^2 at a peak voltage of 0.478 V [32]. According to Yaqoob et al. (2020) obtained high values of PD shown in the research are due to the metallic electrodes used due to the good electrical properties they have, so the current losses are few in the energy generation process [33].

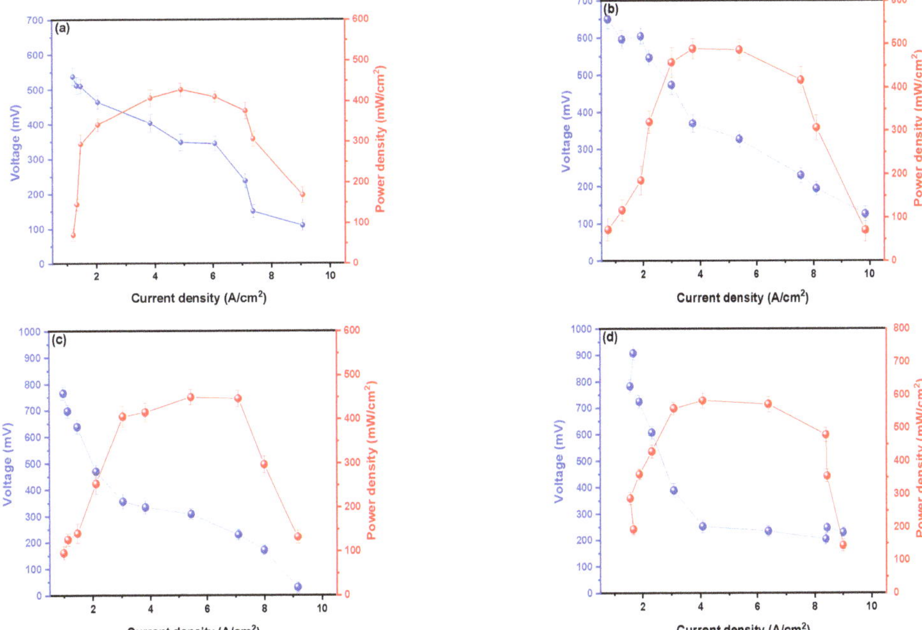

Figure 3. Values of the power densities as a function of the current density of the microbial fuel cells at (**a**) 0, (**b**) 5, (**c**) 10, and (**d**) 20% sucrose.

Figure 4 shows the absorbance spectrum of the compounds present in the different substrates (0%, 5%, 10% and 20% sucrose), observing the most intense peak at 3289 cm^{-1} belonging to the N-H stretch, O-H groups, phenols, and carboxylate acids, while peaks 2904 and 2848 cm^{-1} are associated with the C-H stretching of alkanes, aldehydes and ketones. In the same way, the peak of 1756 cm^{-1} belongs to alkane C-H stretching, while 1660 is associated with C=C stretching, N-H primary amine, C=N stretching and amide stretch. The 1545 cm^{-1} peak indicates the presence of alkane C-H stretching, alkene C=C stretching, C=N stretching, primary and secondary amine C-N stretching and amide; and finally, the peaks at 1255 and 1030 cm^{-1} alkane C-H stretching, alkene C=C stretching, C=N stretching, primary and secondary amine C-N stretching and amide [34–36]. It has been shown that the high content of phenols releases large amounts of electrons which travel through the external circuit to the cathode electrode, thus generating a higher electrical current output [37,38].

Table 1 shows the regions sequenced and analyzed in the BLAST program in which an identity percentage of 99.32% was obtained, which corresponds to the Achromobacter xylosoxidans species, 99.93% to the Acinetobacter bereziniae species, and with 100.0% to the species Stenotrophomonas maltophilia. Figure 5 shows the dedongram, which was built using the MEGA program, which relates and groups sequences of species [39]. These bacteria are ubiquitous, they are found in soil, water, air, plants and animals. They transfer electrons to the anode via external loop carrier proteins, such as cytochrome c, or via membrane appendages called [40,41] nanowires. An essential factor in the production of electric current is the formation of biofilms on the anode electrode. This consists of two types of microorganisms, fermentative and electrogenic. Where the former hydrolyzes organic compounds and the metabolites they secrete and are used as substrates for electrogenic bacteria to generate electrons, protons and CO_2 through oxidative processes [42]. Figure 6 shows the electricity generation process through microbial fuel cells, where MFCs with 5, 10 and 20% sucrose connected in series were used; managing to generate a voltage of 2.09 V, enough to turn on a red LED bulb. This shows that papaya residues have great values for the generation of bioelectricity. Recent research has shown the importance of other residues in other changes, which leads to the sustainability of these types of products [43,44].

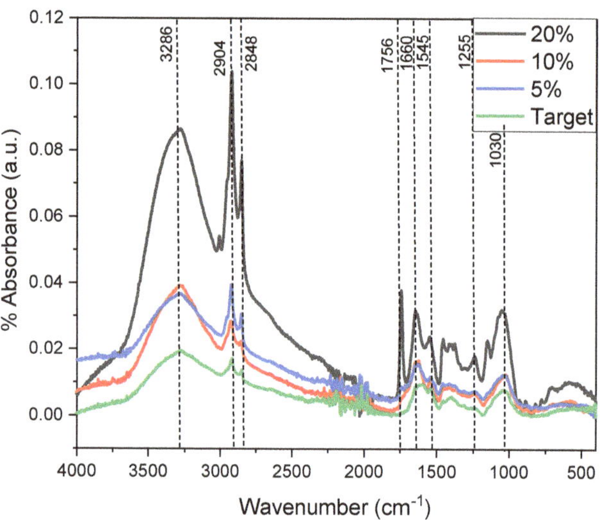

Figure 4. FTIR spectrophotometry of the papaya residues with saccharose.

Table 1. BLAST characterization of the rDNA sequence of bacteria isolated from the MFCs anode plate.

BLAST Characterization	Consensus Sequence Length (nt)	% Maximum Identity	Accession Number	Phylogeny
Achromobacter xylosoxidans	1451	99.32%	CP053617.1	Cellular organisms; Bacterium; Proteobacteria; Betaproteobacteria; burkholderials; Alcaligenaceae; Achromo-bacter
Acinetobacter bereziniae	1468	99.93%	CP018259.1	Cellular organisms; Bacteria; Proteobacteria; Gammaproteobacteria; Pseudomonadales; Moraxellaceae; Acinetobacter
Stenotrophomonas maltophilia	1477	100.00%	NR_041577.1	Cellular organisms; Bacteria; Proteobacteria; Gammaproteobacteria; Xanthomonadales; Xanthomonadaceae; Stenotrophomonas; Stenotrophomonas maltophilia group

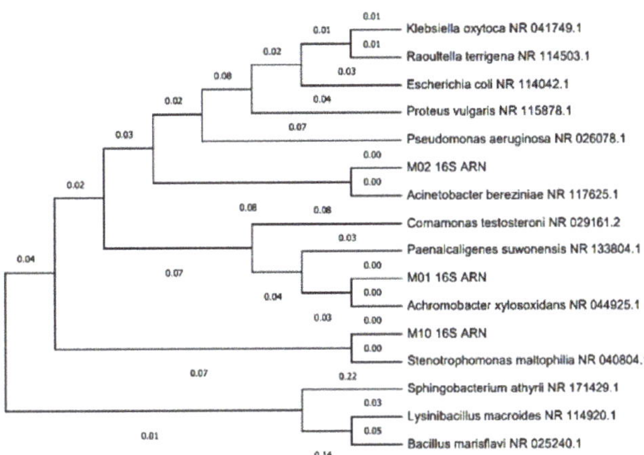

Figure 5. Dendrogram of bacterial clusters isolated from the MFCs anode plate.

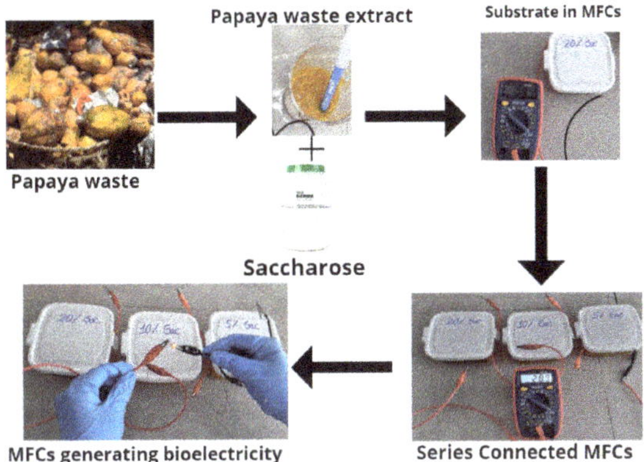

Figure 6. Electricity production in the MFCs.

3. Materials and Methods

3.1. Fabrication of Single-Chamber Microbial Fuel Cells

For the chambers of the microbial fuel cells (three in total), 400 cm^3 polyethylene terephthalate cubic containers were used, to which an 18 cm^2 hole was made on one of the faces in which the electrode was placed, cathodic (Zinc, Zn), while the anodic electrode (Copper, Cu) was placed inside the container; both electrodes were joined by means of an external circuit with a resistance of 100 Ω. As a proton exchange membrane, 10 mL of the solution obtained from 6 g of KCl and 14 g of agar in 400 mL of H$_2$O were used (see Figure 7). While the preparation of sucrose was carried out at 0 (target), 5, 10, and 20%, for this a 50% sucrose stock solution and papaya residue extract were used, with the final working volume being 200 mL.

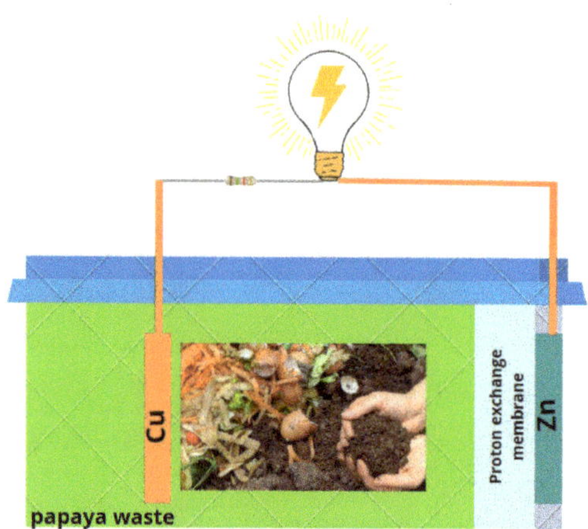

Figure 7. Schematization of single-chamber microbial fuel cells.

3.2. Collection of Papaya Waste

Three decomposing papayas (approximately 5 kg) were collected from La Hermelinda market, Trujillo, Peru. Which were collected in hermetic bags and transferred to the laboratory for use where they were washed three times with distilled water to remove any type of impurities (sand, dust or insects). These wastes were ground in an extractor (Labtron, LDO-B10-USA) until obtaining homogeneity throughout the substrate and then stored in a 1000 mL bottle at 20 ± 2 °C until used in microbial fuel cells.

3.3. Characterization of Microbial Fuel Cells

The electrical parameters of current, voltage, power density and current density were measured using a multimeter (Prasek Premium PR-85, Chicago, IL, USA) using the method described by Rojas-Flores et al. (2021), whose external resistances were 10 ± 0.2, 40 ± 2.3, 50 ± 2.7, 100 ± 3.2, 300 ± 6.2, 390 ± 7.2, 560 ± 10, 680 ± 12.3, 820 ± 14.5, $1000 \pm 20.5\ \Omega$ [45]. While the internal resistance was found using the energy sensor (Vernier- ± 30 V & ± 1000 mA, USA). Likewise, the values of pH and electrical conductivity were monitored with a pH-meter (110 Series Oakton, Chicago, IL, USA) and a conductivity meter (CD-4301, Chicago, IL, USA) during the 30 days of operation. The initial and final transmittance values were measured by FTIR (Thermo Scientific IS50, Chicago, IL, USA).

3.4. Molecular Identification of Microorganisms by Sequencing the 16S rRNA Genes

Molecular identification was carried out by the Analysis and Research Center of the "Biodes Laboratories". From pure or axenic cultures of bacteria, which were based on DNA extraction using the CTAB extraction method, which were analyzed molecularly by amplification of the 16S rRNA gene [46]. The genetic sequences were evaluated with the bioinformatic program MEGA-X to generate consensus sequences and develop phylogenetic trees. The identification of the microbial species was carried out using the Gen-Bank databases and the programs Nucleotide Blast (Basic Local Alignment Search Tool) and EzBio-Cloud [47,48]. The molecular analysis was analyzed only from the MFC with papaya waste with 20% sucrose.

4. Conclusions

Bioelectricity was successfully generated using papaya waste with sucrose in different percentages (0%, 5%, 10%, and 20%) as fuel through laboratory-scale microbial fuel cells using zinc and copper as electrodes. The cell that obtained the best electrical parameters was the one that contained the highest percentage of sucrose (20%), managing to generate an electrical voltage and current of 0.955 V and 5.079 mA, respectively, with an optimal operating pH of 4.98 on the fifteenth day. Likewise, the internal resistance of the cells decreased as sucrose increased, with the maximum internal resistance being 0.044306 ± 0.0014 KΩ and the minimum being 0.1952 ± 0.00214 KΩ belonging to the cells with 0 and 20% sucrose, respectively. Thus, it was also observed that the maximum power density was 583.09 mW/cm^2 at a current density of 407.13 A/cm^2 with a peak voltage of 910.94 mV, belonging to the cell with 20% sucrose. Finally, the absorbance peaks demonstrate the presence of phenols, which gives indications of the high values of current and voltage. Being able to identify 99.32, 99.93 and 100% of the species Achromobacter xylosoxidans, Acinetobacter bereziniae and Stenotrophomonas maltophilia, respectively, from the anode electrode of the MFCs with 20% sucrose. For future work, replicas (at least three) of each MFC should be made and, using the optimal pH values (4.98) found in this research, standardize the pH, as well as cover the metal electrodes with some chemical compound that is not harmful for the species of microorganisms found (Achromobacter xylosoxidans, Acinetobacter bereziniae and Stenotrophomonas maltophilia species) on the substrates to improve the efficiency of microbial fuel cells.

Author Contributions: Conceptualization, S.R.-F.; methodology, S.M.B. and C.-C.L.; software, S.R.-F.; validation, A.-S.L. and F.D.; formal analysis, S.R.-F. and M.D.L.C.-N.; investigation S.R.-F.; data curation, M.D.L.C.-N. and C.-C.L.; writing—original draft preparation, D.D.-N. and G.C.M.; writing—review and editing, S.R.-F. and G.C.M.; project administration, S.R.-F. All authors have read and agreed to the published version of the manuscript.

Funding: This research received no external funding.

Institutional Review Board Statement: Not applicable.

Informed Consent Statement: Not applicable.

Data Availability Statement: Not applicable.

Conflicts of Interest: The authors declare no conflict of interest.

Sample Availability: Not available.

References

1. Kim, B.; Jang, N.; Lee, M.; Jang, J.K.; Chang, I.S. Microbial fuel cell driven mineral rich wastewater treatment process for circular economy by creating virtuous cycles. *Bioresour. Technol.* **2021**, *320*, 124254. [CrossRef] [PubMed]
2. Rojas-Flores, S.; De La Cruz-Noriega, M.; Nazario-Naveda, R.; Benites, S.M.; Delfín-Narciso, D.; Rojas-Villacorta, W.; Romero, C.V. Bioelectricity through microbial fuel cells using avocado waste. *Energy Rep.* **2022**, *8*, 376–382. [CrossRef]
3. Pandit, S.; Savla, N.; Sonawane, J.M.; Sani, A.M.; Gupta, P.K.; Mathuriya, A.S.; Rai, A.K.; Jadhav, D.A.; Jung, S.P.; Prasad, R. Agricultural Waste and Wastewater as Feedstock for Bioelectricity Generation Using Microbial Fuel Cells: Recent Advances. *Fermentation* **2021**, *7*, 169. [CrossRef]
4. Segundo, R.-F.; Magaly, D.L.C.-N.; Benites, S.M.; Daniel, D.-N.; Angelats-Silva, L.; Díaz, F.; Luis, C.-C.; Fernanda, S.-P. Increase in Electrical Parameters Using Sucrose in Tomato Waste. *Fermentation* **2022**, *8*, 335. [CrossRef]
5. Montalván, G.; Lizeth, D. Efecto de los ácidos acético y cítrico para control de antracnosis (*Colletotrichum* sp.) en poscosecha de papaya (*Carica papaya* L.). Bachelor's Thesis, Universidad Central del Ecuador, Quito, Ecuador, 2018.
6. Cornejo-Condori, G.B.; Lima-Medina, I.; Bravo-Portocarrero, R.Y.; Barzola-Tito, K.; Casa-Coila, V.H. Nematodos asociados a la papaya andina (*Carica pubescens* L.), en el distrito Sandia, Puno, Perú. *BIOAGRO* **2021**, *33*, 191–203. [CrossRef]
7. Kung, C.A.L.L.; Panduro, S.K.D.; Sangama, E.D. Mermelada a base de papaya enriquecida con pulpa de camu camu. *J. Agro-Ind. Sci.* **2021**, *3*, 55–61. [CrossRef]
8. Rojas-Flores, S.; Benites, S.; De La Cruz-Noriega, M.; Cabanillas-Chirinos, L.; Valdiviezo-Dominguez, F.; Álvarez, M.Q.; Vega-Ybañez, V.; Angelats-Silva, L. Bioelectricity Production from Blueberry Waste. *Processes* **2021**, *9*, 1301. [CrossRef]
9. Ghazali, N.F.; Mahmood, N.A.N.; Abu Bakar, N.F.; Ibrahim, K.A. Temperature dependence of power generation of empty fruit bunch (EFB) based microbial fuel cell. *Malays. J. Fundam. Appl. Sci.* **2019**, *15*, 489–491. [CrossRef]
10. Slate, A.J.; Whitehead, K.A.; Brownson, D.A.; Banks, C.E. Microbial fuel cells: An overview of current technology. *Renew. Sustain. Energy Rev.* **2019**, *101*, 60–81. [CrossRef]
11. Ma, F.; Yin, Y.; Pang, S.; Liu, J.; Chen, W. A Data-Driven Based Framework of Model Optimization and Neural Network Modeling for Microbial Fuel Cells. *IEEE Access* **2019**, *7*, 162036–162049. [CrossRef]
12. Flores, S.R.; Pérez-Delgado, O.; Naveda-Renny, N.; Benites, S.M.; De La Cruz–Noriega, M.; Narciso, D.A.D. Generation of Bioelectricity Using Molasses as Fuel in Microbial Fuel Cells. *Environ. Res. Eng. Manag.* **2022**, *78*, 19–27. [CrossRef]

13. Abbassi, R.; Yadav, A.K. Introduction to microbial fuel cells: Challenges and opportunities. *Integr. Microb. Fuel Cells Wastewater Treat.* **2020**, *2020*, 3–27. [CrossRef]
14. Al Lawati, M.J.; Jafary, T.; Baawain, M.S.; Al-Mamun, A. A mini review on biofouling on air cathode of single chamber microbial fuel cell; prevention and mitigation strategies. *Biocatal. Agric. Biotechnol.* **2019**, *22*, 101370. [CrossRef]
15. Uddin, M.J.; Jeong, Y.-K.; Lee, W. Microbial fuel cells for bioelectricity generation through reduction of hexavalent chromium in wastewater: A review. *Int. J. Hydrogen Energy* **2021**, *46*, 11458–11481. [CrossRef]
16. Prathiba, S.; Kumar, P.S.; Vo, D.-V.N. Recent advancements in microbial fuel cells: A review on its electron transfer mechanisms, microbial community, types of substrates and design for bio-electrochemical treatment. *Chemosphere* **2022**, *286*, 131856. [CrossRef]
17. Khandaker, S.; Das, S.; Hossain, T.; Islam, A.; Miah, M.R.; Awual, R. Sustainable approach for wastewater treatment using microbial fuel cells and green energy generation—A comprehensive review. *J. Mol. Liq.* **2021**, *344*, 117795. [CrossRef]
18. Li, F.; An, X.; Wu, D.; Xu, J.; Chen, Y.; Li, W.; Cao, Y.; Guo, X.; Lin, X.; Li, C.; et al. Engineering Microbial Consortia for High-Performance Cellulosic Hydrolyzates-Fed Microbial Fuel Cells. *Front. Microbiol.* **2019**, *10*, 409. [CrossRef]
19. Christwardana, M.; Frattini, D.; Accardo, G.; Yoon, S.P.; Kwon, Y. Optimization of glucose concentration and glucose/yeast ratio in yeast microbial fuel cell using response surface methodology approach. *J. Power Sources* **2018**, *402*, 402–412. [CrossRef]
20. Yaqoob, A.A.; Ibrahim, M.N.M.; Yaakop, A.S.; Ahmad, A. Application of microbial fuel cells energized by oil palm trunk sap (OPTS) to remove the toxic metal from synthetic wastewater with generation of electricity. *Appl. Nanosci.* **2021**, *11*, 1949–1961. [CrossRef]
21. Mbugua, J.K.; Mbui, D.N.; Mwaniki, J.; Mwaura, F.; Sheriff, S. Influence of Substrate Proximate Properties on Voltage Production in Microbial Fuel Cells. *J. Sustain. Bioenergy Syst.* **2020**, *10*, 43–51. [CrossRef]
22. Kalagbor Ihesinachi, A.; Akpotayire Stephen, I. Electricity Generation from Waste Tropical Fruits—Watermelon (Citrullus lanatus) and Paw-paw (Carica papaya) using Single Chamber Microbial Fuel Cells. *Int. J. Energy Inf. Commun.* **2020**, *11*, 11–20. [CrossRef]
23. Pathak, P.D.; Mandavgane, S.A.; Kulkarni, B.D. Waste to Wealth: A Case Study of Papaya Peel. *Waste Biomass Valorization* **2018**, *10*, 1755–1766. [CrossRef]
24. Utami, L.; Yenti, E. View of Produksi Energi Listrik dari Limbah Kulit Pepaya (Carica papaya) Menggunakan Teknologi Microbial Fuel Cells. *Al-Kimia* **2018**, *6*, 56–62. Available online: https://journal3.uin-alauddin.ac.id/index.php/al-kimia/article/view/4681/pdf (accessed on 21 July 2022).
25. Rojas-Flores, S.; Noriega, M.D.L.C.; Benites, S.M.; Gonzales, G.A.; Salinas, A.S.; Palacios, F.S. Generation of bioelectricity from fruit waste. *Energy Rep.* **2020**, *6*, 37–42. [CrossRef]
26. Rojas Flores, S.; Naveda, R.N.; Paredes, E.A.; Orbegoso, J.A.; Céspedes, T.C.; Salvatierra, A.R.; Rodríguez, M.S. Agricultural wastes for electricity generation using microbial fuel cells. *Open Biotechnol. J.* **2020**, *14*, 52–58. [CrossRef]
27. Kumar, P.; Nagarajan, A.; Uchil, P. Analysis of Cell Viability by the Lactate Dehydrogenase Assay. *Cold Spring Harb. Protoc.* **2018**, *18*. [CrossRef]
28. Yoon, K.-J.; Ringeling, F.R.; Vissers, C.; Jacob, F.; Pokrass, M.; Jimenez-Cyrus, D.; Su, Y.; Kim, N.-S.; Zhu, Y.; Zheng, L.; et al. Temporal Control of Mammalian Cortical Neurogenesis by m6A Methylation. *Cell* **2017**, *171*, 877–889.e17. [CrossRef]
29. Fujimura, S.; Kamitori, K.; Kamei, I.; Nagamine, M.; Miyoshi, K.; Inoue, K. Performance of stacked microbial fuel cells with barley–shochu waste. *J. Biosci. Bioeng.* **2022**, *133*, 467–473. [CrossRef]
30. Buñay, A.; Miguel, L. Modelado y Simulación del Proceso de Generación de Bio-Electricidad en una Celda Microbiana (MFC) con los Sustratos: Glucosa y Lixiviados. Bachelor's Thesis, Escuela Superior Politécnica de Chimborazo, Riobamba, Ecudaor, 2019.
31. Santiago, B.; Rojas-Flores, S.; De La Cruz Noriega, M.; Cabanillas-Chirinos, L.; Otiniano, N.M.; Silva-Palacios, F.; Luis, A.S. Bioelectricity from Saccharomyces cerevisiae yeast through low-cost microbial fuel cells. In Proceedings of the 18th LACCEI International Multi-Conference for Engineering, Education, and Technology: Engineering, Integration, and Alliances for a Sustainable Development, Virtual, 27–31 July 2020; pp. 27–31.
32. Leiva, E.; Leiva-Aravena, E.; Rodríguez, C.; Serrano, J.; Vargas, I. Arsenic removal mediated by acidic pH neutralization and iron precipitation in microbial fuel cells. *Sci. Total Environ.* **2018**, *645*, 471–481. [CrossRef]
33. Pietrelli, A.; Bavasso, I.; Lovecchio, N.; Ferrara, V.; Allard, B. MFCs as biosensor, bioreactor and bioremediator. In Proceedings of the IEEE 8th International Workshop on Advances in Sensors and Interfaces (IWASI), Otranto, Italy, 13–14 June 2019; pp. 302–306.
34. Ueda, M.; Tojo, S.; Chosa, T.; Uchigasaki, M. Decomposition characteristics of propionate when changing the electrode material, external resistance and reactor temperature of microbial fuel cells. *Int. J. Hydrogen Energy* **2022**, *47*, 2783–2793. [CrossRef]
35. Yang, X.-L.; Wang, Q.; Li, T.; Xu, H.; Song, H.-L. Antibiotic removal and antibiotic resistance genes fate by regulating bioelectrochemical characteristics in microbial fuel cells. *Bioresour. Technol.* **2022**, *348*, 126752. [CrossRef] [PubMed]
36. Mohamed, S.N.; Hiraman, P.A.; Muthukumar, K.; Jayabalan, T. Bioelectricity production from kitchen wastewater using microbial fuel cell with photosynthetic algal cathode. *Bioresour. Technol.* **2020**, *295*, 122226. [CrossRef] [PubMed]
37. Kondaveeti, S.; Mohanakrishna, G.; Kumar, A.; Lai, C.; Lee, J.-K.; Kalia, V.C. Exploitation of Citrus Peel Extract as a Feedstock for Power Generation in Microbial Fuel Cell (MFC). *Indian J. Microbiol.* **2019**, *59*, 476–481. [CrossRef] [PubMed]
38. Yaqoob, A.A.; Ibrahim, M.N.M.; Rodríguez-Couto, S. Development and modification of materials to build cost-effective anodes for microbial fuel cells (MFCs): An overview. *Biochem. Eng. J.* **2020**, *164*, 107779. [CrossRef]
39. Sharma, N.; Khajuria, Y.; Singh, V.K.; Kumar, S.; Lee, Y.; Rai, P.K.; Singh, V.K. Study of Molecular and Elemental Changes in Nematode-infested Roots in Papaya Plant Using FTIR, LIBS and WDXRF Spectroscopy. *At. Spectrosc.* **2020**, *41*, 110–118. [CrossRef]

40. Ahlawat, J.; Kumar, V.; Gopinath, P. *Carica papaya* loaded poly (vinyl alcohol)-gelatin nanofibrous scaffold for potential application in wound dressing. *Mater. Sci. Eng. C* **2019**, *103*, 109834. [CrossRef]
41. Kale, R.; Barwar, S.; Kane, P.; More, S. Green synthesis of silver nanoparticles using papaya seed and its characterization. *Int. J. Res. Appl. Sci. Eng. Technol.* **2018**, *6*, 168–174. [CrossRef]
42. Hedbavna, P.; Rolfe, S.A.; Huang, W.E.; Thornton, S.F. Biodegradation of phenolic compounds and their metabolites in contaminated groundwater using microbial fuel cells. *Bioresour. Technol.* **2016**, *200*, 426–434. [CrossRef]
43. Shen, J.; Du, Z.; Li, J.; Cheng, F. Co-metabolism for enhanced phenol degradation and bioelectricity generation in microbial fuel cell. *Bioelectrochemistry* **2020**, *134*, 107527. [CrossRef]
44. Hall, B.G. Building Phylogenetic Trees from Molecular Data with MEGA. *Mol. Biol. Evol.* **2013**, *30*, 1229–1235. [CrossRef]
45. Tangarife García, N.S. *Control Biológico, la Nueva era de la Agricultura*; Universidad de Ciencias Aplicadas y Ambientales: Bogota, Colombia, 2021.
46. Romo, D.M.R.; Gutiérrez, N.H.H.; Pazos, J.O.R.; Figueroa, L.V.P.; Ordóñez, L.A.O. Bacterial diversity in the Cr (VI) reducing biocathode of a Microbial Fuel Cell with salt bridge. *Rev. Argent. Microbiol.* **2019**, *51*, 110–118.
47. Umar, M.F.; Abbas, S.Z.; Ibrahim, M.N.M.; Ismail, N.; Rafatullah, M. Insights into Advancements and Electrons Transfer Mechanisms of Electrogens in Benthic Microbial Fuel Cells. *Membranes* **2020**, *10*, 205. [CrossRef]
48. Wang, X.; Li, C.; Lam, C.H.; Subramanian, K.; Qin, Z.-H.; Mou, J.-H.; Jin, M.; Chopra, S.S.; Singh, V.; Ok, Y.S.; et al. Emerging waste valorisation techniques to moderate the hazardous impacts, and their path towards sustainability. *J. Hazard. Mater.* **2022**, *423*, 127023. [CrossRef]

molecules

Article

Mass Balance and Compositional Analysis of Biomass Outputs from Cacao Fruits

Marisol Vergara-Mendoza, Genny R. Martínez, Cristian Blanco-Tirado and Marianny Y. Combariza *

Escuela de Química, Universidad Industrial de Santander, Bucaramanga 680002, Colombia;
marisolvergara14@yahoo.es (M.V.-M.); genmarba@gmail.com (G.R.M.); cblancot@uis.edu.co (C.B.-T.)
* Correspondence: marianny@uis.edu.co

Abstract: The global chocolate value chain is based exclusively on cacao beans (CBs). With few exceptions, most CBs traded worldwide are produced under a linear economy model, where only 8 to 10% of the biomass ends up in chocolate-related products. This contribution reports the mass balance and composition dynamics of cacao fruit biomass outputs throughout one full year of the crop cycle. This information is relevant because future biorefinery developments and the efficient use of cacao fruits will depend on reliable, robust, and time-dependent compositional and mass balance data. Cacao husk (CH), beans (CBs), and placenta (CP) constitute, as dry weight, 8.92 ± 0.90 wt %, 8.87 ± 0.52 wt %, and 0.57 ± 0.05 wt % of the cacao fruit, respectively, while moisture makes up most of the biomass weight (71.6 ± 2.29 wt %). CH and CP are solid lignocellulosic outputs. Interestingly, the highest cellulose and lignin contents in CH coincide with cacao's primary harvest season (October to January). CB contains carbohydrates, fats, protein, ash, and phenolic compounds. The total polyphenol content in CBs is time-dependent, reaching maxima values during the harvest seasons. In addition, the fruit contains 4.13 ± 0.80 wt % of CME, a sugar- and nutrient-rich liquid output, with an average of 20 wt % of simple sugars (glucose, fructose, and sucrose), in addition to minerals (mainly K and Ca) and proteins. The total carbohydrate content in CME changes dramatically throughout the year, with a minimum of 10 wt % from August to January and a maximum of 29 wt % in March.

Keywords: cacao fruit; cacao husk; cacao mucilage exudate; cacao placenta; cacao beans; valorization

Citation: Vergara-Mendoza, M.; Martínez, G.R.; Blanco-Tirado, C.; Combariza, M.Y. Mass Balance and Compositional Analysis of Biomass Outputs from Cacao Fruits. *Molecules* **2022**, *27*, 3717. https://doi.org/10.3390/molecules27123717

Academic Editors: Mohamad Nasir Mohamad Ibrahim, Patricia Graciela Vázquez and Mohd Hazwan Hussin

Received: 7 April 2022
Accepted: 13 May 2022
Published: 9 June 2022

Publisher's Note: MDPI stays neutral with regard to jurisdictional claims in published maps and institutional affiliations.

1. Introduction

Cacao is an important crop in the equatorial regions of Latin America, Western Africa, and Southeast Asia, providing income to more than 4.5 million families worldwide. According to the International Cocoa Organization (ICCO), world gross cacao bean (CB) production in 2019 was 4.75×10^6 tons [1]. The main cacao-producing countries are Côte d'Ivoire, Ghana, Indonesia, Brazil, Cameroon, Nigeria, Ecuador, Peru, and the Dominican Republic. In South America, more than 90% of CBs come from Brazil, Ecuador, Peru, Colombia, and Mexico [2]. In 2019, Colombia produced 5.97×10^4 tons of CBs, an increase of 4.9% with respect to 2018. The Santander region in Colombia contributes 42% of the country´s CBs [3]. As with many other agricultural commodities, CB production generates abundant residual biomass. Typically, only 8 to 10% of the biomass contained in the cacao fruit ends up as cacao beans, while the remaining biomass is discarded as waste. Figure 1 shows the traditional approach to cacao fruit processing worldwide. Clearly, the current process is aimed (with a few exceptions) exclusively at obtaining fermented and dry cacao beans. Considering the scheme in Figure 1 as the standard process for cacao fruit, the residual biomass from cacao crops worldwide could easily reach up to 4.5×10^7 tonnes/year (2019).

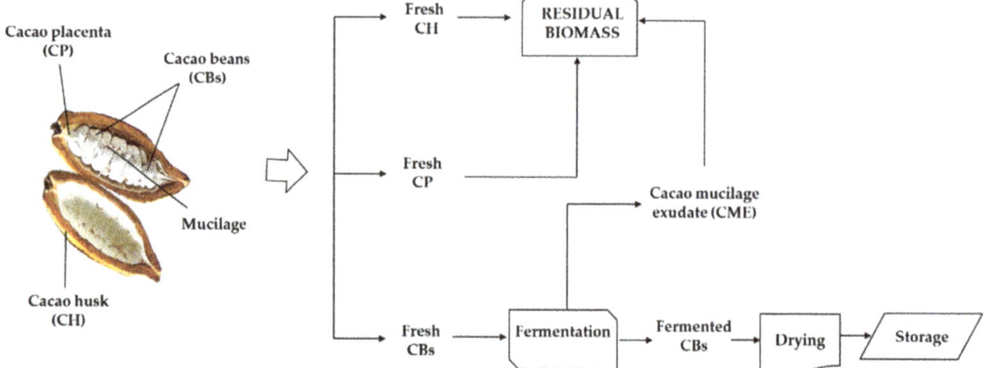

Figure 1. Traditional scheme for cacao fruit processing in tropical cacao-producing countries.

The cacao fruit (Figure 1) has a dense shell, or cacao husk (CH), protecting the seeds. Once harvested, the fruit is opened, the husk is discarded, and the seeds (covered by a white mucilage) are stored and fermented in heaps or wooden crates to produce cacao beans. Literature reports from 2000 to 2021 [4–7] show a steady increase in the number of publications related to the use of some residual biomass outputs from cacao bean production. The scientific interest in residual biomass usage is fueled by the need to increase the cacao crop's circularity, strengthen the cacao value chain, increase the economic profitability of the crop for producers, and reduce the environmental impacts.

There are many uses for cacao residual biomass, particularly cacao husk (CH), involving energy generation, biomaterials extraction, the isolation of active ingredients, catalyst support, human food, animal feed, and biocomposite synthesis, among others. For instance, cacao husk (CH) anaerobic digestion and fermentation result in biogas production [8–10]. Energetic potential analysis showed that gasification is better than pyrolysis or combustion for CH processing [11]. Along the same lines, the CH potential for electricity generation in Uganda was evaluated, reporting a potential energy generation of 24×10^6 MWh for the 9.03×10^3 tonnes of CH produced in Uganda in 2018 [12].

Materials production is another alternative for utilizing cacao residual biomass. For instance, controlled CH pyrolysis can be used for activated carbon production. Some authors reported that the yields and textural properties of the material improve by using acid (H_3PO_4) or basic (KOH) treatments [13]. An activated carbon surface area of over $1300 \text{ m}^2/\text{g}$ using de-ashed CH as the raw material has been reported [14]. Activated carbon derived from CH exhibited an adsorption performance comparable to that of commercial activated carbon for the removal of Congo Red and residual drugs (diclofenac) from aqueous solutions [15,16]. Similarly, there are reports of catalyst and nanoparticle synthesis using CH; for example, the production of Neem seed oil methyl esters, using CH ash as a catalyst, or silver nanoparticle synthesis using CH extract [17,18]. Additionally, researchers reported the use of the pectins in CH as a nutraceutical and functional pharmaceutical excipients [19]. CH has also been tested as a filler in a bioplastic-based composite [20] and a synthetic PP composite [21]. The tensile strength, flexural strength, and modulus of thermoplastic polyurethane (TPU) were improved by the addition of CH fibers [22]. In the field of animal nutrition, CH can be a substitute for traditional ingredients, like replacing maize as a dietary staple for raising rabbits [23] or pigs [24].

However, there is little information in the literature regarding the uses of biomass outputs from cacao fruit other than CH, such as cacao placenta (CP) or cacao mucilage exudate/sweatings (CME). Cacao mucilage exudate (CME), in particular, is generally lost as a lixiviate during cacao bean fermentation. However, it is occasionally consumed as a fresh drink or is used for manufacturing syrups, jams, marmalades, and alcoholic beverages [25–32]. Likewise, the technological applications of CME include ethanol production by fermentation [33]. Recently, our group reported the use of CME as culture media for bacterial cellulose biosynthesis [34]. Finally, the cacao placenta (CP), a tissue that holds the cacao beans together inside the pod (Figure 1), has been used in the production of beverages and nutritional bars [35].

Cacao bean prices, like any other commodity in the world, are controlled globally by the interaction between supply and demand chains. However, global markets are changing, and chocolate consumers are driven by additional factors such as fair trade, direct trade, and reduced environmental impacts (directly associated with crop circularity and the emission of pollutants during beans processing). The cacao fruit has a lot to offer besides its beans, and with cacao production and consumption on the rise, there is abundant residual biomass from the crop that can be used to strengthen the crop value chain, particularly in the producers' countries. However, despite the many possible uses for residual biomass from cacao, currently, there is a systematic lack of information in terms of compositional and mass balance data for the whole cacao fruit, let alone data regarding the seasonal variations in these fundamental crop characteristics. Most of the compositional data available in the current literature relates to small samples of cacao fruit collected exclusively during the harvest months. Knowing the seasonal compositional and mass balance changes in biomass outputs from cacao fruit processing is of fundamental importance to planning for efficient usage of the residual biomass from fermented beans production. In addition, future biorefinery approaches and final usage methods for cacao fruit biomass will be determined by its abundance and composition.

Thus, we hypothesize that it is possible to observe cacao fruit composition and mass-balance dynamics by analyzing representative one-tonne samples, collected monthly for one year. Initially, we separated the traditional biomass outputs of cacao fruit processing (as seen in Figure 1) and determined their abundance, as percentages by weight of the fruit. Next, we collected, processed, and stored all cacao fruit biomass outputs for a detailed mass balance and compositional analysis. CH, CP, CBs, and a new biomass output labeled cacao mucilage exudate (CME) were studied. In the traditional cacao-processing scheme, only a low percentage of the cacao fruit biomass, represented in dry/fermented CBs, is used in the chocolate industry. The remaining percentage, represented in dry CH and CP, and CME are considered to be residual biomass. These residues could be used as raw materials to feed other processes to produce value-added products, to strengthen the cacao value chain in the producers' countries.

2. Results and Discussion

2.1. Cacao Fruit Mass Balance

The traditional process for cacao bean production is manually intensive and starts with cacao fruit harvesting, followed by fruit-opening and the extraction of the fresh cacao beans (CBs), as shown in Figure 1. All these processes are performed on-site (on-farm). Typically, the empty fruit husk (CH) and the placenta (CP) are discarded and left to decompose in the field. In the traditional process, only fresh CBs undergo further treatment, in a step called fermentation.

Figure 2 shows the seasonal percentage by weight variation of cacao beans (CBs), cacao placenta (CP), and cacao fruit husk (CH), of twelve cacao fruit loads (one tonne each), gathered over the months of February to January, using the traditional cacao fruit-processing scheme (Figure 1). Table S1 of the Supplementary Information contains detailed wt % data for fresh CH, CP, and CBs in every cacao-fruit load processed. The percentage by weight values reported in Figure 2 correspond to the fresh (wet) by-products with respect to the fresh fruit weight. Fresh CH is by far the most abundant cacao fruit byproduct, with a total average of 66.96 ± 2.83 wt %, followed by fresh CBs, with 24.55 ± 1.51 wt %, and CP, with 2.58 ± 0.22 wt %. The losses (5.93 ± 3.15 wt %) correspond to missing CH pieces and spoiled fruits and beans. These percentages are similar to those reported for cacao fruit from Bundibugyo District in Uganda (clone not specified) [8], for cacao fruit samples from Ghana (clone not specified) [36], and for cacao fruit from plantations located in Perak, Malaysia (clone not specified) [37]. The percentage by weight of the fresh by-products did not change significantly for cacao fruit loads gathered over the course of one year, as seen in Figure 2.

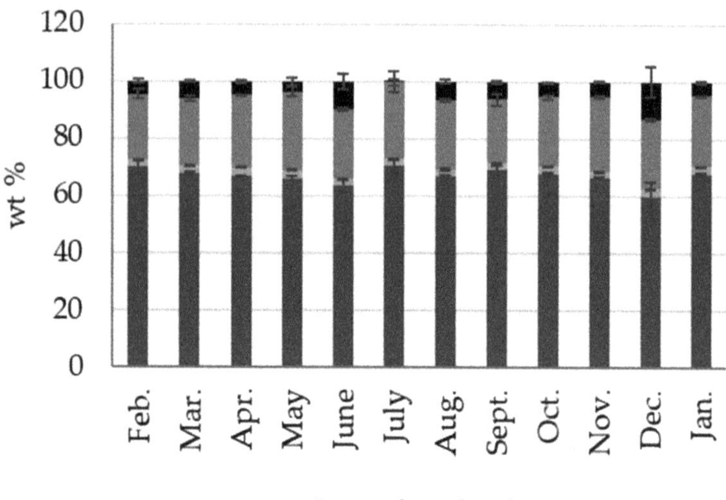

Figure 2. Seasonal variations in the percentage weight of fresh cacao fruit biomass outputs, cacao husk (CH), cacao placenta (CP), and cacao beans (CBs), derived from the traditional approach of cacao fruit processing (first stage, Figure 1).

For the cacao fruit compositional and mass balance analyses, we used a two-stage procedure shown in Figure 3. The first stage involved opening the cacao fruit and separating the fresh cacao beans (CBs) and the placenta (CP) from the husk (CH), as discussed above. This first process represents the traditional approach to cacao fruit usage (Also shown in Figure 1). In the second stage, the CP and the CH were ground, dried, and stored for further compositional analysis (Figure 3). Once the cacao fruit components were separated and their fresh percentage by weight measured, we subjected the materials to different procedures, as shown in Figure 3 (stage 2). Fresh CBs were fermented for seven days, then the cacao mucilage exudate (CME) lixiviated from the process was collected. Fresh CP and CH were ground and dried. Figure 4 shows the seasonal variation in percentage by weight of dried cacao husk (CH), dried cacao placenta (CP), liquid cacao mucilage exudate (CME), and fermented and dried cacao beans (CBs). Table S2 of the Supplementary Information contains detailed wt % data for dried CH, CP, CBs, and liquid CME in every cacao fruit load processed. Dried CB and CH exhibit similar average percentages by weight of 8.92 ± 0.90 wt % and 8.87 ± 0.52 wt %, respectively (see Table S2). Dried CP amounts to a small contribution of 0.57 ± 0.05 wt %, as well as CME, with 4.13 ± 0.80 wt %. These results show that four different biomass outputs (dried CH, dried CP, dried CBs, and CME) can be derived from the cacao fruit. Each type of biomass has a particular composition and potential use, as discussed in the following sections.

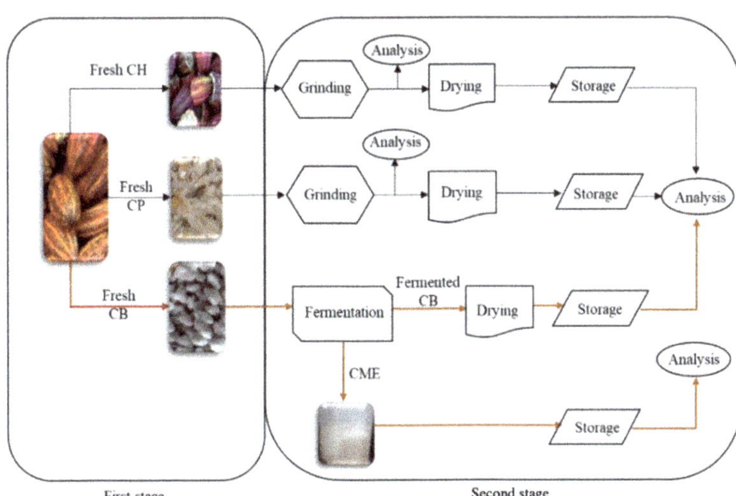

Figure 3. General scheme of cacao-fruit processing for this work.

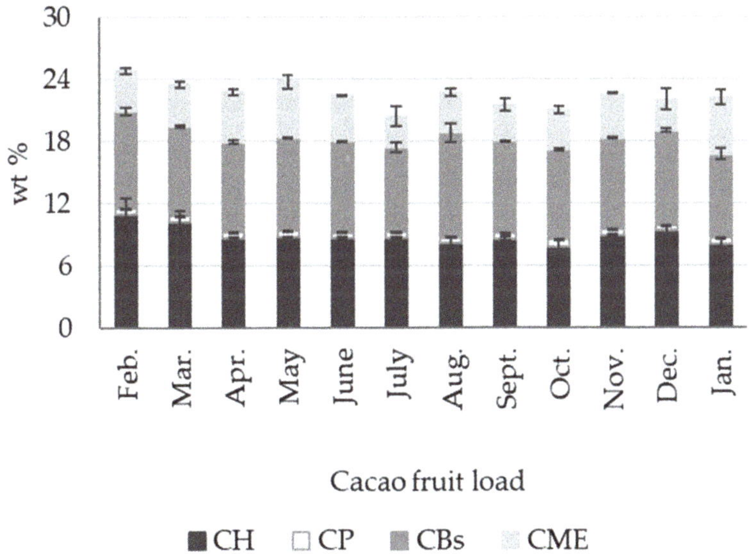

Figure 4. Seasonal variations in the percentage weight of cacao fruit biomass outputs, dried cacao husk (CH), dried cacao placenta (CP), dried cacao beans (CBs), and liquid cacao mucilage exudate (CME) derived from the second stage of cacao fruit processing (Figure 3). Note: the remaining percentage corresponds to moisture.

The mass balance for the first and second stages of cacao fruit processing is shown in Figure 5. The values reported correspond to the actual weights measured while processing the load collected in November (Nov.), the main cacao fruit harvest season in Colombia. Figure S1 of the Supplementary Information also contains mass balance information for the first and second stages of cacao fruit processing, with averaged wt % data for all biomass outputs. Starting with one tonne of the fresh cacao fruit, we collected 66.3 kg of fresh CH (66.33 wt %), 26.4 kg of fresh CP (2.64 wt %), and 256.9 kg of fresh CBs (25.69 wt %). These materials are high in moisture; after the second stage, which involves grinding (CH, CP), drying (CH, CP), and fermenting (CBs), their mass is reduced dramatically. At this point, 89.4 kg of dried CH, 87.7 kg of dry/fermented CBs, and 5.7 kg of dried CP are the solid materials resulting from the second stage. In addition, a liquid by-product, cacao mucilage exudate CME, can be recovered during the second stage. The CME (43.1 kg) is produced as a lixiviate during the first hours of CB fermentation. Overall, the moisture content in the solid materials (CH, CBs, and CP) present in one tonne of fresh cacao fruit (Nov.) was 720.7 kg. It is worth noting that the traditional way of cacao fruit processing (first stage, Figure 4) results in only 8.77 wt % of useful product, in the form of fermented/dry cacao beans. Ideally, other byproducts of cacao fruit processing, such as CH, CP, and CME, corresponding to an additional 13.82 wt % of the fruit, could be potentially used as raw materials for value-added products. However, the use of these residues depends on availability and the compositional data, which are relatively scarce in the scientific literature. We provide average compositional data for the main biomass outputs, along with compositional changes over one year, in the following sections.

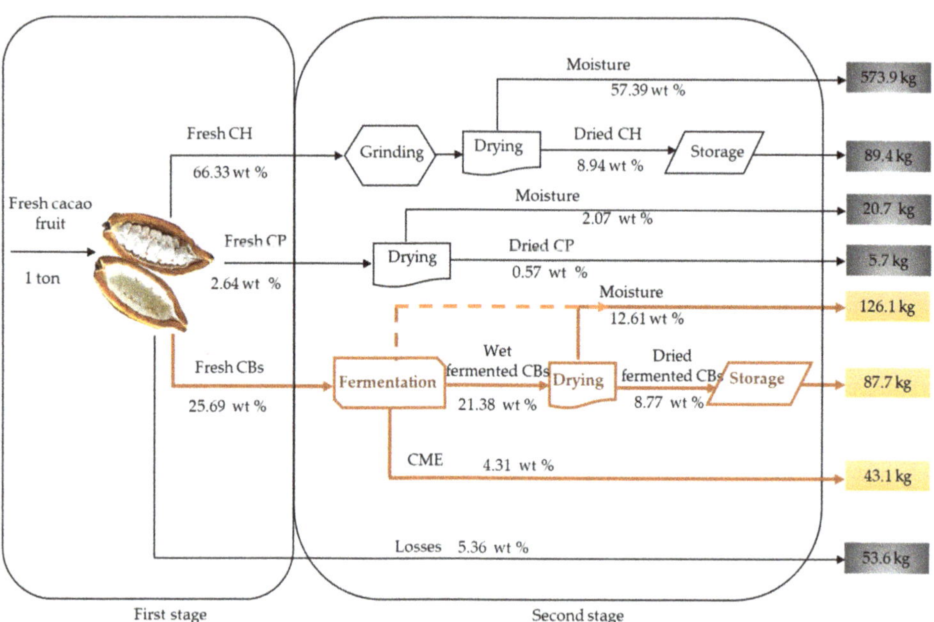

Figure 5. Mass balance for 1 tonne of fresh cacao fruit during the first and second stages of processing (load: Nov.).

2.2. Compositional Analysis

2.2.1. Cacao Fruit Husk (CH)

The cacao fruit husk (CH) functions as a protective pod for the seeds. Thus, CH exhibits a complex chemical structure involving supportive tissue of high mechanical resistance and high water-holding capability. Table 1 shows the average proximate analysis of fresh cacao husk (CH) clone CCN 51, corresponding to twelve loads (one tonne each) of cacao fruit collected over one year. Table S3 of the Supplementary Information also contains detailed wt % CH proximate analysis data for each cacao fruit load processed. The proximate analysis provides a broad classification of components in biomass, which is information of fundamental importance, mainly for animal feed evaluation and biomass energy use. CH is mainly composed of water (84.62 ± 1.97 wt %) and solids (15.38 ± 1.97 wt %). In the literature, the values reported for ash (9.1 wt %) and crude fiber (30.93) [38] are slightly lower than those found in this work (Table 1). Additionally, the reported protein contents of 5.9 wt % [38] are comparable to the 5.44 wt % found in this work. Another study in the literature [39] shows lower values for protein content of 2.42 ± 0.37 wt %, along with 0.93 ± 0.34 for crude fat and 87.06 ± 0.58 moisture, with a comparable value of 10.65 wt % for the ash.

Table 1. Average proximate analysis of fresh cacao husk (CH) from clone CCN51 cacao fruit.

Component.	Percentage by Weight (wt %)
Moisture	84.62 ± 1.97
Total solids	15.38 ± 1.97
Crude fat *	0.29 ± 0.08
Crude protein *	5.44 ± 0.80
Crude fiber *	30.93 ± 2.03
Ash *	10.65 ± 1.62
Total nitrogen *	0.88 ± 0.13
Nitrogen-free extract (NFE) *	52.69 ± 2.50

* As a percentage of the total solids.

The protein, fat, and crude fiber content in CH makes it attractive as animal feed. For example, ruminants fed with CH from West Java, Indonesia (clone not specified) gained more weight than control ruminants [40]. CH-fed rabbits (20% of their food intake) from Ghana exhibited increased weight compared to corn-fed control animals [9]. In addition, replacing maize with CH (variety not specified) in the diet of *Oreochromis niloticus* (tilapia) reduces its culture costs without significantly affecting the fish's survival rates and weight [41].

Nowadays, non-edible lignocellulosic materials, readily available as residues from agro-industrial processes, are being considered as alternatives to renewable chemicals and fuels to replace oil-based products. Table 2 compares previous works on cacao husk composition in terms of structural carbohydrates (cellulose and hemicellulose) and lignin, with the average values for twelve cacao loads, as measured in this work. Structural carbohydrates and lignin, collectively, account for up to 78 wt % of the total solids in cacao husk (CH). The average cellulose content for all loads in terms of CH (25.64 ± 3.49 wt %) is lower than that found in various agricultural biomass outputs, such as rice husk (37.1 wt %), sugarcane bagasse (32–44 wt %), cassava peels (37.9 wt %), corn cob (38.27 wt %) and sweet sorghum bagasse (45 wt %). Likewise, the hemicellulose content in CH (19.96 ± 2.42 wt %) is lower than the values found in nutshells and cassava peels, which are 25–30 and 23.9 wt %, respectively. Lignin content, on the other hand, is higher in CH (32.73 ± 5.15 wt %) than that in spent coffee grounds (23.5 wt %), rice husk (24.1 wt %), sugarcane bagasse (19–24 wt %), cassava peels (7.5 wt %), corn cob (7.16 wt %), and sweet sorghum bagasse (21 wt %) [42–47]. Table S4 of the Supplementary Information contains detailed compositional information on various agricultural biomasses for comparison purposes.

Table 2. Structural carbohydrates and lignin in cacao husk (CH) from cacao fruit samples of various origins and this work.

Cellulose	Hemicellulose	Lignin	References
	(wt %)		
31.7 ± 0.1	27.0 ± 0.1	21.7 ± 0.1	[10]
30.79	21.09	25.55	[48]
18.42	10.04	12.06	[49]
35	10	14.6	[50]
30.41 ± 0.20	11.97 ± 3.17	33.96 ± 1.9	[47]
35.4 ± 0.33	37 ± 0.5	14.7 ± 0.35	[42]
41.92 ± 0.09	35.26 ± 0.05	0.95 ± 0.04	[51]
26.15 ± 3	12.7 ± 56	21.16 ± 2.6	[52]
25.64 ± 3.49	19.96 ± 2.42	32.73 ± 5.15	This work *

* As a percentage of the average total solids value from Table 3.

Table 3. Average proximate analysis for sun-dried CH from cacao fruit samples of various origins.

Moisture	Ash	Crude Fat	Crude Protein	Crude Fiber	Reference
		wt %			
12.5	12.3	VNR	VNR	VNR	[12]
VNR	11.44 ± 0.41	0.93 ± 0.34	2.42 ± 0.37	VNR	[39]
10.5	9	1.5	2.1	VNR	[54]
VNR	VNR	2.5	8.4	32.3	[40]
13	13	0.6	8	50	[23]
VNR	9.1–10.1	VNR	5.9–7.6	22.6–32.5	[50]
VNR	VNR	1.2–10	5.9–9.1	22.6–35.7	[36]
10.04 ± 0.3	12.67 ± 0.14	VNR	VNR	33.6 ± 0.15	[55]
6.72 ± 0.17	8.32 ± 0.7	2.24 ± 0.1	4.22 ± 0.07	VNR	[11]
8.5 ± 0.6	6.7 ± 0.2	1.5 ± 0.13	8.6 ± 0.9	36.6 ± 0.01	[53]
16.1	13.5	VNR	VNR	VNR	[56]
VNR	9.07 ± 0.04	VNR	9.1 ± 1.7	35.7 ± 0.9	[52]
9.86 ± 0.75	10.61 ± 1.39	1.61 ± 0.86	5.90 ± 0.91	31.91 ± 2.98	This work

VNR: Value not reported.

The average cellulose content in Colombian CH from the CCN 51 clone (25.64 ± 3.49 wt %) is lower than the values reported for CH from Nigeria, in the Ile-Ife region [10]; Brazil, in the Bahia area [48]; Indonesia, in Sumatra province (clone not specified) [50]; and Malaysia, in Jabatan (clone not specified) [42]. The cellulose contents in CH of 41.92 wt % for Peruvian cacao samples (clone not specified) are the highest values found in the literature so far [51]. We found average hemicellulose values of 19.96 ± 2.42 wt % for CCN 51 CH, higher than the values of 10.0, 10.4, and 12 wt % reported for two Colombian and one Ghanaian CH samples [49,50,52]. In contrast, the average hemicellulose values in this work are lower than the 37 ± 0.5 and 35.26 ± 0.05 wt % reported for the Malaysian and Nigerian CH samples [44,51]. On the other hand, the average lignin content of 32.73 ± 5.15 wt % in the CH of clone CCN51 measured in this study is at the high end of the range of values reported by other authors (Table 2). The high lignin content in CCN51 CH could be a characteristic of the clone. However, the literature reporting cacao fruit composition, cited in this work and used to compare our findings, does not include information on the type of material examined. Additionally, to the best of our knowledge, the information currently reported in the literature is derived from grab-sampling of the biomass, understood as samples reflecting the composition at that point in time when the sample was collected. In contrast, in this contribution, we follow changes in cacao fruit composition over time.

The structural biopolymers in CH showed variations over the one-year period considered in this work, as seen in Figure 6. The cellulose contents in CH ranged from 22 to 35 wt % and hemicellulose ranged from 14 to 24 wt %. Lignin content, on the other hand, exhibited significant variability, with values ranging from 25 to 39 wt %. Table S3 of the Supplementary Information contains detailed wt % compositional data for the structural biopolymers and lignin in fresh CH, as fractions of the total solids, for each cacao fruit load. The cellulose, hemicellulose, and lignin contents in cacao fruit can be influenced by factors such as location, tree age, and growth conditions. In Colombia, the cacao crop generally has two harvest seasons per year [2]. The highest productivity season ranges from October to January, while an additional, lower-productivity season runs from April to June. We observed increased cellulose and lignin contents in CH during the October–January harvest season.

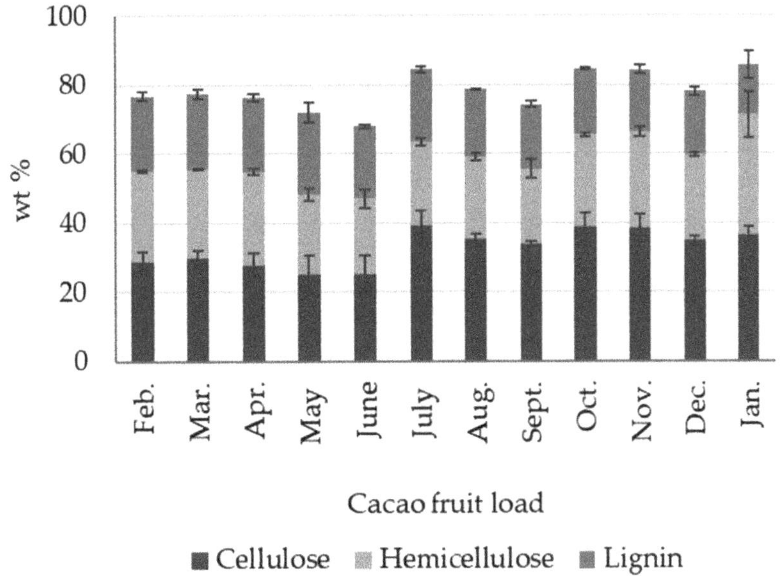

Figure 6. Seasonal variations in structural carbohydrates (cellulose and hemicellulose) and lignin content in CH as fractions of the total solids in fresh CH, from Colombian CCN51 cacao fruit.

Following the second-stage processing methodology (Figures 3 and 5) and as reported in Section 3, fresh CH was ground and sun-dried. Table 3 compares the composition of sun-dried CH, an important biomass output readily generated in cacao farms and used as an energy source in cacao-producing countries. Moisture content in sun-dried CH affects the combustion efficiency of the material. For sun-dried CH moisture contents in the literature, reports range from 6 to 16 wt %, depending on drying time, ambient temperature, relative humidity conditions, and perhaps the cacao plant variety. However, the literature reports normally do not include information regarding the cacao clone or the compositional changes over time. We found average moisture contents of 9.87 ± 1.40 wt % in sun-dried CH (CCN51). Likewise, ash content is an important parameter to estimate biomass behavior when burned, and as an input in models to describe ash transport and deposition during combustion. High ash contents also affect the production of granulated fuel. The average ash content in sun-dried CH (CCN 51) corresponds to 10.61 ± 1.39 wt %, a value within the ranges of 6 and 13 wt % reported for CH in the literature (Table 3). The lowest ash content in Table 3 (6.7 ± 0.2 wt %) was reported for CH samples from Bahia, Brazil (clone not specified) [53]. The average crude fat content in sun-dried CH (1.61 ± 0.86 wt %) lies within the values of 0.6 and 2.24 wt % reported by various authors, as shown in Table 3. The amount of protein of 5.90 ± 0.91 wt % in CH (Colombian CCN 51) is also within the range of published values but is lower than the 8.6 ± 0.9 wt % and 9.1 ± 1.7 wt % reported in other studies [52,53]. Regarding the average crude fiber, the results of this work lie within the values reported by other researchers, as seen in Table 3.

The proximal analysis for sun-dried CH belonging to individual cacao loads shows no significant seasonal variations, as seen in Figure 6 (Table S5). Only in some exceptional cases is there a significant change. For instance, in load 12 (January), the moisture percentage was low (6.62 wt %) and the protein content was high (8.08 wt %) compared to the average of the values. Likewise, load 2 (March) showed a low fat content (0.57 wt %) and a high percentage of crude fiber (37.92 wt %). Sun-dried CH calorific values ranged from 13.22 (Mar) to 14.62 (Oct), with an average of 13.69 ± 0.43 MJ kg $^{-1}$ (Table S5, Figure 7). The direct combustion of sun-dried CH is common for energy generation in cacao-producing countries. Compared to other residual biomass sources, such as soybean cake, rapeseed, cotton cake, potato peel, apricot bagasse, and peach bagasse, with values from 15.41 to 19.52 MJ kg $^{-1}$ [57], the calorific value of sun-dried CH is slightly lower, with an average value of 13.69 ± 0.43 MJ kg $^{-1}$.

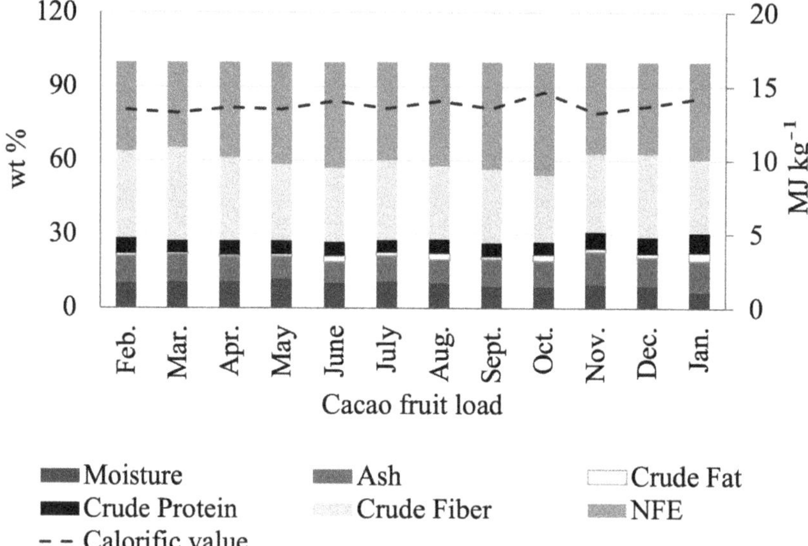

Figure 7. Seasonal variations in the proximate analysis of sun-dried cacao husk (CH) from Colombian CCN51 cacao fruit.

According to the data in Tables 1 and 2 and Figure 5, CH can be cataloged as a lignocellulosic output and, as such, can be processed via physical, chemical, or biological methods to produce added-value byproducts. For instance, the enzymatic or chemical hydrolysis of cellulose in CH can produce sugars, which, in turn, could be used to obtain biofuels such as ethanol, hydrogen, and biobutanol, as well as organic acids through fermentation [58]. Likewise, cellulose could be used for advanced materials production, nanocrystals, and nanofibers. The hemicellulose in CH could be used as a source of probiotics since the main components of this fraction include heteropolymers like xylan, glucuronoxylan, arabinoxylan, xyloglucans, and galactomannans [53,59]. A recent review discussed low- and high-value applications for CH. Among the former, the authors included fertilizer, soil amendments, animal feed, activated carbon, and soap, while the latter involve paper-making, biofuel, dietary fiber, and antioxidant isolation [60].

2.2.2. Cacao Beans (CBs)

CBs are the seeds of the *Theobroma cacao* tree and are the main product of the cacao crop. Table 4 shows the average proximate analysis for dried and fermented cacao beans

(CB) from cacao fruit samples of various origins. Colombian CCN51 dried and fermented CBs contain mainly fat (54.41 ± 0.92 wt %), carbohydrates (27.02 ± 0.98 wt %), and protein (12.10 ± 0.44 wt %); these values are within the ranges reported for Ghanaian and Ecuadorian CBs (Table 4). The protein content in Colombian CCN-51 CBs is within the ranges of 10–15 wt % reported for CBs from Venezuela (Trinitarian variety), Ecuador, Ghana (hybrid clones), Peru, Madagascar, Dominican Republic, and the Ivory Coast [61–64]. The fat content (54.41 ± 0.49) in Colombian CBs is similar to reports of 56–58 wt % for fermented and dried CBs (Forastero variety) in West Africa [63] and is higher than the values of 43.46 ± 0.25 and 41.93 ± 0.13 wt % found in the Ecuadorian and Ghanaian CB samples [61,62]. The carbohydrate content in Colombian CBs (27.02 ± 0.98 wt %) is similar to the reports of 23.1 ± 0.54 wt % in Ghanaian Forastero [63] and lower than the values of 33.79 ± 0.24 and 36.58 ± 0.15 wt % reported for the Ecuadorian and Ghanaian CB samples (clone not specified) [62], as seen in Table 4.

Table 4. Average proximate analysis for dried and fermented cacao beans (CB) from cacao fruit samples of various origins.

Moisture	Ash	Crude Fiber	Crude Protein	Crude Fat	Carbohydrates	Total Polyphenols (mg	Reference
			wt %			Gallic Acid/g)	
4.3 ± 0.09	2.3 ± 0.04	VNR	18.2 ± 0.13	52.2 ± 0.1	23.1 ± 0.54	VNR	[63]
6.22 ± 0.2	2.84 ± 0.04	VNR	12.25 ± 0.16	42.7 ± 0.6	VNR	VNR	[61]
5.95 ± 0.04	4.03 ± 0.01	VNR	12.79 ± 0.03	VNR	33.78 ± 0.02	VNR	[62] *
5.11 ± 0.01	3.56 ± 0.02	VNR	12.82 ± 0.01	VNR	36.58 ± 0.01	VNR	[62] +
3.96 ± 0.50	2.51 ± 0.17	3.20 ± 0.90	12.10 ± 0.44	54.41 ± 0.92	24.00 ± 2.24	47.31 ± 8.03	This work

VNR: Value not reported. * Ecuador, + Ghana.

Cacao beans are rich in antioxidants and are, thus, considered a functional food. Many scientific studies demonstrate the link between improved cardiovascular health and the consumption of antioxidant-rich chocolate [65]. Antioxidants in CBs exist mainly as polyphenols (flavonoids), which are responsible for the characteristic bitter taste of raw cacao seeds. The total polyphenols in Colombian CBs (47.31 ± 8.03 mg GAE/g) are in the low end of the range, according to a review of antioxidant content in CBs from various origins. For instance, the polyphenols content in Ecuadorian CCN51 CBs ranged from 84 mg GAE/g to 91 mg/g (catechin equivalents), while the Criollo clones from the Dominican Republic and Peru exhibited lower polyphenol contents of 40 and 50 mg GAE/g [66]. However, the polyphenol content in Colombian CBs was within the reported values of 34 to 60 mg/g, for a series of CB samples from the Ivory Coast (Forastero variety), Colombia (Amazon variety), Equatorial Guinea (Amazon Forastero variety), Ecuador (Amazon-Trinitario-Canelo, Amazon hybrid), Venezuela (Trinitario variety), Peru (Criollo variety), and the Dominican Republic (Criollo variety) [67].

Figure 8 and Table S6 show the seasonal variations in CB composition. There are no significant differences in moisture (2.94 % wt–4.74 % wt), ash content (2.40 wt %–2.93 wt %), fat (52.77 wt %–55.36 wt %), protein (11.34 wt %–12.67 wt %), total carbohydrates (22.39 wt %–26.93 wt %), and calorific value (26.51 MJ kg^{-1}–27.42 MJ kg^{-1}) for the various CB loads studied. However, parameters such as crude fiber and total polyphenols (Table S6) in CBs exhibited significant changes over time. For instance, the total polyphenol content reached maxima values during the harvest seasons, with contents ranging from 49.97 to 61.68 mg GAE/g from January to April, and 52.12 to 57.58 mg GAE/g in November and December. The total polyphenols in CCN51 CBs reached minima values from May to September (36.21 to 42.52 mg GAE/g).

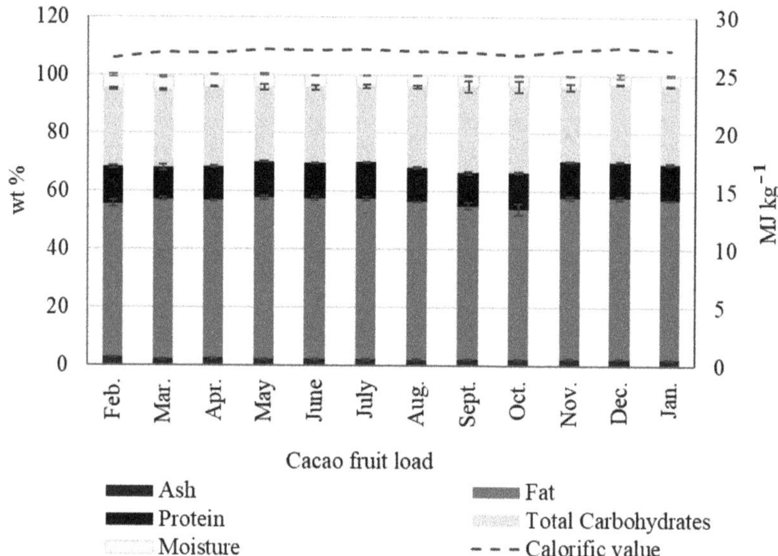

Figure 8. Seasonal variations in the proximate analysis of fermented and dried cacao beans (CB) from Colombian CCN51 cacao fruit.

We believe that compositional information is fundamental to making informed decisions in cacao crop management. For CBs, for instance, knowing when the beans reach maximum antioxidant values could determine the fate of the product, such as antioxidant extraction vs. roasting, allowing producers to increase their profitability and product portfolio.

2.2.3. Cacao Placenta (CP)

The CP is a fibrous supportive tissue that holds the CBs together inside the pod. According to Tables S1 and S2 (Supplementary Information), fresh CP corresponds to $2.58 \pm 0.22\%$ of the total fruit weight. The material mainly contains water; after drying, the total solid content in CP is 0.57 ± 0.05 wt %. Perhaps, due to its scarceness, CP has received little attention in the scientific literature. Table 5 shows an average CP moisture content of 78.36 wt % for CCN51 fruit, as in reports of 80.55 ± 0.42 wt % and 80.925 ± 1.06 wt % for Côte d'Ivoire cacao fruit (Forastero variety) [68] and Ecuadorian cacao fruit (CCN51 clone) [69]. We found low ash percentages in CP (0.54 wt %), in contrast with the 9.34 wt % reported for Côte d'Ivoire cacao fruit [68]. The protein content (1.69 wt %) is within the values reported by other authors. The total nitrogen in CP (0.27 wt %) is lower than 1.005 ± 0.134 wt %, as reported by the authors of [69].

Table 5. Average proximate analysis for fresh CP from cacao fruit samples of various origins.

Moisture	Ash	Protein	Total Nitrogen	Reference
		wt %		
80.55 ± 0.42	9.34 ± 0.89	5.12 ± 0.02	VNR	[68]
80.925 ± 1.067	5.560 ± 0.424	VNR	1.005 ± 0.134	[69]
VNR	1.28 ± 0.051	1.38 ± 0.028	VNR	[30]
78.36	0.54	1.69	0.27	This work

VNR: Value not reported.

Lately, CP has attracted some attention as a source of nutrients, antioxidants, and bioactive compounds. For instance, one study has found that fermentation increased the antioxidant activity of CP, while the levels of sugars, tannins, and flavonoids decreased [69];

likewise, the nutritional value of dehydrated CP makes it an ingredient in nutritional bars [31] or for the production of an alcoholic drink and nectar made from CP from Ecuadorian cacao fruit samples (CCN51 clone) [30]. In the field of animal nutrition, previous studies reported the use of dried CP from Ecuadorian cacao in chicken feed formulations [70] and quantities of CP flour (CPF) for pig feeding [71].

2.2.4. Cacao Mucilage Exudate (CME)

Cacao mucilage exudate or sweatings (CME), a sweet liquid that seeps from the white pulp surrounding the cacao bean, is typically lost during the cacao bean fermentation process. CME comprises up to 4.13 ± 0.80 wt % of the cacao fruit's biomass (see Table S2) and is mainly composed of water and carbohydrates. Table 6 compares the average proximate analysis for CME from cacao fruit samples of various origins and from this work. The average moisture content of Colombian CCN51 CME (82.68 ± 6.09 wt %) is similar to the values of 85.86 ± 0.09 wt % reported for samples from Bahia, Brazil (clone not specified) [72], and 82.5 % and 80.5 % for the Ecuadorian samples (Nacional and CCN-51 clones) [73]. The protein content in the Colombian CCN51 CME samples (0.31 ± 0.09 wt %) is lower than (1.2 ± 0.49 wt %) [69], (5.41 wt %) from the Ecuador cacao fruit (CCN-51 clone) [74], and (5.56 ± 0.1 wt % and 5.47 ± 0.12 wt %) from the Taura and Cone Ecuadorian samples [11]. The carbohydrate content in CCN51 CME has an average value of 17.18 ± 6.44 wt %, lower than the values of 68.35 ± 0.16 and 67.99 ± 0.14 for Cone and Taura Ecuadorian samples [11]. CME contains small amounts of polyphenols (0.55 ± 0.45 mg gallic acid/g), a characteristic that has not previously been reported. Table S7 of the Supplementary Information includes the detailed proximate analysis of Colombian CCN51 CME samples collected over one year.

Table 6. Average proximate analysis for CME from cacao fruit samples of various origins and this work.

Moisture	Ash	Protein	Total Carbo-hydrates	Total Polyphenols (mg	Reference
		wt %		Gallic Acid/g)	
85.86 ± 0.09	0.59 ± 0.15	1.20 ± 0.49	11.80 ± 0.09	VNR	[72]
77.34 [a]	2.91	5.41	VNR	VNR	[73]
82.5 [b]	VNR	0.87	VNR	VNR	[73]
80.5	VNR	0.38	VNR	VNR	[74]
VNR [c]	7.51 ± 0.14	5.47 ± 0.12	68.35 ± 0.16	VNR	[11]
VNR [d]	7.68 ± 0.18	5.56 ± 0.10	67.99 ± 0.14	VNR	[11]
82.67 ± 6.09	0.45 ± 0.01	0.31 ± 0.09	17.18 ± 6.44	0.55 ± 0.45	This work

VNR: Value not reported. [a] Nacional, [b] CCN-51, [c] Cone, [d] Taura.

Figure 9 and Tables S7 and S8 show the seasonal variations in CME composition. There are significant differences in moisture, with higher values observed toward the end of the year from August to January (88.6 wt %) and lower values during February and March (71.5 wt %). The ash (0.27 wt % to 0.61 wt %) and protein contents (0.17 wt % to 0.46 wt %) in CME are low. In contrast, the total carbohydrate content in CME changes dramatically during the year, with a minimum of 10 wt % during August and January and a maximum of 29 wt % in March. Interestingly, the first three loads of the year showed the highest total carbohydrate content and calorific values, and the lowest moisture content of all samples.

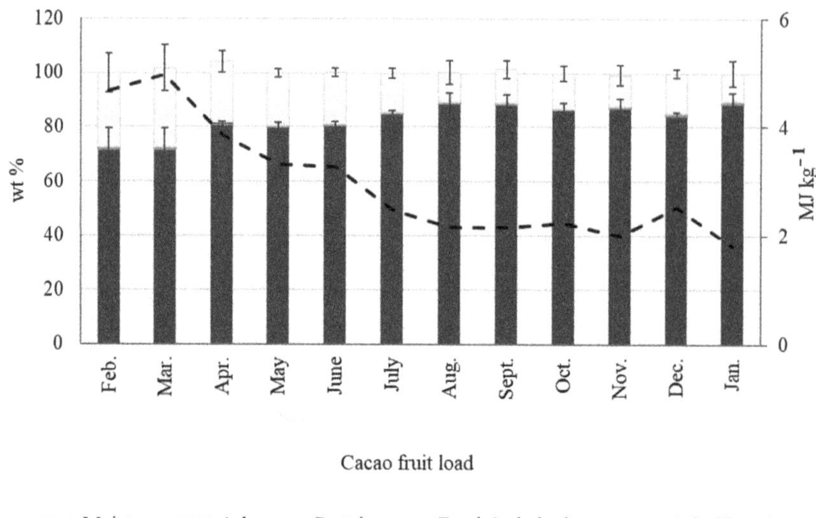

Figure 9. Seasonal variations in the proximate analysis of CME from Colombian CCN51 cacao fruit.

CME's high sugar content hints at a valuable biomass output, with potential uses as human food or as a carbon source in industrial processes. The average sugar values in CME from CCN51 cacao fruit correspond to 75.54 ± 9.54 g L^{-1} for fructose, 67.15 ± 8.36 g L^{-1} for glucose, and 14.08 ± 8.89 g L^{-1} for saccharose. The glucose and fructose contents are higher than the values of 45.8 ± 1.2 g L^{-1} and 32.5 ± 0.3 g L^{-1} [72]; however, the glucose content is much lower than the 214.2 ± 6.2 g L^{-1} reported for South Cote D'Ivoire samples (clone not specified) [25].

There are significant differences between individual sugars in CME from different cacao fruit loads, as seen in Figure 10. The saccharose content ranged from 6.84 g L^{-1} to 33.7 g L^{-1}, while the glucose and fructose contents stretched from 54.3 g L^{-1} to 80.89 g L^{-1} and from 59.3 g L^{-1} to 84.43 g L^{-1}, respectively. The content of the most complex sugar in CME (saccharose) increases, at the expense of a decreased content of glucose and fructose during the main harvest (October–December), which is also the dry season in the region. However, from October to December, we also registered the lowest total carbohydrate content in CME. Sugar concentration in CME depends on many factors, such as fruit ripeness, rainfall, crop conditions, the time of year, and cacao variety, among others. Sugar accumulation can be enhanced by more prolonged exposure of the fruit to the sun [75]. A Pearson's correlation coefficient (r) of 0.63 suggests a direct correlation between the sugar content in CME and rainfall in the area. The CME extracted from cacao fruit loads processed during November, December, and January exhibit the lowest sugar content, matching up with the dry season (low rainfall). On the other hand, during the rainy season (March to October—an abnormally long rainy season), we observed the highest total sugar content. Table S8 of the Supplementary Information contains detailed seasonal compositional information for CME.

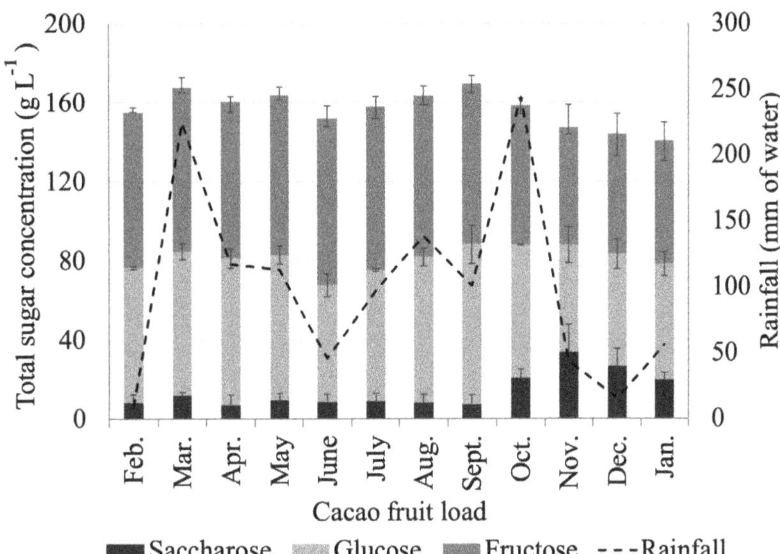

Figure 10. Seasonal variations in CME sugar content from Colombian CCN51 cacao fruit.

CME also contains trace minerals in the form of calcium, sodium, potassium, and aluminum. These micronutrients are relevant nowadays because of their proven benefits to the human body. Sodium, for instance, an essential mineral that allows maintaining the water balance in the body and adequate blood pressure, has an average concentration of 1.71 ± 0.078 ppm in CME. In contrast, cacao beans exhibit higher Na contents, ranging from 3 to 32 ppm for samples from various origins [76]. Potassium, which is also fundamental for water balance and cardiovascular health [77], has an average concentration of 2413.77 ± 194.89 ppm in CME. The high potassium content in CME makes it a natural source of the mineral, rivaling traditional high-potassium foods such as potatoes, bananas, apricots, tomatoes, carrots, passion fruit, and many others. In contrast, reports dealing with the multi-elemental analysis of cacao beans from worldwide samples show K contents of around 10–13 ppm, suggesting a selective accumulation of this nutrient in the mucilage surrounding the beans [76]. Interestingly, Na and K contents increased in processed cacao bean byproducts, such as cacao mass and cocoa [78].

The literature reports several uses for CME. For instance, the use of CME for making marmalade, ending up with a product with similar organoleptic properties as apricot marmalade and a nutritional value comparable with some tropical fruits [25], following jelly formulations from CME, extracted from national varieties and CCN-51 [73]. Likewise, jam was made with 67.14° Brix using CME, cacao placenta, and added cane sugar [25]. A cacao beverage containing CME and a liqueur was extracted from Trinidad and Tobago cacao beans (hybrid clone) [26]. CME's uses are also reported in patents. For instance, one application suggested a method for obtaining a syrup from CME and cacao mucilage from CBs that was suitable for food preparations as a flavoring or texturizing agent [27].

The high sugar content in CME makes it an ideal carbon source for biotechnological applications. For instance, one study used CME and *Saccharomyces cerevisiae* to produce a fermented drink with 22.4% of ethanol [32]. In the same way, the production of ethanol from CME fermentation with *Saccharomyces cerevisiae*, with a maximum yield of 0.073 g g^{-1} and volumetric productivity of 0.168 g L^{-1} h^{-1} was achieved [72]. In another work, the authors reached a maximum concentration of 13.8 g L^{-1} of ethanol, the production of 0.5 g of ethanol g glucose $^{-1}$, and a productivity of 0.25 g L^{-1} h^{-1} of ethanol from CME fermented with *Pichia kudriavzevii* [33]. Elsewhere, CME from cacao fruit was sourced from the Selemadeg Barat District, Bali Province (clone not specified) to produce vinegar [79]. The authors found that a single-phase fermentation, plus additional alcohol, resulted in a high acetic and propionic acid product. Finally, more recently, our group demonstrated the feasibility of using CME for bacterial cellulose (BC) production [34]. We measured maximum bacterial cellulose yields and achieved production rates of 15.57 g L^{-1} and 0.041 g L^{-1} h^{-1}, which were replicated at laboratory and pilot-plant scales. This observation suggests an industrial scenario, wherein BC production from CME is a real possibility.

3. Materials and Experimental Methods

3.1. Area of Study

Ripe cacao fruits were harvested from trees of the CCN51 clone ranging from 6 to 10 years of age. The cacao plantation was situated in the rural area of San Vicente de Chucurí (latitude: 6°52′59″ N, longitude: 73°25′1″ W) at the heart of the main cacao-producing region in the country (Santander—Colombia). In 2018, San Vicente de Chucurí contributed 28 wt % to Santander's cacao production [80].

3.2. Cacao Fruit Collection and Treatment

Ripe cacao fruit loads, of approximately one tonne each (1000 kg), were brought to the lab monthly (February–January of 2018–2019) for a total number of twelve loads (12,000 kg). For the cacao fruit compositional and mass balance analyses, we devised a two-stage procedure. The first stage involved opening the cacao fruit and separating the fresh cacao beans (CBs) and the placenta (CP) from the husk (CH), as seen in Figure 3. This first process represents the traditional approach to cacao fruit usage. In the second stage, the CP and the CH were ground, dried, and stored for further compositional analysis (Figure 3). After CB extraction, the CH and CP were ground down to a particle size of 0.1–2 cm, using a TRAPP TRF 300 mill (Trapp, Brazil), and were then sun-dried for 5–6 days. The CH was turned over every 4 h to facilitate water evaporation.

Fresh CBs were placed in a double-wall stainless-steel container fitted with an inner mesh that allowed the collection of the cacao mucilage exudate (CME) designed and built by our research group. The same container was used to perform the fermentation process for seven days. The CME and CBs were stored for further analysis. All materials (wet and dry) were weighed to determine the mass balances. Table 7 shows the equations used to determine the percentages in terms of wet and dry weight for the cacao fruit byproducts, as reported throughout the manuscript.

Table 7. Equations used to determine the percentages by weight of cacao fruit outputs.

Percentages	Equation	Definitions
Wet weight	$\text{CH wt \%} = \frac{M_{FCH}}{M_{FCF}}$ $\text{CP wt \%} = \frac{M_{FCP}}{M_{FCF}}$ $\text{CB wt \%} = \frac{M_{FCBs}}{M_{FCF}}$ $\text{Fermented CBs wt \%} = \text{CBs wt \%} - \text{CME wt \%}$	M_{FCH} mass of fresh CH M_{FCF} mass of fresh cacao fruit M_{FCP} mass of fresh CP M_{FCBs} mass of fresh CBs
Dry weight	$\text{Dried CH wt \%} = \frac{M_{DCH}}{M_{FCF}}$ $\text{Dried CP wt \%} = \frac{M_{DCP}}{M_{FCF}}$ $\text{Dried fermented CBs wt \%} = \frac{M_{DFCBs}}{M_{FCF}}$	M_{DCH} mass of dried CH M_{DCP} mass of dried CP M_{DFCBs} mass of dried fermented CBs
Moisture content (MC)	$\text{CH}_{MC}\text{ \%} = \frac{(M_{FCH} - M_{DCH})}{M_{FCH}}$ $\text{CP}_{MC}\text{ \%} = \frac{(M_{FCP} - M_{DCP})}{M_{FCP}}$ $\text{CB}_{MC}\text{ \%} = \frac{(M_{FCBs} - M_{DCB})}{M_{FCB}}$ $\text{CME}_{MC}\text{ \%} = \frac{(M_{FCME} - M_{DCME})}{M_{FCME}}$	M_{FCH} mass of fresh CH M_{DCH} mass of dried CH M_{FCP} mass of fresh CP M_{DCP} mass of dried CP M_{FCBs} mass of fresh CBs M_{DCP} mass of dried CP M_{FCME} mass of fresh CME M_{DCME} mass of dried CME

The percentage weight and compositional information reported throughout the manuscript correspond to the average values and standard deviations from twelve measurements. Statistics analysis was performed, using the average and standard deviation functions, in an Excel spreadsheet. The Pearson coefficient was also calculated, using the correlation function, in an Excel spreadsheet.

Table 7 illustrates the equations used to determine the percentages in wet and dry weight for the cacao fruit byproducts.

3.3. Compositional Analyses

Table 8 shows the analysis and reference methods used to characterize the components isolated from the cacao fruit: cacao husk (CH), cacao placenta (CP), cacao mucilage exudate (CME), and cacao beans (CBs).

Table 8. Chemical characterization of the various cacao fruit components.

Analysis	Reference Method	CH	CP	CME	CBs
Moisture	AOAC 931.04 [81]	X		X	X
	AOAC 925.10 [82]		X		
Ash	AOAC 972.15 [83]	X		X	X
	AOAC7,009/84-94205/90		X		
Protein	AOAC 970.22 [84]	X		X	X
	AOAC 2001.11 [85]		X		
Crude fiber	AOAC 930.20a [86]	X	X		X
Fat	AOAC 920.75a [87]	X	X		X
Total carbohydrates	By difference: 100—(% Ash)—(% Total Fat)—(% Moisture)—(% Protein)—(% Fiber)	X	X		X
Calorific value	By equation: (% T Carbohydrate × 4 Kcal/g) + (% Protein × 4 Kcal/g) + (% Total Fat × 9 Kcal/g)	X	X		X
Total polyphenols	Standard Methods 5530 B, D [88]			X	X
Glucose	AOAC 925.36 [89]			X	
Fructose	AOAC 925.36			X	
Sucrose	AOAC 925.36			X	
Total soluble solids °Brix	AOAC 931.12 [90]			X	
pH 24.2 (°C)	AOAC 960.19 [91]			X	
Calcium	AOAC 985.35 [92]			X	
Potassium	AOAC 985.35			X	
Sodium	AOAC 985.35			X	
Aluminum	Emission mode			X	
Total nitrogen	Standard Methods 4500 N [93]	X	X	X	
Total solids	Standard Methods 2540B [94]	X			
Holocellulose	Jayme-Wise Method [95]	X			
Cellulose	Kurschner and Hoffer Method [96]	X			
Hemicellulose	By difference: % Holocellulose—% Cellulose	X			
Lignin	Klason Method [97]	X			

4. Conclusions

1. We identified five distinctive biomass outputs from cacao fruit: cacao husk (CH), cacao beans (CBs), cacao placenta (CP), and cacao mucilage exudate (CME). CH, CBs, and CP are solid lignocellulosic outputs that comprise, in terms of dry weight, 8.92 ± 0.90 wt %, 8.87 ± 0.52 wt %, and 0.57 ± 0.05 wt % of the cacao fruit weight, respectively. Moisture, on the other hand, constitutes most of the biomass weight (71.6 ± 2.29 wt %). Cellulose and lignin contents in CH are time-dependent, reaching maximum values during the crop's primary harvest season (October–January).

2. Dried CH is mostly used as an energy source in the cacao-producing regions of the world. We found no significant changes in CH calorific values during the crops' yearly cycle, with an average of 13.69 ± 0.43 MJ kg^{-1}. This value is similar to the calorific value content in other residual biomass outputs, such as rice straw/husk, soybean cake, potato peels, rapeseed, sugarcane bagasse, and cotton cake.

3. As a lignocellulosic output, CH can potentially be processed via physical, chemical, or biological methods to produce added-value byproducts, such as simple sugars, ethanol, hydrogen, biobutanol, and volatile organic acids, among others. Likewise, the structural biopolymers in CH, such as cellulose, can be the precursors of high-performance materials, such as nanocrystals and nanofibers. The hemicelluloses in CH, which are rich in heteropolymers like xylan, glucuronoxylan, arabinoxylan, xyloglucans, and galactomannans, could become a potential source of probiotics such as xylo-oligosaccharides.

4. CB contains carbohydrates, fats, protein, ash, and phenolic compounds. The contents of these materials in Colombian CCN51 CBs do not change significantly during the yearly crop cycle and are within the range of CBs from different geographical sources. Interestingly, the total polyphenol content in CBs is time-dependent, reaching maxima values during the harvest seasons. For instance, from January to April, CBs exhibit 49.97 to 61.68 mg GAE/g, and, from November to December, 52.12 to 57.58 mg GAE/g. The total polyphenols in CCN51 CBs reach minimum values of 36.21 to 42.52 mg GAE/g from May to September.

5. Cacao mucilage exudate (CME) is a liquid biomass output that is equivalent to 4.13 ± 0.80 wt % of the cacao fruit. CME is rich in simple sugars (glucose, fructose, and saccharose) and minerals (K), with an average of 20 wt % of total carbohydrates. Interestingly, the total carbohydrate content in CME changes dramatically during the year, with a minimum of 10 wt % from August to January and a maximum of 29 wt % in March. Likewise, we observed a positive correlation between sugar content in CME and rainfall, with the highest sugar content in CME being measured during April when rainfall was at its highest.

6. CME uses include fermentation to produce alcohol and concentration to produce syrups and jams. However, the high nutrient content of CME makes it an ideal culture media for biotechnological applications, particularly biopolymer production, as demonstrated recently by our group.

Supplementary Materials: The following supporting information can be downloaded at: https://www.mdpi.com/article/10.3390/molecules27123717/s1, Figure S1: Supplementary Information with mass balance for the first and second stages of cacao fruit processing. with averaged wt % data for all biomass outputs; Table S1: Traditional fresh cacao fruit biomass outputs: cacao husk (CH), cacao placenta (CP), and cacao beans (CBs), derived from the first stage of cacao fruit processing; Table S2: Potential cacao fruit biomass outputs: dried cacao husk (CH), dried cacao placenta (CP), dried cacao beans (CBs), and cacao mucilage exudate (CME) derived from the second stage of cacao fruit processing; Table S3: Composition analysis by load for fresh CH; Table S4: Composition of various agricultural lignocellulosic biomass outputs; Table S5: Proximate composition of sun-dried CH; Table S6: Physicochemical composition of CBs; Table S7: Seasonal variations in the proximate analysis of CME from CCN51 cacao fruit; Table S8: Physicochemical analysis of CME.

Author Contributions: M.V.-M.: Investigation, methodology, validation, visualization, formal analysis, writing—original draft. G.R.M.: Investigation, methodology, validation. C.B.-T.: Conceptualization, resources, funding acquisition, project administration. M.Y.C.: Conceptualization, methodology, visualization, formal analysis, supervision, data curation, writing—review and editing, funding acquisition. All authors have read and agreed to the published version of the manuscript.

Funding: This research was funded by the Colombian Science and Technology Ministry—Minciencias Grant 251-208 postdoctoral fellowship and the SGR-Gobernación de Santander, Grant BPIN 2016000100046.

Institutional Review Board Statement: Not applicable.

Informed Consent Statement: Not applicable.

Data Availability Statement: The data presented in this study are available in supplementary material.

Acknowledgments: We thank Guatiguará Technological Park and the Central Research Laboratory Facility at Universidad Industrial de Santander for infrastructural support. MVM acknowledges a postdoctoral fellowship from the Colombian Science and Technology Ministry—Minciencias—Grant 251-208. This work was supported by the SGR-Gobernación de Santander, Grant BPIN 2016000100046.

Conflicts of Interest: The authors declare no conflict of interest.

Sample Availability: Samples of the compounds are available from the authors.

References

1. ICCO International Cocoa Organization. Quarterly Bulletin of Cacao Statistics, Issue No. 2—Volume XLVI. 2020. Available online: https://www.icco.org/icco-documentation/quarterly-bulletin-of-cocoa-statistics/ (accessed on 5 March 2021).
2. Arvelo, M.; González, D.; Delgado, T.; Maroto, S.; Montoya, P. *Estado Actual Sobre la Producción, el Comercio y Cultivo del Cacao en América*; Instituto Interamericano de Cooperación para la Agricultura, Fundación Colegio de Postgraduados en Ciencias Agrícolas. C.R., IICA: San José, CA, USA, 2017.
3. Fedecacao-Federación Nacional de Cacaoteros. Boletín de Prensa. Así Quedó el Ranking de Producción de Cacao en Colombia. Publicado 18/021/2020. 2020. Available online: http://www.fedecacao.com.co/portal/index.php/es/2015-04-23-20-00-33/1193-boletin-de-prensa-asi-quedo-el-ranking-de-produccion-de-cacao-en-colombia/ (accessed on 12 May 2021).
4. Vásquez, Z.S.; de Carvalho Neto, D.P.; Pereira, G.V.M.; Vandenberghe, L.P.S.; de Oliveira, P.Z.; Tiburcio, P.B.; Rogez, H.L.G.; Góes Neto, A.; Soccol, C.R. Biotechnological approaches for cocoa waste management: A review. *Waste Manag.* **2019**, *90*, 72–83. [CrossRef] [PubMed]
5. Ouattara, L.Y.; Kouassi, E.K.A.; Soro, D.; Soro, Y.; Yao, K.B.; Adouby, K.; Drogui, A.P.; Tyagi, D.R.; Aina, P.M. Cocoa pod husks as potential sources of renewable high-value-added products: A review of current valorizations and future prospects. *BioResources* **2021**, *16*, 1988–2020. [CrossRef]
6. Mendoza-Meneses, C.J.; Feregrino-Pérez, A.A.; Gutiérrez-Antonio, C. Potential Use of Industrial Cocoa Waste in Biofuel Production. *J. Chem.* **2021**, *2021*, 3388067. [CrossRef]
7. Indiarto, R.; Raihani, Z.; Dewi, M.; Zsahra, A. A Review of Innovation in Cocoa Bean Processing By-Products. *Int. J.* **2021**, *9*, 1162–1169. [CrossRef]
8. Dahunsi, S.O.; Osueke, C.O.; Olayanju, T.M.A.; Lawal, A. Co-digestion of *Theobroma cacao* (Cocoa) pod husk and poultry manure for energy generation: Effects of pretreatment methods. *Bioresour. Technol.* **2019**, *283*, 229–241. [CrossRef]
9. Antwi, E.; Engler, N.; Nelles, M.; Schuch, A. Anaerobic digestion and the effect of hydrothermal pretreatment on the biogas yield of cocoa pods residues. *Waste Manag.* **2019**, *88*, 131–140. [CrossRef]
10. Dahunsi, S.O.; Adesulu-Dahunsi, A.T.; Izebere, J.O. Cleaner energy through liquefaction of Cocoa (*Theobroma cacao*) pod husk: Pretreatment and process optimization. *J. Clean. Prod.* **2019**, *226*, 578–588. [CrossRef]
11. Martínez, R.; Torres, P.; Meneses, M.A.; Figueroa, J.G.; Pérez-Álvarez, J.A.; ViudaMatos, M. Chemical, technological and in vitro antioxidant properties of cocoa (*Theobroma cacao* L.) co-products. *Food Res. Int.* **2012**, *49*, 39–45. [CrossRef]
12. Kilama, G.; Lating, P.O.; Byaruhanga, J.; Biira, S. Quantification and characterization of cocoa pod husks for electricity generation in Uganda. *Energy Sustain. Soc.* **2019**, *9*, 22. [CrossRef]
13. Villota, S.M.; Lei, H.W.; Villota, E.; Qian, M.; Lavarias, J.; Taylan, V.; Agulto, I.; Mateo, W.; Valentin, M.; Denson, M. Microwave-assisted activation of waste cocoa pod husk by H_3PO_4 and KOH-comparative insight into textural properties and pore development. *ACS Omega* **2019**, *44*, 7088–7095. [CrossRef]
14. Tsai, W.T.; Jiang, T.J.; Lin, Y.Q. Conversion of de-ashed cocoa pod husk into high-surface-area microporous carbon materials by CO_2 physical activation. *J. Mater. Cycles Waste Manag.* **2019**, *21*, 308–314. [CrossRef]
15. Olakunle, M.O.; Inyinbor, A.A.; Dada, A.O.; Bello, O.S. Combating dye pollution using cocoa pod husks: A sustainable approach. *Int. J. Sustain. Eng.* **2018**, *11*, 4–15. [CrossRef]
16. De Luna, M.D.G.; Murniati Budianta, W.; Rivera, K.K.P.; Arazo, R.O. Removal of sodium diclofenac from aqueous solution by adsorbents derived from cocoa pod husks. *J. Environ. Chem. Eng.* **2017**, *2*, 1465–1474. [CrossRef]
17. Betiku, E.; Etim, A.O.; Pereao, O.; Ojumu, T.V. Two-step conversion of neem (*Azadirachta indica*) seed oil into fatty methyl esters using a heterogeneous biomass-based catalyst: An example of cocoa pod husk. *Energy Fuels* **2017**, *31*, 6182–6193. [CrossRef]
18. Lateef, A.; Azeez, M.A.; Asafa, T.B.; Yekeen, T.A.; Akinboro, A.; Oladipo, I.C.; Azeez, L.; Ojo, S.A.; Gueguim-Kana, E.B.; Beukes, L.S. Cocoa pod husk extract-mediated biosynthesis of silver nanoparticles: Its antimicrobial, antioxidant and larvicidal activities. *J. Nanostruct. Chem.* **2016**, *6*, 159–216. [CrossRef]
19. Adi-Dako, O.; Ofori-Kwakye, K.; El Boakye-Gyasi, M.; Bekoe, S.O.; Okyem, S. In Vitro Evaluation of Cocoa Pod Husk Pectin as a Carrier for Chronodelivery of Hydrocortisone Intended for Adrenal Insufficiency. *J. Drug Deliv.* **2018**, *2017*, 828405. [CrossRef]
20. Lubis, M.; Harahap, M.B.; Ginting, M.H.S.; Maysarah, S. The effect of ethylene glycol as plasticizer against mechanical properties of bioplastic originated from jackfruit seed starch and cocoa pod husk. *Nusant. Biosci.* **2018**, *10*, 76–80. [CrossRef]

21. Chun, K.S.; Husseinsyah, S.; Yeng, C.M. Effect of green coupling agent from waste oil fatty acid on the properties of polypropylene/cacao pod husk composites. *Polym. Bull.* **2016**, *73*, 3465–3484. [CrossRef]
22. El-Shekeil, Y.A.; Sapuan, S.M.; Algrafi, M.W. Effect of fiber loading on mechanical and morphological properties of cocoa pod husk fibers reinforced thermoplastic polyurethane composites. *Mater. Des.* **2014**, *64*, 330–333. [CrossRef]
23. Esong, R.N.; Etchu, K.A.; Bayemi, P.H.; Tan, P.V. Effects of the dietary replacement of maize with sun-dried cacao pods on the performance of growing rabbits. *Trop. Anim. Health Prod.* **2015**, *47*, 1411–1416. [CrossRef]
24. Oddoye, E.O.K.; Rhule, S.W.A.; Agyente-Badu, K.; Anchirinah, V.; Owusu Ansah, F. Fresh cocoa pod husk as an ingredient in the diets of growing pigs. *Sci. Res. Essays* **2010**, *5*, 1141–1144.
25. Anvoh, K.Y.B.; Zoro, B.A.; Gnakr, D. Production and characterization of juice from mucilage of cacao beans and its transformation into marmalade. *Pak. J. Nutr.* **2009**, *8*, 129–133. [CrossRef]
26. Badrie, N.; Escalante, M.; Bekele, F.L. Production and quality characterization of pulp from cacao beans from Trinidad: Effects of varying levels of pulp on value-added carbonated cacao beverages. In Proceedings of the CFCS Meeting, 49th Annual Meeting, Caribbean Food Crops Society 49™ Annual Meeting, 30 June–6 July 2013. [CrossRef]
27. Toth, J.; Lopata, J.; Schweizer, C.; Pachard, S. Method for Production and Use of Syrup Derived from the Fruit Pulp of the Cacao Pod. WO 2017/044610 A1, 16 March 2017.
28. Dias, D.R.; Schwan, R.F.; Freire, E.S.; Serôdio, R.D.S. Elaboration of a fruit wine from cocoa (*Theobroma cacao* L.) pulp. *Int. J. Food Sci. Technol.* **2007**, *42*, 319–329. [CrossRef]
29. Puerari, C.; Teixeira-Magalhães, K.; Freitas-Schwan, R. New cocoa pulp-based kefir beverages: Microbiological, chemical composition and sensory analysis. *Food Res. Int.* **2012**, *48*, 634–640. [CrossRef]
30. Quimbita, F.; Rodriguez, P.; Vera, E. Uso del exudado y placenta del cacao para la obtención de subproductos. *Rev. Tecnol. ESPOL—RTE* **2013**, *26*, 8–15.
31. Segura, J.; García, M.I.; Tigre, A.; Dominguez, V.; Barragán, U.; Guamán, J.; Ramón, R.; Bayas-Morejón, F. Elaboration of an alcoholic drink from the aerobic fermentation of mucilage of cacao (*Theobroma cacao* L.). *Res. J. Pharm. Biol. Chem. Sci.* **2018**, *9*, 1375–1380.
32. Nguyen, T.T.; Nguyen, T.A.; Ho, T.T.; Nguyen, T.T. A study of wine fermentation from mucilage of cacao beans (*Theobroma cacao* L.). *Dalat Univ. J. Sci.* **2016**, *6*, 387–397.
33. Romero-Cortés, T.; Cuervo-Parra, J.A.; Robles-Olvera, V.J.; Rangel-Cortes, E.; López-Pérez, P.A. Experimental and kinetic production of ethanol using mucilage juice residues from cacao processing. *Int. J. Chem. React. Eng.* **2018**, *16*. [CrossRef]
34. Saavedra-Sanabria, O.L.; Durán, D.; Cabezas, J.; Hernández, I.; Blanco-Tirado Combariza, M.Y. Cellulose biosynthesis using simple sugars available in residual cacao mucilage exudate. *Carbohydr. Polym.* **2021**, *274*, 118645. [CrossRef]
35. Morejón-Lucio, R.; Vera-Chang, J.; Vallejo-Torres, C.; Morales-Rodríguez, W.; Díaz-Ocampo, R.; Alvarez-Aspiazu, A. Valor nutricional de la placenta deshidratada de cacao (*Theobroma cacao* L.) nacional, para la elaboración de barras nutricionales. *Revista* **2018**, *5*, 57–62.
36. Oddoye, E.O.K.; Agyente-Badu, C.K.; Gyedu-Akoto, E. Cocoa and Its By-Products: Identification and Utilization. In *Chocolate in Health and Nutrition*; Watson, R., Preedy, V., Zibadi, S., Eds.; (Nutrition and Health); Humana Press: Totowa, NJ, USA, 2013; Volume 7. [CrossRef]
37. Chun, K.S.; Husseinsyah, S.; Osman, H. Modified Cocoa Pod Husk-Filled Polypropylene Composites by Using Methacrylic Acid. *BioResources* **2013**, *8*, 3260–3275. [CrossRef]
38. Campos-Vega, R.; Karen, H.; Nieto-Figueroa, K.H.; Oomah, D. Cacao (*Theobroma cacao* L.) pod husk: Renewable source of bioactive Compounds. *Trends Food Sci. Technol.* **2018**, *81*, 172–184. [CrossRef]
39. Nguyen, V.T.; Nguyen, N.H. Proximate composition, extraction, and purification of theobromine from Cacao pod husk (*Theobroma Cacao* L.). *Technologies* **2017**, *5*, 14. [CrossRef]
40. Laconi, E.B.; Jayanegara, A. Improving nutritional quality of cocoa pod (*Theobroma cacao*) through chemical and biological treatments for ruminant feeding: In vitro and in vivo evaluation. *Asian Australas. J. Anim. Sci.* **2015**, *28*, 343–350. [CrossRef]
41. Ashade, O.O.; Osineye, O.M. Effect of replacing maize with cocoa pod husk in the nutrition of *Oreochromis niloticus*. In Proceedings of the 25th Annual Conference of the Fisheries Society of Nigeria (FISON), Lagos, Nigeria, 25–29 October 2010; pp. 504–511. [CrossRef]
42. Daud, Z.; Mohd Kassim, A.S.; Aripin, A.S.; Awang, H.; Mohd Hatta, M.Z. Chemical composition and morphological of cocoa pod husks and cassava peels for pulp and paper production. *Aust. J. Basic Appl. Sci.* **2013**, *7*, 406–411.
43. Kalita, E.; Nath, B.K.; Deb, P.; Agan, F.; Islam, M.R.; Saikia, K. High quality fluorescent cellulose nanofibers from endemic rice husk: Isolation and characterization. *Carbohydr. Polym.* **2015**, *122*, 308–313. [CrossRef]
44. Karp, S.G.; Woiciechowski, A.L.; Soccol, V.T.; Soccol, C.R. Pretreatment Strategies for Delignification of Sugarcane Bagasse: A Review. *Braz. Arch. Biol. Technol.* **2013**, *56*, 679–689. [CrossRef]
45. Kim, M.; Day, D.F. Composition of sugar cane, energy cane, and sweet sorghum suitable for ethanol production at Louisiana sugar mills. *J. Ind. Microbiol. Biotechnol.* **2011**, *38*, 803–807. [CrossRef]
46. Kumar, A.K.; Sharma, S. Recent updates on different methods of pretreatment of lignocellulosic feedstocks: A review. *Bioresour. Bioprocess.* **2017**, *4*, 7. [CrossRef]
47. Titiloye, J.O.; Bakar, M.S.A.; Odetoye, T.E. Thermochemical characterization of agricultural wastes from West Africa. *Ind. Crops Prod.* **2013**, *47*, 199–203. [CrossRef]

48. Santana, N.B.; Dias, J.C.T.; Rezende, R.P.; Franco, M.; Oliveira, L.K.S.; Souza, L.O. Production of xylitol and bio-detoxification of cocoa pod husk hemicellulose hydrolysate by Candida boidinii XM02G. *PLoS ONE* **2018**, *13*, e0195206. [CrossRef] [PubMed]

49. Ward-Doria, M.; Arzuaga-Garrido, J.; Ojeda, K.A.; Sánchez, E. Production of biogas from acid and alkaline pretreated cocoa pod husk (*Theobroma cacao* L.). *Int. J. ChemTech Res. IJCRGG* **2016**, *9*, 252–260.

50. Mansur, D.; Tago, T.; Masuda, T.; Abimanyu, H. Conversion of cacao pod husks by pyrolysis and catalytic reaction to produce useful chemicals. *Biomass Bioenergy* **2014**, *66*, 275–285. [CrossRef]

51. Adeyi, O. Proximate composition of some agricultural wastes in Nigeria and their potential use in activated carbon production. *J. Appl. Sci. Environ. Manag.* **2010**, *14*, 55–58. [CrossRef]

52. Alemawor, F.; Dzogbefia, V.P.; Oddoy, E.O.K.; Oldham, J.H. Enzyme cocktail for enhancing poultry utilization of cocoa pod husk. *Sci. Res. Essay* **2009**, *4*, 555–559. Available online: http://www.academicjournals.org/SRE (accessed on 12 May 2021).

53. Vriesmann, L.C.; de Mello Castanho Amboni, R.D.; de Oliveira Petkowicz, C.L. Cacao pod husks (*Theobroma cacao* L.): Composition and hot-water-soluble pectins. *Ind. Crops Prod.* **2011**, *34*, 1173–1181. [CrossRef]

54. Chun, K.S.; Husseinsyah, S. Agrowaste-based composites from cocoa pod husk and polypropylene: Effect of filler content and chemical treatment. *J. Thermoplast. Compos. Mater.* **2016**, *29*, 1332–1351. [CrossRef]

55. Cruz, G.; Pirilä, M.; Huuhtanen, M.; Carrión, L.; Alvarenga, E.; Keiski, R.L. Production of activated carbon from cacao (*Theobroma cacao*) pod husk. *J. Civ. Environ. Eng.* **2012**, *2*, 1–6. [CrossRef]

56. Syamsiro, M.; Saptoadi, H.; Tambunan, B.H.; Pambudi, N.A. A preliminary study on use of cacao pod husk as a renewable source of energy in Indonesia. *Energy Sustain. Dev.* **2012**, *16*, 74–77. [CrossRef]

57. Erol, M.; Haykiri-Acma, H.; Küçükbayrak, S. Calorific value estimation of biomass from their proximate analyses data. *Renew. Energy* **2010**, *35*, 170–173. [CrossRef]

58. Robak, K.; Balcerek, M. Review of second-generation bioethanol production from residual biomass. *Food Technol. Biotechnol.* **2018**, *56*, 174–187. [CrossRef] [PubMed]

59. Serra, B.J.; Aragay, B.M. Composition of dietary fibre in cocoa husk. *Z. Für Lebensm. Und—Forsch. A* **1988**, *207*, 105–109. [CrossRef]

60. Lu, F.; Rodriguez-Garcia, J.; Van Damme, I.; Westwood, N.J.; Shaw, L.; Robinson, J.S.; Warren, G.; Chatzifragkou, A.; Mason, S.M.; Gomez, L.; et al. Valorisation strategies for cocoa pod husk and its fractions. *Curr. Opin. Green Sustain. Chem.* **2018**, *14*, 80–88. [CrossRef]

61. Sandoval, A.J.; Barreiro, J.A.; De Sousa, A.; Valera, D.; López, J.V.; Müller, J.A. Composition and thermogravimetric characterization of components of Venezuelan fermented and dry Trinitario cocoa beans (*Theobroma cacao* L.): Whole beans, peeled beans and shells. *Rev. Téc. Ing. Univ. Zulia* **2019**, *42*, 39–47. [CrossRef]

62. Torres-Moreno, M.; Torrescasana, E.; Salas-Salvadó, J.; Blanch, C. Nutritional composition and fatty acids profile in cocoa beans and chocolates with different geographical origin and processing conditions. *Food Chem.* **2015**, *166*, 125–132. [CrossRef]

63. Afoakwa, E.O.; Quao, J.; Takrama, J.; Budu, A.; Sand Saalia, F.K. Chemical composition and physical quality characteristics of Ghanaian cocoa beans as affected by pulp pre-conditioning and fermentation. *J. Food Sci. Technol.* **2013**, *50*, 1097–1105. [CrossRef]

64. Bertazzo, A.; Comai, S.; Brunato, I.; Zancato, M.; Costa, C.V. The content of protein and non-protein (free and protein-bound) tryptophan in *Theobroma cacao* beans. *Food Chem.* **2011**, *124*, 93–96. [CrossRef]

65. Valussi, M.; Minto, C. Cacao as a Globalised Functional Food: Review on Cardiovascular Effects of Chocolate Consumption. *Open Agric. J.* **2016**, *10*, 36–51. [CrossRef]

66. Oracz, J.; Zyzelewicz, D.; Nebesny, E. The Content of Polyphenolic Compounds in Cocoa Beans (*Theobroma cacao* L.), Depending on Variety, Growing Region, and Processing Operations: A Review. *Crit. Rev. Food Sci. Nutr.* **2015**, *55*, 1176–1192. [CrossRef]

67. Tomas-Barberan, F.A.; Cienfuegos-Jovellanos, E.; Marín, A.; Muguerza, B.; Gil-Izquierdo, A.; Cerda, B.; Zafrilla, P.; Morillas, J.; Mulero, J.; Ibarra, A.; et al. A new process to develop a cacao powder with higher flavonoid monomer content and enhanced bioavailability in healthy humans. *J. Agric. Food Chem.* **2007**, *55*, 3926–3935. [CrossRef]

68. Goude, K.A.; Adingra, K.M.D.; Gbotognon, O.J.; Kouadio, E.J.P. Biochemical characterization, nutritional and antioxidant potentials of cocoa placenta (*Theobroma Cacao* L.). *Ann. Food Sci. Technol.* **2019**, *20*, 603–613.

69. Delgado-Gutiérrez, N.A. Plan de Manejo Integral de Residuos Derivados de la Extracción de la Pulpa de Cacao en la Hacienda Bellavista, Luz de América, Provincia de Azuay-Ecuador. Ph.D. Thesis, Universidad de Cuenca, Cuenca, Ecuador, 2018.

70. Sánchez-Prieto, V.; Ahmed-El, S.; Yépez-Anchundia, M.; Mosquera, C.; Arizaga, R.; Cadena, N. Elaboración de alimento balanceado para pollo broiler a base de subproductos de cacao (cáscara, cascarilla y placenta) Elaboration of balanced feed for chicken broiler based on cocoa by-products (shell, husk and placenta). *Espirales Rev. Multidiscip. De Investig.* **2018**, *2*, 1–13. Available online: https://www.researchgate.net/publication/323583608 (accessed on 12 May 2021).

71. Damsere, J.D.; Boateng, M.; Amoah, K.O.; Frimpong, Y.O.; Okai, D.B. The response of pigs to diets containing varying levels of cocoa placenta meal (CPM) supplemented with an exogenous enzyme complex. *Afr. J. Agric. Res.* **2020**, *15*, 538–545. [CrossRef]

72. Leite, P.B.; Machado, W.M.; Guimarães, A.G.; Mafra de Carvalho, G.B.; Magalhães-Guedes, K.T.; Druzian, J.I. Cocoa's residual honey: Physicochemical characterization and potential as a fermentative substrate by *Saccharomyces cerevisiae*. *Sci. World J.* **2019**, *7*, 5698089. [CrossRef] [PubMed]

73. Vallejo-Torres, C.A.; Díaz-Ocampo, R.; Morales-Rodríguez, W.; Soria-Velasco, R.; Vera-Chang, J.F.; Baren-Cedeño, C. Utilización del mucílago de cacao, tipo nacional y trinitario, en la obtención de jalea. *Espamciencia* **2016**, *7*, 51–58.

74. Arteaga-Estrella, Y. Estudio del desperdicio del mucilago de cacao en el cantón naranjal (provincia del Guayas). *ECA Sinergia* **2013**, *4*, 49–59.

75. Cubillos-Bojacá, A.F.; García-Muñoz, M.C.; Calvo-Salamanca, A.M.; Carvajal-Rojas, G.H.; Tarazona-Díaz, M.P. Study of the physical and chemical changes during the maturation of three cocoa clones, EET8, CCN51, and ICS60. *J. Sci. Food Agric.* **2019**, *99*, 5910–5917. [CrossRef]
76. Bertoldi, D.; Barbero, A.; Camin, F.; Caligiani, A.; Larcher, R. Multielemental fingerprinting and geographic traceability of Theobroma cacao beans and cocoa products. *Food Control* **2016**, *65*, 46–53. [CrossRef]
77. Soetan, K.O.; Olaiya, C.O.; Oyewole, O.E. The importance of mineral elements for humans, domestic animals and plants: A review. *Afr. J. Food Sci.* **2010**, *4*, 200–222.
78. Kruszewski, B.; Obiedziński, M.W. Multivariate analysis of essential elements in raw cocoa and processed T chocolate mass materials from three different manufacturers. *LWT—Food Sci. Technol.* **2018**, *98*, 113–123. [CrossRef]
79. Ganda-Putra, G.P.; Wartini, N.M.; Trisna Darmayanti, L.P. Characteristics of Cocoa Vinegar from Pulp Liquids Fermentation by Various Methods. In Proceedings of the of the 2nd International Conference on Biosciences and Medical Engineering (ICBME2019) AIP Conf. Proc. Bali, Indonesia, 11–12 April 2019. [CrossRef]
80. Ministerio de Agricultura y Desarrollo. 2017. Available online: https://www.agronet.gov.co/Documents/SANTANDER_2017.pdf/ (accessed on 26 June 2019).
81. *AOAC 931.04-1931*; Loss on Drying (Moisture) in Cacao Products. AOAC International: Rockville, MD, USA, 2015.
82. *AOAC 925.10-1925*; Solids (Total) and Loss on Drying (Moisture). AOAC International: Rockville, MD, USA, 2014.
83. *AOAC 972.15-1974*; Ash of Cacao Products. AOAC International: Rockville, MD, USA, 2015.
84. *AOAC 970.22-1970*; Nitrogen (Total) in Cacao Products. AOAC International: Rockville, MD, USA, 2015.
85. *AOAC 2001.11-2005*; Protein (Crude) in Animal Feed, Forage (Plant). AOAC International: Rockville, MD, USA, 2015.
86. *AOAC 930.20-1930*; Crude Fiber in Cacao Products. AOAC International: Rockville, MD, USA, 2015.
87. *AOAC 920.75-1920*; Procedure for Separation of Fat in Cacao Products. AOAC International: Rockville, MD, USA, 2015.
88. Standard Methods Committee of the American Public Health Association, American Water Works Association, and Water Environment Federation. *5530 phenols In Standard Methods for the Examination of Water and Wastewater*; Lipps, W.C., Baxter, T.E., Braun-Howland, E., Eds.; APHA Press: Washington, DC, USA, 2017. [CrossRef]
89. *AOAC 925.36-1925*; Sugars (Reducing) in Fruits and Fruit Products. AOAC International: Rockville, MD, USA, 2020.
90. *AOAC 931.12-1931(1996)*; Calomel in Ointments. Titrimetric Method. AOAC International: Rockville, MD, USA, 2015.
91. *AOAC 960.19-1960*; pH of Wines. AOAC International: Rockville, MD, USA, 2015.
92. *AOAC 985.35-1988*; Minerals in Infant Formula, Enteral Products. AOAC International: Rockville, MD, USA, 2014.
93. Standard Methods Committee of the American Public Health Association, American Water Works Association, and Water Environment Federation. *4500-norg nitrogen (organic) In Standard Methods for the Examination of Water and Wastewater*; Lipps, W.C., Baxter, T.E., Braun-Howland, E., Eds.; APHA Press: Washington, DC, USA, 2018. [CrossRef]
94. Standard Methods Committee of the American Public Health Association, American Water Works Association, and Water Environment Federation. 2540 solids In *Standard Methods for the Examination of Water and Wastewater*; Lipps, W.C., Baxter, T.E., Braun-Howland, E., Eds.; APHA Press: Washington, DC, USA, 2018. [CrossRef]
95. Green, J.W. *Methods in Carbohydrate Chemistry*; Whistler, R.L., Ed.; Academic Press: New York, NY, USA, 1963; Volume 3, pp. 9–21.
96. Kurschner, K.; Hoffer, A. Cellulose and cellulose derivative: Fresenius. *J. Anal. Chem.* **1993**, *92*, 145–154.
97. Dence, C.W. The determination of lignin. In *Methods in Lignin Chemistry*; Lin, S.Y., Dence, C.W., Eds.; Springer: Heidelberg, Germany, 1992; pp. 33–61.

 *molecules*

Article

Starch-Silane Structure and Its Influence on the Hydrophobic Properties of Paper

Tomasz Nowak [1,2], Bartłomiej Mazela [1], Konrad Olejnik [3], Barbara Peplińska [4] and Waldemar Perdoch [1,*]

[1] Faculty of Forestry and Wood Technology, Poznań University of Life Sciences, Wojska Polskiego 28, 60-637 Poznan, Poland; tomasz.nowak@poskladani.pl (T.N.); bartlomiej.mazela@up.poznan.pl (B.M.)

[2] POSkładani.pl A.T. Nowak spółka jawna, ul. Władysława Nehringa 8 lok. 1, 60-247 Poznan, Poland

[3] Centre of Papermaking and Printing, Lodz University of Technology, Wolczanska 221, 93-005 Lodz, Poland; konrad.olejnik@p.lodz.pl

[4] NanoBioMedical Centre, Adam Mickiewicz University, Umultowska 85, 61-614 Poznan, Poland; barp@amu.edu.pl

* Correspondence: waldemar.perdoch@up.poznan.pl

Abstract: Starch is an inexpensive, easily accessible, and widespread natural polymer. Due to its properties and availability, this polysaccharide is an attractive precursor for sustainable products. Considering its exploitation in adhesives and coatings, the major drawback of starch is its high affinity towards water. This study aims to explain the influence of the silane-starch coating on the hydrophobic properties of paper. The analysis of the organosilicon modified starch properties showed an enhanced hydrophobic behavior, suggesting higher durability for the coatings. Molecules of silanes with short aliphatic carbon chains were easily embedded in the starch structure. Longer side chains of silanes were primarily localized on the surface of the starch structure. The best hydrophobic properties were obtained for the paper coated with the composition based on starch and methyltrimethoxysilane. This coating also improved the bursting resistance and compressive strength of the tested paper. A static contact angle higher than 115° was achieved. PDA analysis confirmed the examined material exhibited high barrier properties towards water. The results extend the knowledge of the interaction of silane compositions in the presence of starch.

Keywords: silylated starch; hydrophobic agent; starch hydrophobization; paper coating; cellulose

Citation: Nowak, T.; Mazela, B.; Olejnik, K.; Peplińska, B.; Perdoch, W. Starch-Silane Structure and Its Influence on the Hydrophobic Properties of Paper. *Molecules* **2022**, *27*, 3136. https://doi.org/10.3390/molecules27103136

Academic Editors: Mohamad Nasir Mohamad Ibrahim, Patricia Graciela Vázquez and Mohd Hazwan Hussin

Received: 29 March 2022
Accepted: 11 May 2022
Published: 13 May 2022

Publisher's Note: MDPI stays neutral with regard to jurisdictional claims in published maps and institutional affiliations.

1. Introduction

Starch, along with cellulose and lignin, is perceived as one of the most attractive biopolymers due to its properties, availability, and low price. Starch is a polysaccharide of plant origin. Sources include roots (e.g., sweet potatoes, tapioca), tubers (e.g., potatoes), stems (e.g., sago palms), cereal grains (e.g., corn, rice, wheat, barley, oat, sorghum), and legumes (e.g., peas and beans), among others [1]. Starch has thermoplastic properties, is biodegradable, is soluble in hot water, and does not create technological obstacles in the recycling process [2]. However, poor mechanical properties, high moisture sorption, and brittleness are disadvantages that limit its industrial use as a stand-alone material [3,4]. Nevertheless, starch, subjected to an appropriate chemical treatment, possesses film-forming properties [2,5] that are used, for example, to control the absorbency of the paper surface. As a result, it improves the printing properties [6].

However, available methods of starch usage in the industry do not guarantee sufficient hydrophobization of paper products used in high moisture conditions, especially transport packaging [7]. Spiridon et al. [8] modified starch microparticles (TA-SM) with tartaric acid. They used the dry preparation technique in which the TA-SM microparticles were introduced as fillers within glycerol plasticized-corn starch. The modified starch was used as a coating agent. The water resistance and thermal stability were slightly improved through the addition of a large amount of cellulose due to the inter-component H-bonding between components. The evaluation of the mechanical properties revealed a significant increase in the tensile strength of the composites with increasing cellulose content [6].

The use of the starch properties to create colloidal solutions in hot water and its film-forming properties to evenly distribute compounds with hydrophobic characteristics, such as organosilicon compounds on the surface of biodegradable materials, appears to be a sustainable method of hydrophobization [9–11]. The search for synergy and compensation of the flaws between biopolymers and organic compounds seems to be the right direction for developing new industrial technologies [12]. Organosilicon compounds are used with great success in the hydrophobization of wood [13,14]. In hydrolysis, alcoholysis, condensation reactions, radical reactions, and sol-gel type reactions, bonds between hydroxyl groups in cellulose and organic substituents of alkoxysilanes are created [15–18]. These bonds are permanent and make the material significantly resistant to water activity. In terms of the chemical structure, starch is a polymer similar to cellulose. It is built out of repeating glucose units, combined with α-1,4-glycosidic (amylose) bonds and fractions of amylopectins that have branches, due to the presence of α-1,6-glycosidic bonds [19]. Its main difference from cellulose is the orientation of the glucose rings in the chain of amylose, and this seemingly minor difference is why starch, unlike cellulose, dissolves in hot water, creating a colloidal solution. The viscosity of the solution can be controlled by changing the concentration, which makes it easier to distribute it on a carrier in the form of a thin film [13,20].

Surface sizing and creating a film on a cellulose carrier is one of the most common operations used in the paper industry [19,21]. In terms of internal sizing to increase starch retention in paper, it is advisable to use cationic starch, which is characterized by a higher affinity to cellulose fibers [22]. Starch solutions can also be modified by adding various substances for papermaking purposes [19,21,22]. Furthermore, the binding and film-forming properties of starch create, for example, the possibility of applying alkoxysilanes on the surface of paper products. The hydroxyl groups of the glucose repeating units give the possibility of creating starch-silane bonds first, and subsequently, after applying on the paper substrate, starch–silane–cellulose bonds are created. That way, a consolidated hydrophobic layer is created.

There are known methods of paper hydrophobization using alkoxysilanes with substituents (e.g., isocyanate [23,24] or chlorine [25]), which guarantee the creation of silane-cellulose bonds. Several solutions in the literature present organosilicon compounds as a hydrophobizing agent for starch. The solution is known from the Glittenberg reports [22], in which a powdery starch was mixed with alkali silanes in the presence of alkali metal salts. In its description, granular starch was first treated with an aqueous solution of alkali metal salts of alkali silicates, such as sodium methyl-silicate, and subsequently air-dried at room temperature. The effect of the starch modification was the increase in the hydrophobicity of the dried granules. In addition, the new granulated silicone starch had a neutral pH, and in the form of a dry powder, it had mobility and flow properties similar to fluids. Another beneficial feature of the product was that it was resistant to water activity at a temperature up to 50 °C. Adding modified starch to water at the temperature of its gelling (or close to it) resulted in the ability to create smooth pastes and dispersions. According to the invention, modified starch had increased durability properties.

The invention described by Satterly [26] presented the method of obtaining starch in a powdery form that has increased hydrophobic properties. Hydrophobization was reached as a result of adding water-soluble silicones. The hydrophobic mixture was prepared at a temperature lower than the temperature of starch gelatinization (<60 °C). Amort et al. [27] described the modification of starch by using silanes, which was conducted in the presence of alkali metal hydroxides or alkali metal aluminates to silane hydrolyzates. According to the authors, the modified starch was characterized by better processing properties than a corresponding unmodified starch. Chen et al. [28] examined starch-silane (γ-methacryloxypropyl trimethoxysilane) systems to describe their adhesive properties. The glue samples prepared using trimethoxies γ-methacryloxypropyl silane as a cross-linking agent had increased stability during storage and increased shear resistance of the weld. Furthermore, the addition of silanes improved the plastic properties of the weld. Wei et al. [29] described a nanocrystalline starch that was subjected to modification through organosilicon compounds to increase the hydrophobic properties of the final product. The measured contact angle of the modified nanocrystalline starch was three times higher than the contact angle measured for the unmodified starch. Two-stage modification of starch using vinyltrimethoxysilane and a copolymer of methyl methacrylate (MMA), butyl acrylate (BA), and 2,2,2-trifluoroethyl methacrylate (3FMA) significantly changed the properties of the material. The latex foils obtained showed an increased hydrophobicity, lower free surface energy, and higher thermal stability than the unmodified starch [30].

Jariyasakoolroj and Chirachanchai [31] modified starch with silanes and polylactic acid. Their study demonstrated the creation of permanent bonds between components of the reaction mixture, and the created products had an increased level of crystallinity, lower glass transition temperature, and a slight increase in the tensile strength. In the study by Sandrine et al. [32], they described the influence of chemical modification on the properties of hemp-starch composites subjected to an alkaline treatment using silanes. Those actions increased the mechanical properties of the produced composites, especially their rigidity. Ganicz et al. [33,34] showed a method of paper hydrophilization through water emulsion of triethoxymethylsilane as a paper coating. The authors examined the effects of the applied coatings on the paper's tensile strength, tear index, roughness, air permeance, and brightness.

The methods described above demonstrate the great potential of starch as a natural polymer, providing excellent opportunities within the scope of its chemical modification, especially in the range of conferring hydrophobic properties. However, the mentioned solutions do not apply to phenomena and interactions occurring between silanes and starch. In the literature, there is still a lack of complex studies regarding the influence of modified starch on the properties of cellulose material. For this reason, our study aims to examine the physicochemical interactions using Scanning Electron Microscopy (SEM) with Energy Dispersive X-Ray (EDX) analysis and hydrophobic properties (contact angle, Penetration Dynamics Analysis—PDA) between starch and organosilicon compounds in their application on a paper as a cellulose matrix. This study aims to explain the influence of the structure of the silane-starch coating on the hydrophobic properties of paper.

2. Results and Discussion

2.1. Water Penetration Dynamics of Modified Starch Applied on a Paper

The liquid–paper interaction is a very complex phenomenon. This process can be divided into two main stages: wetting and penetration. During the wetting stage, the water contact with the paper surface is hampered due to an air film adhering to the surface and air presence inside the surface pores. After the wetting stage, water finally begins to enter the paper structure and displaces air from the pores. The internal specific surface area of paper and its thickness increase, and micro-bubbles of air are observed within the adsorbed water. These air micro-bubbles are scattering centers for ultrasonic radiation; thus, a lower intensity signal is transmitted through the sample. The ultrasonic signal decreases faster as the liquid penetrates faster into the paper. The results of the measurements of the dynamics of water penetration into the structure of the paper samples with the coatings applied and the reference paper (without the coating) are presented in Table 1 and Figure 1a,b. The process of wetting and penetrating liquids is described by several parameters. Parameter "t95" correlates to the wetting rate of the surface and it corresponds to the size and structure of the pores at the paper surface. Parameter "Max" is the information about the surface resistance against the water. The wetting stage lasts until time "Max". The water penetration begins after time "Max" [35,36]. This is the indicator of hydrophobicity and the sizing degree of the paper surface. Moreover, the slope of the curve beyond the Max value can be considered the penetration speed ($\Delta I/\Delta t$) [35,36]. In the presented research, the calculations were performed for two times: 0.2 s and 3 s. This was because the reference paper and the starch coated paper had such a high water absorbency that after 3 s, the water absorption process was finished. Therefore, in order to show the dynamics of water absorption, it was necessary to also use a time shorter than 0.5 s. On the other hand, papers coated with a hydrophobization agent absorbed water much slower. In their case, performing a calculation for a time base of 3 s allowed for better presentation of the water absorption rate. The results in Table 1 show that the reference paper and the paper with only a starch based coating showed no wettability resistance. The wetting period described by the Max parameter was 0 s, and the values of $\Delta I / \Delta t$ (water absorption speed) during 0.2 s were very high. For the samples with coatings based on various starch-silane compositions, the lowest absorption rate in 0.2 s was recorded for the paper with the methyltrimethoxysilane (MTMS) coating. In this case, the value was -9.6, which means that the wetting barrier has not yet been overcome for this paper. This is confirmed by the value of the Max parameter, which was 0.4 s. The values of $\Delta I / \Delta t$ calculated for a time of 3 s confirmed that for the reference paper and paper with only a starch coating, the absorption process has already been completed. On the other hand, paper with a starch and MTMS coating was characterized by the lowest water absorption rate (i.e., the highest hydrophobicity).

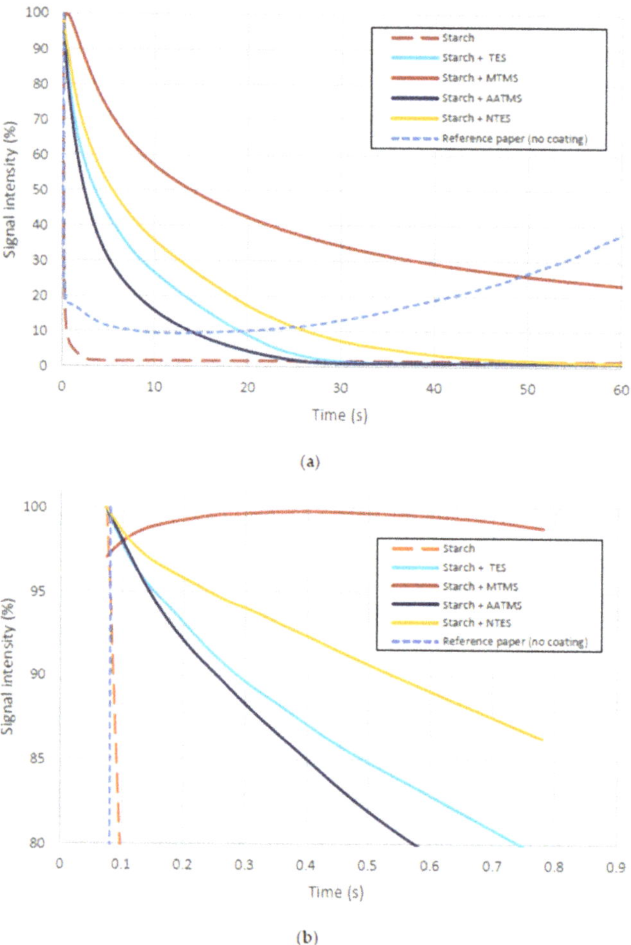

Figure 1. Analysis of PDA—the intensity of ultrasonic waves as a function of time for modified starch applied on a paper; (**a**) 0–60 s; and (**b**) 0–0.8 s.

Table 1. Results of the measurements of water penetration into the structure of the tested paper samples. The coefficient of variation for all cases ranged from 6.4 to 11.7%.

	Reference Paper (Uncoated)	Paper Coated with Starch	Paper Coated with Starch + AATMS	Paper Coated with Starch + NTES	Paper Coated with Starch + TES	Paper Coated with Starch + MTMS
MAX	0.0	0.0800	0.0780	0.0777	0.0777	0.4003
t95	0.0	0.0817	0.1837	0.2583	0.1537	1.3593
$\Delta I/\Delta t$ (0.2 s)	253.7	303.1	13.7	18.6	11.2	−9.6
$\Delta I/\Delta t$ (3 s)	1.6	3.9	36.3	9.8	30.4	5.6
Water uptake (Cobb60), g/m^2	154.5	135.8	133.5	119.1	128.2	21.5

The results presented in the table are confirmed by the curves presented in Figure 1a,b. The curve shape for the reference paper (uncoated) is typical for highly hydrophilic materials. The increase in the signal after 10 s of the measurement means that water penetrated the material structure quickly and evenly and the structure of the material began to swell (Figure 1a). Different materials suppress ultrasonic waves to varying degrees depending on their sorption properties. Water sorption occurs the fastest for starch applied on paper and the slowest for the samples coated with the addition of tetraethoxysilane tetraethoxysilane (TES), n-octyltriethoxysilane (NTES), and MTMS. This observation indicates that MTMS provided the best hydrophobization effect against water for paper samples with a cellulose surface. As was stated for the results from Table 1, the results from the PDA analysis indicate that in the initial phase of contact with water (i.e., between 0 up to 1 s), the examined variants of paper samples exhibited various wetting properties (Figure 1b). The rapid decrease in the signal for the uncoated paper in the initial wetting phase indicates the lack of a hydrophobic barrier on the surface and its rapid wetting (Figure 1b). The paper coated with starch modified through MTMS displayed a decreased wettability and permeability towards water, which was reflected in the increase in the intensity of the ultrasonic waves in the initial phase of the experiment (before the Max time) and the slower decrease after the time Max. The water uptake test (Cobb60 method) confirms the high effectiveness of the starch + MTMS composition. Compared to paper without any coating (Reference paper), the absorbency of starch + MTMS coated paper was more than seven times lower. The high absorbency of the other tested samples indicates that coating compositions containing AATMS, NTES or TES are not suitable as hydrophobic agents in papermaking applications.

2.2. Water Repellency Effect of Modified Starch Applied on a Paper

The study of the water-barrier properties of the investigated paper samples was supported by the measurement of the contact angle (Figure 2). For the reference paper, the contact angle was not measurable due to the immediate penetration of water into the material structure. The four organosilicone modified starch (OMS) act as hydrophobizing agents but to a different extent, and in particular, MTMS and NTES enhance the contact angle over 100°, while *N*-(2-Aminoethyl)-3-aminopropyltrimethoxy silane (AATMS) is much less effective (63°). These results (Figure 2) suggest that the chemistry of MTMS and NTES comes from the hydrophobic alkyl chain. A chain that includes an amine structure, which is hydrophilic, does not protect a paper against water. The hydrophilic and hydrophobic nature of the alkyl part of silane probably influences the final results of the hydrophobic properties. Despite the low barrier properties demonstrated in the PDA study, starch applied on the paper slightly increased the static contact angle parameter (51°).

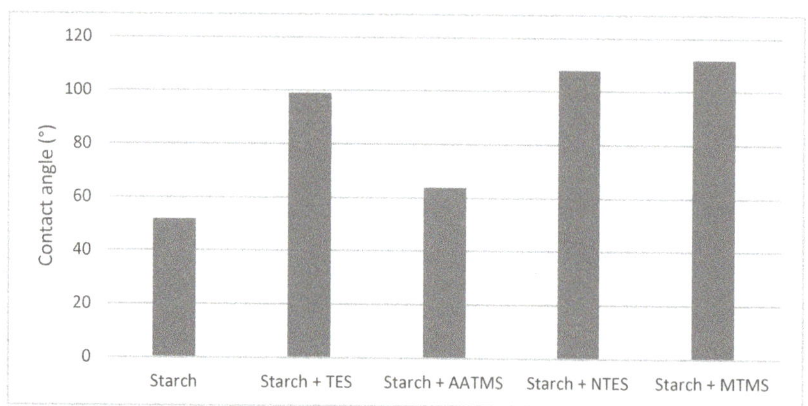

Figure 2. The static contact angle for modified starch applied on a paper.

2.3. Effect of the Starch-MTMS Coating on the Selected Strength Properties of Paper Sheets

Among all the starch and silane-based coating compositions tested, the best barrier effects were obtained for the coating containing MTMS. To investigate the effect of this coating on the mechanical properties of paper, proper tests were carried out. Typical strength properties that are typically determined for paper packaging materials were selected for this purpose: bursting strength and compressive strength. Figure 3 shows the comparison of the bursting resistance of a reference paper (uncoated), a paper with only a starch-based coating, and a paper with a coating containing starch and MTMS. The obtained results indicate that the application of the starch on the surface of the paper increased its bursting resistance. The coating based on the starch and MTMS mixture resulted in a reduction in the value of the bursting pressure by about 10 kPa in relation to the paper with the coating based on only starch. Despite the presence of MTMS, the bursting resistance was higher by approx. 60 kPa compared to the reference paper (without any coating). This means that the coating consisting of starch and MTMS still exhibits the properties that improve the bursting strength of paper.

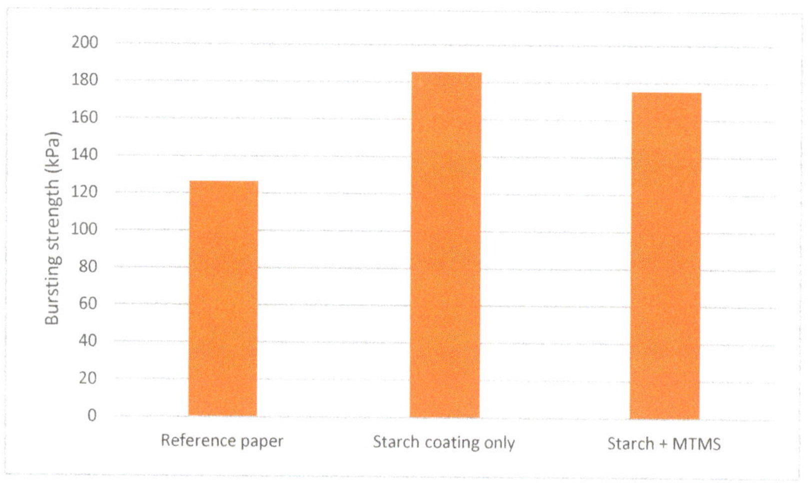

Figure 3. Comparison of the bursting strength for a reference paper, paper coated only with starch, and paper coated with a starch + MTMS composition.

The results presented in Figure 4 show the compressive strength of the tested paper samples. In this case, the measurements were conducted for both directions of the paper (MD—Machine Direction and CD—Cross Direction). The obtained results confirm that the coating based on starch and MTMS does not adversely affect the strength properties of the tested paper.

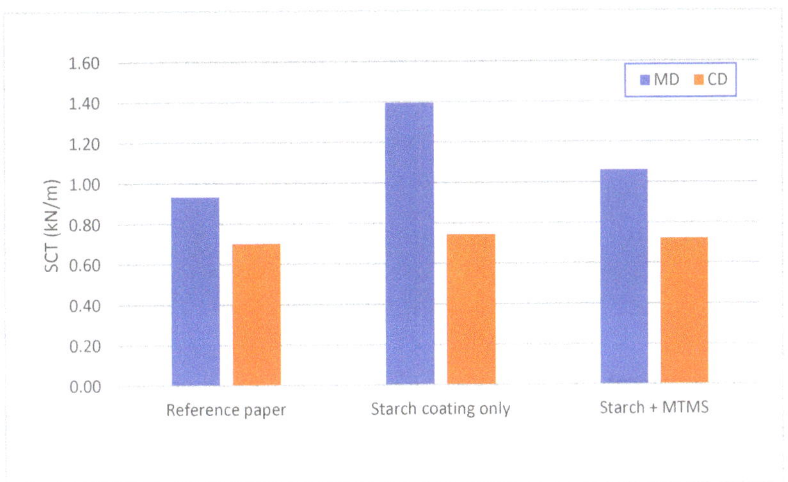

Figure 4. Comparison of the compression strength for a reference paper, paper coated with only starch, and paper coated with a starch + MTMS composition.

2.4. Morphology Modification of OMS

The size of the starch granules subjected to gelatinization and modification varied and ranged between 4–30 μm (Figure 5). Starch subjected to thermal treatment in an aqueous solution (gelatinization process) and subsequently frozen out has a strongly porous structure (Figure 6a,b). It is worth noting that the examined material subjected to the physical treatment mentioned above presents a very different appearance of the surface structure and inside structure (i.e., cross-sectional area) of the sample. Those differences (Figure 7) are most likely the result of changes taking place in the starch gruel structure.

Figure 5. SEM micrograph of native starch granules ×600.

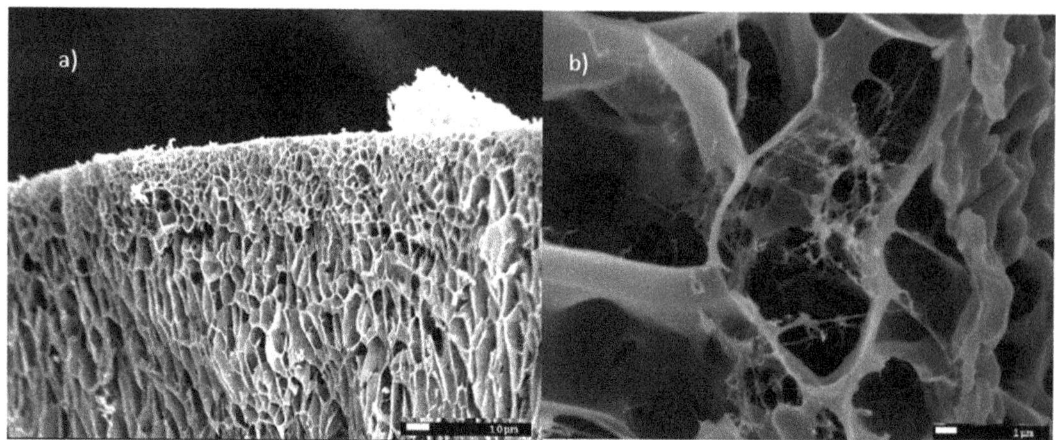

Figure 6. Cryo-SEM micrographs of an aqueous solution of starch (**a**) sample structure ×500; and (**b**) sample structure ×5000.

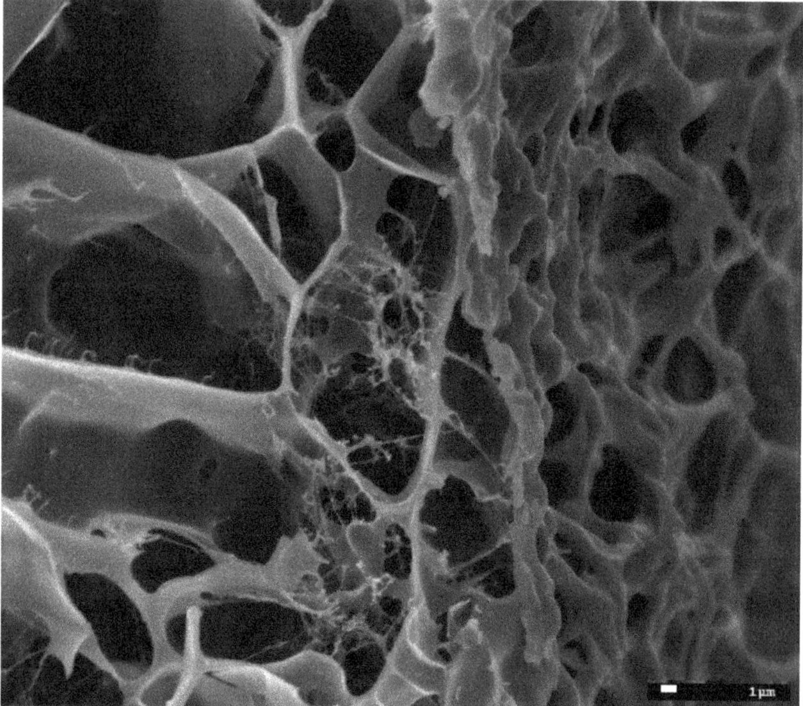

Figure 7. Cryo-SEM micrographs of an aqueous solution of starch ×3000—differences of the inside and surface structures.

Figure 8a–d summarizes the images of starch modified through MTMS. Similar to the case of unmodified starch gruel, a pronounced difference can be observed on the surface and inside the structure of the silane-modified material (Figure 8a). The microscopic image of starch modified by MTMS (Figure 8b) is similar to that of unmodified starch (Figures 6 and 7). The examined material had a significantly higher stability than the structure of unmodified starch because the 10 kV energy used during the analysis allowed for the microscopic images with a magnification of 20,000 and 50,000 to be acquired without destroying the sample (Figure 8c,d). In comparison, the unmodified starch gruel degraded at 10 kV, i.e., during the analyses performed at a magnification of 5000. In Figure 6b, a cross-linking of the starch chains can be observed, at the end of which spherical structures are visible that contain a very high silicon content (according to EDX analysis), i.e., 2.35% by weight (1.27 atomic %). Organosilicon compounds were also embedded in the starch chain structure, which could be observed as distinct brighter spherical areas of the SEM micrographs of OMS through MTMS (Figure 8c,d). The silicon concentration in those areas was two times higher than in the remaining fragments of the cross-linked starch gruel. Following the EDX analysis, the mean content of MTMS in the gruel structure was 1.8% by weight (silicon is 1% of the weight in the structure of MTMS-modified starches). The most crucial information of the analysis of the structure of starch modified through MTMS was the confirmation that the added organosilicon compound had been distributed across all the cross-linked structures of starch. In Figure 9, the map of the silicon distribution in starch modified through MTMS is shown. The presence of silicon atoms precisely matches the structure of the starch gruel.

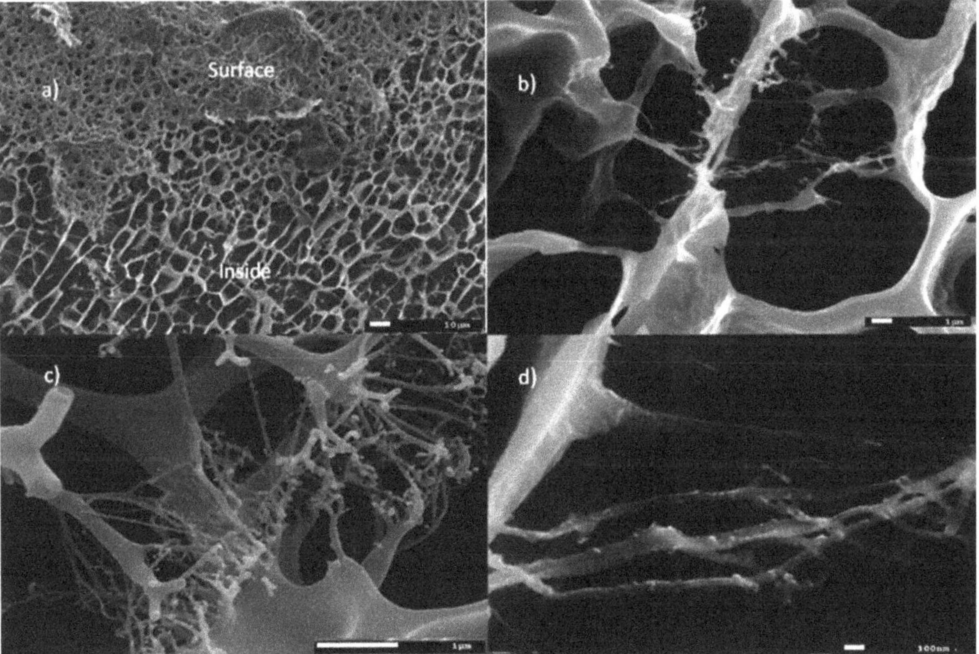

Figure 8. Cryo-SEM micrographs of an aqueous solution of starch modified through MTMS: (**a**) differences of the inside and surface structures ×500; (**b**) internal bonding in a modified starch structure ×5000; (**c**) internal bonding including silane groups ×20,000; and (**d**) internal bonding including silane groups ×50,000.

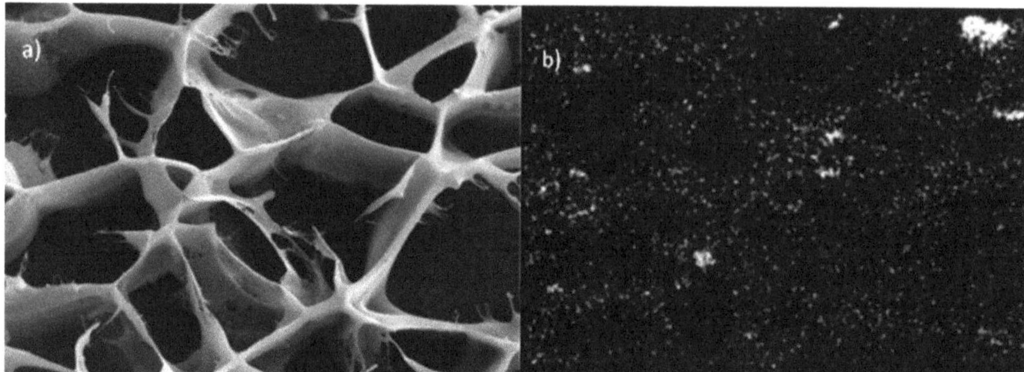

Figure 9. Cryo-SEM micrographs of an aqueous solution of starch modified through MTMS (**a**) and Si atom distribution (EDX mapping) in the starch structure (**b**).

The modification of starch through NTES did not influence the change in the durability of the structure of the starch gruel. Similarly, in the case of unmodified starch, the structure of the examined material was degraded (Figure 10a) during the analysis (10 kV; magnification of 5000). In Figure 10b, an interesting structure was observed on the surface of the examined material. The EDX analysis (Figure 10c) confirmed that the sample surface is covered by silane. The silicon concentration on the surface of the examined material was 9.11% by weight (4.53 atomic %). There were no organosilicon compounds localized in the cross-section of the examined material. Therefore, NTES was not embedded in the starch structure, but only accumulated on its surface. Figure 10b shows the difference between the surface and cross-sectional structure of the examined material. The microscopic images allowed for the observation of characteristic, bright, spherical structures of silane embedded into the starch structure. EDX analysis allowed us to observe that AATMS in the starch structure accumulated mainly on the surface of the examined material. The mean concentration of silicon on the surface of the material was 5.03% by weight (2.69 atomic %), which was two times higher than the concentration of silicon measured in the inside layer of the sample (2.26% by weight; 1.23 atomic %). A significant fact is that more areas are visible with embedded organosilicon compounds in the surface layer than in the areas localized deeper, indicating limited penetration of the silane deep into the modifying polymer (Figure 10b–f).

The distribution of the organosilicon compound in the starch coating seems to be the limiting factor for the water-repellent properties of the paper. Therefore, the impact of the structure of the organosilicon compounds on their location in the starch–silane network should be considered first. From the chemical point of view, the properties of the components used in the research were diverse. MTMS, which hydrolyzes in the water environment, contributed to the uniform distribution of silicon in the structure of the aqueous solution of starch. It resulted in a highly hydrophobic surface formed on the cellulosic materials. The probable cause of the even distribution of silicon was also the small size of the silane molecule itself, the aliphatic chain of which contains only one carbon atom. The use of organosilicon compounds with longer aliphatic chains (NTES) or containing amino groups (AATMS), regardless of their solubility in water, formed a compact structure on the surface of the aqueous solution of starch. Notably, the production of a similar coating on the starch surface was not tantamount to giving the hydrophobic properties of cellulose' materials. The accumulation of starch-containing NTES sparingly soluble in water on the surface of the paper created a water barrier, which was confirmed by the results of the high contact angle and PDA analysis. The use of AATMS, which is soluble in water under the same conditions, did not increase the hydrophobic properties of the material.

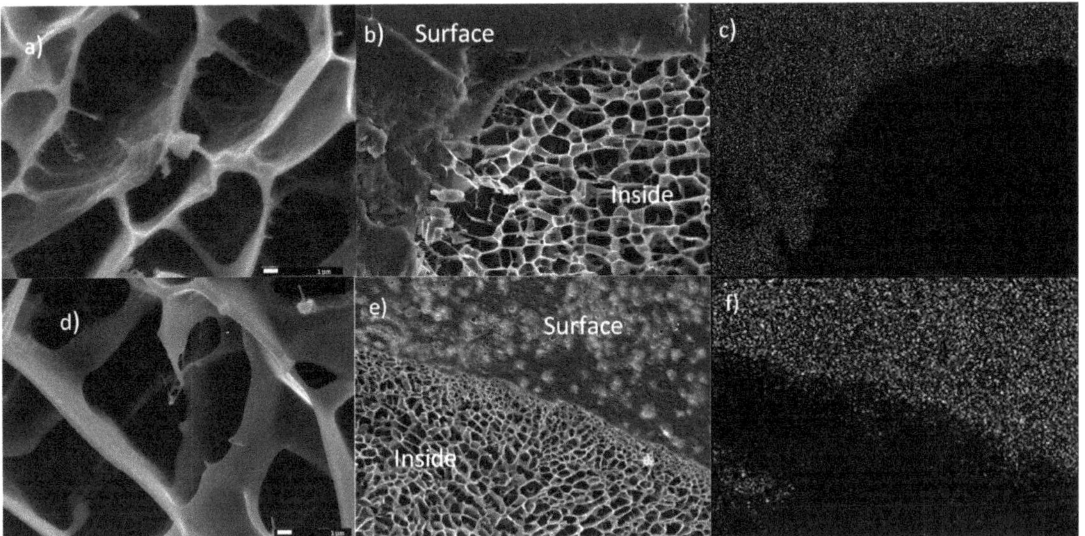

Figure 10. Cryo-SEM micrographs of starch modified through silanes: (**a**) internal bonding in an NTES modi-fied starch structure ×5000; (**b**) differences of the inside and surface structures with starch modified with NTES ×500; (**c**) EDX mapping of the (NTES) silicon distribution; (**d**) internal bonding in an AATMS modified starch structure ×5000; (**e**) differences of the inside and surface structures with starch modified with AATMS ×500; and (**f**) EDX mapping of the (AATMS) silicon distribution.

3. Materials and Methods

3.1. Preparation of the Starch-Silane Hydrophobic Agent

Commercial wheat starch was used in the presented research (C*Flex 20002, Cargill, Incorporated, Minneapolis, MN, USA). The amylose content in the starch was 31.2 ± 0.16 wt%. The lipid content was 0.25 ± 0.02 wt% and the protein content was 0.30 ± 0.01 wt%. An aqueous solution of wheat starch (5% concentration) was stirred at $70 \pm 5\,°C$ in the presence of sodium hydroxide (0.25% w/w) to reduce the temperature of starch gelatinization. It also allowed the starch structure to loosen, as reported by other authors [37–39], to ease the penetration of silanes between the amylose and amylopectin chains. The suspension obtained was cooled to $40 \pm 5\,°C$, and while mixing, organosilicon compounds were added (2.5% w/w). Organosilicons used in the current study are shown in Table 2.

Organosilicons used in the presented research have carbon chains of different lengths: methyltrimethoxysilane (MTMS), n-octyltriethoxysilane (NTES), *N*-(2-Aminoethyl)-3-amino-propyltrimethoxy silane (AATMS), and tetraethoxysilane (TES).

These four formulations were applied as a coating, using $100\ g/m^2$ for cellulose paper (Whatman cellulose no. 3), by dipping it in the organosilicon solution (5 s) and letting it dry at room temperature for 48 h. Morphological and chemical characterization was performed to understand the hydrophobization mechanism.

Table 2. Organosilicons applied to modify starch.

Acronym	Name	CAS Number	Chemical Structure
MTMS	methyltrimethoxysilane	1185-55-3	
NTES	n-octyltriethoxysilane	2943-75-1	
AATMS	N-(2-Aminoethyl)-3-aminopropyltrimethoxy silane	1760-24-3	
TES	Tetraethoxysilane (Tetraethyl orthosilicate)	78-10-4	

3.2. Evaluation of Hydrophobic Properties

3.2.1. Water Penetration Dynamics Analysis

Water penetration dynamics for paper samples coated with the starch-silane coated cellulose was measured using the PDA apparatus, Module S 05 (Emtec Electronic GmbH). The hand sheets were cut into three samples, each with a size of 3 cm × 7 cm. Measurements were performed separately for each sample in demineralized water at 20 °C, following the standard procedure described by the manufacturer [40]. The following parameters were used: ultrasound frequency of 2 MHz, measuring diameter of 35 mm, measurement duration of 60 s. The result was an arithmetic average value calculated from the series of three measurements.

3.2.2. Contact Angle Analysis

A PGX+ Goniometer from Testing Machines Inc. (New Castle, DE, USA) was used for the contact angle measurements. The tests were carried out according to the TAPPI T 458 standard method. The water drop volume was 4 μL.

3.2.3. Water Uptake

Water uptake test was conducted according to ISO 535:2014 (Cobb method, Cobb tester, Lorentzen & Wettre, Kista, Sweden). Time of the test was 60 s.

3.3. Mechanical Properties of the Paper Samples

Compressive strength (SCT test) was conducted according to ISO 9895:2008 standard (Instron 5564 machine, High Wycombe, UK). Bursting strength was conducted according to ISO 2758:2001 (Mullen Burst Machine, Lorentzen & Wettre, Kista, Sweden).

3.4. SEM-EDX Analysis of the Modified Starch Coating

The surface morphology of the samples was examined by the SEM method with cryogenic preparation system (Cryo-SEM), which enables direct investigation of the sample in the vitrified state. Cryo-scanning electron microscopy effectively observes wet samples without causing drying artifacts. It preserves the "natural" internal structure of the sample. This is particularly true for samples in the life sciences, where structures will retain their original conformation only in their fully hydrated state. Images were taken using the JEOL JSM-7001F TTLS (JEOL Ltd., Tokyo, Japan) scanning electron microscope equipped with the PP3000T cryo-SEM preparation system, which allows the cryo specimen to be prepared, processed, and transferred into the SEM chamber. The samples were cryo-fixed by plunging them into sub-cooled nitrogen (nitrogen slush, temperature of about $-210\,^{\circ}C$) and transferred by the vacuum cryo-transfer shuttle to the preparation chamber mounted onto the SEM. Inside the preparation chamber, at $-185\,^{\circ}C$, the specimen was fractured to expose a fresh surface. Then, it was sublimated and coated with a thin platinum layer. Finally, the sample was loaded under vacuum into the SEM chamber, where it remained frozen during imaging on the cold-stage, cooled by a nitrogen carrier ($-190\,^{\circ}C$). The images of all samples were taken by applying an accelerating voltage of 10 kV and using a secondary electron (SEI) detector. Elemental analysis and cryo-SEM elemental mapping were performed using the energy dispersive microanalysis (EDX) mode of an X-ray-equipped SEM that also applied a voltage of 10 kV.

4. Conclusions

The interaction of water and cellulose coated with starch modified with organosilicon compounds depends primarily on the physical properties and structure of the organosilicon compound itself. Based on the research, it can be concluded that the distribution of silicon in the gelatinized starch network depends primarily on the length of the organic chain in the organosilane molecule. MTMS molecules, which have very short side chains (methyl group), were easily embedded in the starch structure. Silanes containing longer side chains (e.g., AATMS, NTES) were localized mainly on the surface of the starch structure. The distribution of silanes resulting from their size does not translate unequivocally into making the coated materials hydrophobic. The physical properties of the organosilicon compounds primarily determine these properties. Poorly soluble substances such as NTES increase the hydrophobic properties, and water-soluble substances such as AATMS do not improve these properties. An even distribution of silane in starch and the developed permanent MTMS-starch bond indicate that this composition may be considered as a new type of hydrophobic agent for papermaking purposes. Static contact angle analysis and PDA analysis confirmed the high barrier properties of the examined material towards water. NTES containing an octa-carbon side chain in its structure produced a thin film on the starch surface, which also increased the barrier property of the coated paper. Despite producing a thin film on the starch surface, silane containing hydrophilic amine groups in its side chain did not provide barrier properties to the paper.

Author Contributions: Conceptualization, T.N., B.M. and W.P.; methodology, T.N., B.M., K.O., B.P. and W.P.; software, T.N. and B.P.; validation, B.M., K.O. and W.P.; formal analysis, W.P.; investigation, T.N., B.P. and W.P.; resources, T.N., B.M., K.O., B.P. and W.P.; data curation, T.N., B.M., K.O., B.P. and W.P.; writing—original draft preparation, T.N. and W.P.; writing—review and editing, B.M., K.O., B.P. and W.P.; supervision, B.M., K.O. and W.P.; project administration, B.M. and W.P. All authors have read and agreed to the published version of the manuscript.

Funding: This research was funded by National Center for Research and Development, III edition of EEA and Norway grants; The Program 'Applied Research in the frame of Norway Grants 2014-2021/POLNOR 2019 (NOR POLNOR/CellMat4ever/0063/2019-00)" and "Ministry of Science and Higher Education Programme—"Regional Initiative Excellence" 2019–2022, project no. 005/RID/2018/19.

Institutional Review Board Statement: Not applicable.

Informed Consent Statement: Not applicable.

Data Availability Statement: The data presented in this study are available on request from the corresponding author. The data are not publicly available due to University policies.

Conflicts of Interest: The authors declare no conflict of interest.

Sample Availability: Samples of the materials are available on request from the authors.

References

1. Swinkels, J.J.M. Composition and Properties of Commercial Native Starches. *Starch-Stärke* **1985**, *37*, 1–5. [CrossRef]
2. BeMiller, J.N.; Whistler, R.L. *Starch: Chemistry and Technology*; Academic Press: Cambridge, MA, USA, 2009; ISBN 0-08-092655-X.
3. Moriana, R.; Vilaplana, F.; Karlsson, S.; Ribes-Greus, A. Improved Thermo-Mechanical Properties by the Addition of Natural Fibres in Starch-Based Sustainable Biocomposites. *Compos. Part A Appl. Sci. Manuf.* **2011**, *42*, 30–40. [CrossRef]
4. Ribba, L.; Garcia, N.L.; D'Accorso, N.; Goyanes, S. Disadvantages of Starch-Based Materials, Feasible Alternatives in Order to Overcome These Limitations. In *Starch-Based Materials in Food Packaging*; Elsevier: Amsterdam, The Netherlands, 2017; pp. 37–76.
5. Jiménez, A.; Fabra, M.J.; Talens, P.; Chiralt, A. Edible and Biodegradable Starch Films: A Review. *Food Bioprocess Technol.* **2012**, *5*, 2058–2076. [CrossRef]
6. Biricik, Y.; Sonmez, S.; Ozden, O. Effects of Surface Sizing with Starch on Physical Strength Properties of Paper. *Asian J. Chem.* **2011**, *23*, 3151.
7. Larotonda, F.D.S.; Matsui, K.N.; Sobral, P.J.A.; Laurindo, J.B. Hygroscopicity and Water Vapor Permeability of Kraft Paper Impregnated with Starch Acetate. *J. Food Eng.* **2005**, *71*, 394–402. [CrossRef]
8. Spiridon, I.; Teacă, C.-A.; Bodîrlău, R.; Bercea, M. Behavior of Cellulose Reinforced Cross-Linked Starch Composite Films Made with Tartaric Acid Modified Starch Microparticles. *J. Polym. Environ.* **2013**, *21*, 431–440. [CrossRef]
9. Molavi, H.; Behfar, S.; Shariati, M.A.; Kaviani, M.; Atarod, S. A Review on Biodegradable Starch Based Film. *J. Microbiol. Biotechnol. Food Sci.* **2015**, *4*, 456. [CrossRef]
10. Tai, N.L.; Adhikari, R.; Shanks, R.; Adhikari, B. Flexible Starch-Polyurethane Films: Physiochemical Characteristics and Hydrophobicity. *Carbohydr. Polym.* **2017**, *163*, 236–246. [CrossRef]
11. Martinez-Pardo, I.; Shanks, R.A.; Adhikari, B.; Adhikari, R. Thermoplastic Starch-Nanohybrid Films with Polyhedral Oligomeric Silsesquioxane. *Carbohydr. Polym.* **2017**, *173*, 170–177. [CrossRef]
12. Dal, A.B.; Hubbe, M.A. Hydrophobic Copolymers Added with Starch at the Size Press of a Paper Machine: A Review of Findings and Likely Mechanisms. *BioResources* **2021**, *16*, 2138.
13. Donath, S.; Militz, H.; Mai, C. Weathering of Silane Treated Wood. *Holz Roh-Und Werkst.* **2007**, *65*, 35. [CrossRef]
14. Ratajczak, I.; Szentner, K.; Rissmann, I.; Mazela, B.; Hochmanska, P. Treatment Formulation Based on Organosilanes and Plant Oil Blend—Reactivity to Wood and Cellulose. *Wood Res.* **2012**, *57*, 265–270.
15. Xie, Y.; Hill, C.A.; Xiao, Z.; Militz, H.; Mai, C. Silane Coupling Agents Used for Natural Fiber/Polymer Composites: A Review. *Compos. Part A Appl. Sci. Manuf.* **2010**, *41*, 806–819. [CrossRef]
16. Siuda, J.; Perdoch, W.; Mazela, B.; Zborowska, M. Catalyzed Reaction of Cellulose and Lignin with Methyltrimethoxysilane—FT-IR, 13C NMR and 29Si NMR Studies. *Materials* **2019**, *12*, 2006. [CrossRef] [PubMed]
17. Tshabalala, M.A.; Gangstad, J.E. Accelerated Weathering of Wood Surfaces Coated with Multifunctional Alkoxysilanes by Sol-Gel Deposition. *J. Coat. Technol.* **2003**, *75*, 37–43. [CrossRef]
18. Hill, C.A.; Farahani, M.M.; Hale, M.D. The Use of Organo Alkoxysilane Coupling Agents for Wood Preservation. *Holzforschung* **2004**, *58*, 316–325. [CrossRef]
19. Holik, H. *Handbook of Paper and Board*; John Wiley & Sons: Hoboken, NJ, USA, 2006; ISBN 3-527-60833-8.
20. Jonhed, A.; Andersson, C.; Järnström, L. Effects of Film Forming and Hydrophobic Properties of Starches on Surface Sized Packaging Paper. *Packag. Technol. Sci. Int. J.* **2008**, *21*, 123–135. [CrossRef]
21. Hubbe, M.A. Paper's Resistance to Wetting–A Review of Internal Sizing Chemicals and Their Effects. *BioResources* **2007**, *2*, 106–145.
22. Glittenberg, D.; Becker, A. Cationic Starches for Surface Sizing. *Pap. Technol.* **1998**, *39*, 37–41.
23. Cunha, A.G.; Freire, C.S.; Silvestre, A.J.; Neto, C.P.; Gandini, A. Preparation and Characterization of Novel Highly Omniphobic Cellulose Fibers Organic–Inorganic Hybrid Materials. *Carbohydr. Polym.* **2010**, *80*, 1048–1056. [CrossRef]
24. Paquet, O.; Krouit, M.; Bras, J.; Thielemans, W.; Belgacem, M.N. Surface Modification of Cellulose by PCL Grafts. *Acta Mater.* **2010**, *58*, 792–801. [CrossRef]
25. Cunha, A.G.; Freire, C.; Silvestre, A.; Neto, C.P.; Gandini, A.; Belgacem, M.N.; Chaussy, D.; Beneventi, D. Preparation of Highly Hydrophobic and Lipophobic Cellulose Fibers by a Straightforward Gas–Solid Reaction. *J. Colloid Interface Sci.* **2010**, *344*, 588–595. [CrossRef] [PubMed]
26. Satterly, K.P. Method of Rendering Starch Hydrophobic and Free Flowing. U.S. Patent No. 3,071,492, 1 January 1963.
27. Amort, J.; Hanisch, H.; Klapdor, U.; van der Maas, H.; Suerken, H.-P. Method for the Modification of Starch in an Aqueous Medium. U.S. Patent No. 4,540,777, 10 September 1985.
28. Chen, L.; Wang, Y.; Fei, P.; Jin, W.; Xiong, H.; Wang, Z. Enhancing the Performance of Starch-Based Wood Adhesive by Silane Coupling Agent (KH570). *Int. J. Biol. Macromol.* **2017**, *104*, 137–144. [CrossRef] [PubMed]
29. Wei, B.; Sun, B.; Zhang, B.; Long, J.; Chen, L.; Tian, Y. Synthesis, Characterization and Hydrophobicity of Silylated Starch Nanocrystal. *Carbohydr. Polym.* **2016**, *136*, 1203–1208. [CrossRef] [PubMed]
30. Qu, J.; He, L. Synthesis and Properties of Silane-Fluoroacrylate Grafted Starch. *Carbohydr. Polym.* **2013**, *98*, 1056–1064. [CrossRef]
31. Jariyasakoolroj, P.; Chirachanchai, S. Silane Modified Starch for Compatible Reactive Blend with Poly (Lactic Acid). *Carbohydr. Polym.* **2014**, *106*, 255–263. [CrossRef]
32. Sandrine, U.B.; Isabelle, V.; Hoang, M.T.; Chadi, M. Influence of Chemical Modification on Hemp–Starch Concrete. *Constr. Build. Mater.* **2015**, *81*, 208–215. [CrossRef]

33. Ganicz, T.; Olejnik, K.; Rózga-Wijas, K.; Kurjata, J. New Method of Paper Hydrophobization Based on Starch-Cellulose-Siloxane Interactions. *BioResources* **2020**, *15*, 4124–4142. [CrossRef]
34. Ganicz, T.; Rozga-Wijas, K. Siloxane-Starch-Based Hydrophobic Coating for Multiple Recyclable Cellulosic Materials. *Materials* **2021**, *14*, 4977. [CrossRef]
35. Waldner, C.; Hirn, U. Ultrasonic Liquid Penetration Measurement in Thin Sheets—Physical Mechanisms and Interpretation. *Materials* **2020**, *13*, 2754. [CrossRef]
36. Sarah, K.; Ulrich, H. Short Timescale Wetting and Penetration on Porous Sheets Measured with Ultrasound, Direct Absorption and Contact Angle. *RSC Adv.* **2018**, *8*, 12861–12869. [CrossRef] [PubMed]
37. Kolpak, F.J.; Weih, M.; Blackwell, J. Mercerization of Cellulose: 1. Determination of the Structure of Mercerized Cotton. *Polymer* **1978**, *19*, 123–131. [CrossRef]
38. Okano, T.; Sarko, A. Mercerization of Cellulose. II. Alkali–Cellulose Intermediates and a Possible Mercerization Mechanism. *J. Appl. Polym. Sci.* **1985**, *30*, 325–332. [CrossRef]
39. Tondi, G.; Wieland, S.; Wimmer, T.; Schnabel, T.; Petutschnigg, A. Starch-Sugar Synergy in Wood Adhesion Science: Basic Studies and Particleboard Production. *Eur. J. Wood Wood Prod.* **2012**, *70*, 271–278. [CrossRef]
40. Grüner, G. *Emtec Penetration-Dynamics Analyzer*; Emtec Electronic GmbH Materials: Leipzig, Germany, 1996.

molecules

Article

Behaviour of Extractives in Norway Spruce (*Picea abies*) Bark during Pile Storage

Eelis S. Halmemies [1,2,*], Raimo Alén [1], Jarkko Hellström [3], Otto Läspä [4], Juha Nurmi [2], Maija Hujala [5]
and Hanna E. Brännström [2]

[1] Department of Chemistry, University of Jyväskylä, Survontie 9, 40500 Jyväskylä, Finland; raimo.j.alen@jyu.fi
[2] Natural Resources Institute Finland, Teknologiakatu 7, 67100 Kokkola, Finland;
    juhaottonurmi@gmail.com (J.N.); hanna.brannstrom@luke.fi (H.E.B.)
[3] Natural Resources Institute Finland, Tietotie 4, 31600 Jokioinen, Finland; jarkko.hellstrom@luke.fi
[4] School of Engineering and Natural Resources, Oulu University of Applied Sciences, Yliopistonkatu 9,
    90570 Oulu, Finland; otto.laspa@oamk.fi
[5] School of Business and Management, LUT University, Yliopistonkatu 34, 53850 Lappeenranta, Finland;
    maija.hujala@lut.fi
*   Correspondence: eelis.halmemies@gmail.com or ext.eelis.halmemies@luke.fi; Tel.: +358-400630365

Abstract: The current practices regarding the procurement chain of forest industry sidestreams, such as conifer bark, do not always lead to optimal conditions for preserving individual chemical compounds. This study investigates the standard way of storing bark in large piles in an open area. We mainly focus on the degradation of the most essential hydrophilic and hydrophobic extractives and carbohydrates. First, two large $450\,\mathrm{m}^3$ piles of bark from Norway spruce (*Picea abies*) were formed, one of which was covered with snow. The degradation of the bark extractives was monitored for 24 weeks. Samples were taken from the middle, side and top of the pile. Each sample was extracted at 120 °C with both *n*-hexane and water, and the extracts produced were then analysed chromatographically using gas chromatography with flame ionisation or mass selective detection and high-performance liquid chromatography. The carbohydrates were next analysed using acidic hydrolysis and acidic methanolysis, followed by chromatographic separation of the monosaccharides formed and their derivatives. The results showed that the most intensive degradation occurred during the first 4 weeks of storage. The levels of hydrophilic extractives were also found to decrease drastically (69% in normal pile and 73% in snow-covered pile) during storage, whereas the decrease in hydrophobic extractives was relatively stable (15% in normal pile and 8% in snow-covered pile). The top of the piles exhibited the most significant decrease in the total level of extractives (73% in normal and snow-covered pile), whereas the bark in the middle of the pile retained the highest amount of extractives (decreased by 51% in normal pile and 47% in snow-covered pile) after 24-week storage.

Keywords: pile storage; wood extractives; condensed tannins; stilbenes; gas chromatography with mass selective detection (GC-MS); high-performance liquid chromatography (HPLC)

Citation: Halmemies, E.S.; Alén, R.; Hellström, J.; Läspä, O.; Nurmi, J.; Hujala, M.; Brännström, H.E. Behaviour of Extractives in Norway Spruce (*Picea abies*) Bark during Pile Storage. *Molecules* **2022**, *27*, 1186. https://doi.org/10.3390/molecules27041186

Academic Editors: Mohamad Nasir Mohamad Ibrahim, Patricia Graciela Vázquez and Mohd Hazwan Hussin

Received: 7 January 2022
Accepted: 7 February 2022
Published: 10 February 2022

Publisher's Note: MDPI stays neutral with regard to jurisdictional claims in published maps and institutional affiliations.

1. Introduction

Bark contains the great majority of the hydrophilic extractives present in conifers, and it is produced as various forestry sidestreams annually on a massive scale. In 2016, the Finnish forest industry was estimated to produce 7.9 million tons of solid wood-based sidestreams [1]. Despite the high saturation of bark with potentially useful extractable chemicals for valorisation, conifer bark is still mainly used for purposes not directly related to extractives. Bark is primarily used (i) for the production of heat and energy (sometimes in a pelletised form), (ii) for non-energy purposes (e.g., roof material and mould manufacture) and (iii) for landscaping [1].

Among the various groups of bark extractives, tannins and stilbenes, which are categorised as polyphenolic and anti-oxidative compounds, are considered to be of particular interest. Generally, stilbenes (especially resveratrol) and tannins have multiple commercial applications highlighting their protective and health benefits [2,3]. Therefore, extracting these crucial compounds with suitable solvents followed by purification is considered an industrially attractive approach. However, a possible bottleneck of industrial valorisation is its logistics since high-value applications also set equally high requirements for raw materials. Therefore, it stands to reason that practices that best preserve extractives must be applied before the raw material is extracted.

In general, the storage of wood, especially pile storage, can have a considerable impact on its chemical composition [4–9]. Although pile storage of bark is a standard procedure, it may result in significant material losses, leading even to fires. However, it seems practically inevitable that some forms of raw material storage must be used, and finding a solution that does not compromise the quality of the raw material ought to be considered to be of great importance. Storing bark in an intact form on saw logs has already been discussed in previous studies [10,11]. It seems evident that such a form of storage has many advantages, as compared to pile storage, in preserving extractives in bark. This is understandable, as a smaller particle size (as in pile storage) generally exposes the chemical compounds to more degrading factors. Nevertheless, the storage of whole sawlogs may not always be feasible, and for practical reasons, some form of pile storage bark needs to be used instead. Therefore, it is necessary to understand how the pile's internal thermokinetics affect the behaviour and degradation of extractives.

Bark extractives stored in piles are usually attacked both externally and internally [8]. Among the external factors that contribute to degradation are rain, wind and ultraviolet (UV) radiation, as well as heat, which causes evaporation [12–14]. On the other hand, the internal factors include bark-colonising fungi and bacteria and their enzymatic activity, as well as the self-heating of piles as a result of cellular respiration [15–17]. The main changes in extractives are polymerisation/depolymerisation reactions, oxidation reactions, hydrolysis reactions and phenoxy radical photo-degradation reactions [13,18]. In addition, extractives are also lost as a result of leaching (hydrophilic compounds, e.g., tannins and stilbene glycosides) and evaporation (e.g., monoterpenoids) [19,20].

While there are previous studies which aim at providing the overall picture of spruce bark, such as, the study by Krogell et al., to understand how that picture changes over time is also of key importance [21]. In this study, we evaluated the degradation behaviour of the lipophilic and hydrophilic extractives of Norway spruce (*Picea abies*) bark during pile storage over a period of 24 weeks. The main goal was to understand the speed, extent and nature of degradation and whether there is a significant difference between the sampling locations inside each pile (i.e., middle, side and top). We tested the following hypotheses: (i) the extractive content of bark stored in a pile depends on the physical location inside the pile, (ii) covering the bark pile with snow at the beginning of storage can better preserve the bark extractives and (iii) the degradation rate of extractives during pile storage is faster than that of intact bark on saw logs. Overall, the information gathered in this study facilitates the decision-making process regarding the optimisation of storage conditions for the preservation of extractives needed in the manufacture of value-added products.

2. Results and Discussion

2.1. Overview of the Change in the Chemical Composition of Bark during Storage

An overview of the changes in the chemical composition of the bark during storage is presented in Figure 1. In this figure, the gravimetrically determined amounts of total dissolved solids (TDSs) from hot-water and *n*-hexane extracts, the amount of lignin (both acid-soluble and acid-insoluble) and holocellulose as determined by acid hydrolysis and the amount of hemicelluloses and cellulose as determined by acidic methanolysis are presented. Here, the overall changes in the chemical composition are discussed with regard to the storage time, sampling location and pile covering. A more in-depth analysis of the changes within each extractive group is presented in Section 2.3. The exact values of the various compound groups, individual compounds as well as their standard deviations presented in the subsequent figures are available as Supplementary Files (link at the end of the article).

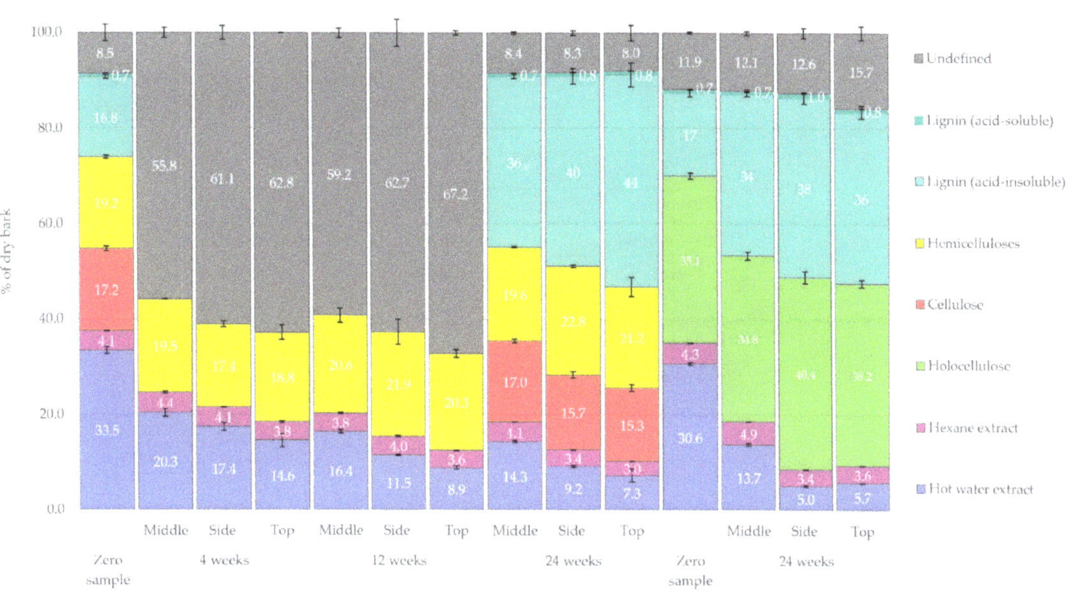

Figure 1. Overall changes in the bark samples' chemical composition during storage as % of dry bark.

2.1.1. Change in Total Dissolved Solids
The Effect of Storage Time

The approximate impact of storage on the relative amounts of chemical compounds in bark was as follows: over 24 weeks of storage, the amount of hydrophilic extractives decreased from 31–34% to 5–14%, the amount of lipophilic extractives changed from 4% to 3–5%, the amount of cellulose decreased slightly from 17% to 15–17%, the amount of hemicelluloses increased slightly from 19% to 20–23%, the amount of acid-insoluble lignin increased from 17% to 34–44%, the amount of acid-soluble lignin (determined by ultraviolet-visible [UV–Vis] spectrometry) increased from 0.7% to 0.7–1.0% and the amount of unidentified compounds changed from 9–12% to 8–16%. The major decrease in hydrophilic extractives agrees with previous storage studies of conifer bark. It has been previously reported that the extractives content in *Pinus sylvestris* chain flailing residue roughly halves during the first 4 weeks of storage, with the most significant changes showing in the hydrophilic fractions [22]. Similarly, Routa et al. studied *Pinus sylvestris* and *Picea abies* bark in pile storage and found that only 56% and 66% of the acetone-soluble extractives remained after eight weeks of storage, respectively [23,24]. Čabalova et al. also reported a significant decrease in *Picea abies* bark extractives extracted by ethanol-toluene mixture (2:1) and a relative increase in lignin and cellulose during 8 months of storage [25]. Compared to our previous study regarding *Picea abies* sawlog bark storage in winter and summer, the difference was noticeable. Although the initial chemical composition in the winter zero samples was very similar, the chemical composition of the 4-week stored piled bark was roughly comparable to that of 24-week stored sawlog bark [10].

Statistical tests revealed that, at the 10% level of significance, the storage time significantly affects the amounts of diterpenoids, unidentified lipophilic compounds, steryl esters, triglycerides, stilbenes, flavonoids, other phenolics, sesquistilbenes, distilbenes, unidentified hydrophilic compounds, proanthocyanidins and the TDSs of the hot-water extracts (Table 1).

Table 1. Results (*p*-values) obtained from testing the statistical differences among the storage duration (0, 4, 12 or 24 weeks), sampling location (middle, side or top) and snow cover (covered or not covered with snow) in terms of the amounts of lipophilic extractives, hydrophilic extractives, condensed tannins (CTs) and total dissolved solids (TDSs). The bold text indicates a statistically significant difference with a *p*-value less than 0.10.

	Storage Time	Sampling Location	Snow Cover
Lipophilic Extractive Groups			
Resin acids	0.280	0.148	**0.018**
Fatty acids	0.313	0.115	0.285
Diterpenoids	**0.058**	0.651	0.157
Sterols	0.236	0.431	0.464
Other lipophilic extractives	0.379	0.166	0.157
Unidentified	**0.022**	0.142	**0.005**
Steryl esters	**0.066**	0.446	0.255
Triglycerides	**<0.001**	0.764	0.200
Hydrophilic Extractive Groups			
Sugars	0.355	**0.078**	0.344
Organic acids	0.527	**0.010**	0.400
Sugar alcohols	0.219	0.192	0.432
Stilbenes	**0.039**	0.670	0.170
Flavonoids	**0.023**	0.430	0.176
Other phenolics	**0.031**	0.404	0.458
Alcohols	**0.076**	0.233	0.319
Lignans	0.124	0.133	0.234
Other hydrophilic extractives	0.795	**0.068**	0.472
Sesquistilbenes	**0.002**	0.862	**n/a**
Distilbenes	**<0.001**	0.805	**n/a**
Unidentified	**0.005**	0.719	0.499
Condensed Tannins			
Total concentration	**0.039**	0.733	0.827
Procyanidins	**0.039**	0.733	0.827
Prodelphinidins	**0.025**	0.424	0.436
DP	**0.039**	1.000	**0.005**
TDSs			
n-Hexane extract	0.288	0.201	0.324
Hot-water extract	**0.006**	0.161	0.364
Biofuel Properties of Stored Bark			
Ash content	0.117	0.233	0.103
Effective heating value	0.280	0.153	**0.024**

Multiple different factors affect the loss of extractives during pile storage. For example, hydrophilic compounds are readily leached by moisture and rainwater, microorganisms rapidly consume some compounds (e.g., sugars), and many extractives are oxidised (e.g., resin acids) or evaporated (e.g., monoterpenoids) [5,26–28]. However, some extractives may be converted via heat and UV-light-induced radical chain reactions to non-extractable polymers (e.g., self-isomerisation and condensation of tannins into phlobaphenes) [20].

The Effect of Sampling Location

The sampling location in the pile (whether from the middle, side or top) appeared to have a systematic and predictable effect on the concentrations of bark components among all storage weeks. Statistical analysis showed that, at the 10% level of significance, the sampling location does not significantly affect the lipophilic extractives. However, a significant statistical result was obtained for the amounts of sugars and organic acids and for the 'other hydrophilic extractives' group (Table 1).

The degradation on the top of the pile was the most pronounced, with less degradation on the side and the most conservative degradation in the middle of the pile. These differences may largely be explained by the complex mechanics of pile storage, which differ in terms of temperature, moisture, ventilation and exposure to external forces depending on the pile formation, pile material (e.g., particle size) and the location in the pile [5,29]. The top of the pile is the part most exposed to both outside influences (e.g., wind, rain and UV light) and the pile's internal activities (steam rising from the pile as a result of self-heating, microbial degradation). Thus, it was not surprising that the top of the pile contained a low concentration of compounds that are easily affected by these factors. Interestingly, after the initial decrease in concentration at weeks 4 and 12, certain extractive groups (e.g., sugars, sugar alcohols and organic acids) experienced an increase only in the middle point of the pile. This observation suggests that the non-volatile hydrophilic extractives from the top of the pile gradually leached downwards, creating a concentrated spot in the middle. A general trend, where the lower one goes in the pile, the higher the concentration of extractives is, could not, however, be confirmed in this study. Routa et al. also looked at the effect of location in bark pile on extractives content in *Pinus sylvestris* and *Picea abies*, but they could not find similar general trends by TDS as were found in this study [23,24]. This difference may be explained by a variety of factors, such as their choice of solvent (pure acetone), difference in extraction method, pile formation and the raw material characteristics.

The Effect of Snow Cover

Minor differences were found between the results of non-covered and snow-covered bark piles. Statistical tests indicated that, at the 10% significance level, snow cover significantly affects the amounts of resin acids and unidentified lipophilic extractives, the degree of polymerisation (DP) of proanthocyanidins and the effective heating value of bark (Table 1).

Notably, the concentrations of hydrophilic TDSs in the snow-covered pile were only slightly low at the beginning and end of storage compared to those in the non-covered pile. The data shown in Figure 2a,b indicate that the snow-covered pile was frozen for 10 days since the beginning of storage, unlike the non-covered pile. This means that the snow cover must have reduced the initial degradation caused by UV light and microbes. However, once the snow melted, additional slow water extraction and consequent leaching of hydrophilic extractives towards the bottom of the pile occurred. The increased moisture also enhanced the conditions for microbial invasion. Overall, although there seemed to be some initial value in covering bark piles with snow, the material losses may have been more significant in the end. Thus, it can be concluded that the hypothesis that covering bark piles with snow can help preserve the bark extractives is invalid (at least when the storage period reaches week 24). Therefore, to study the effect of snow cover on preserving extractives, sampling should be performed before the snow melts. There is evidence that semi-permeable covering of piles can reduce moisture content, temperatures and dry matter losses in forest fuel storage piles [7,9]. However, the impact of such covering during storage on extractives still needs further investigation. Recent study found that thermal drying of *Picea abies* sawmill bark in moderate temperatures will still yield major extractive losses [30].

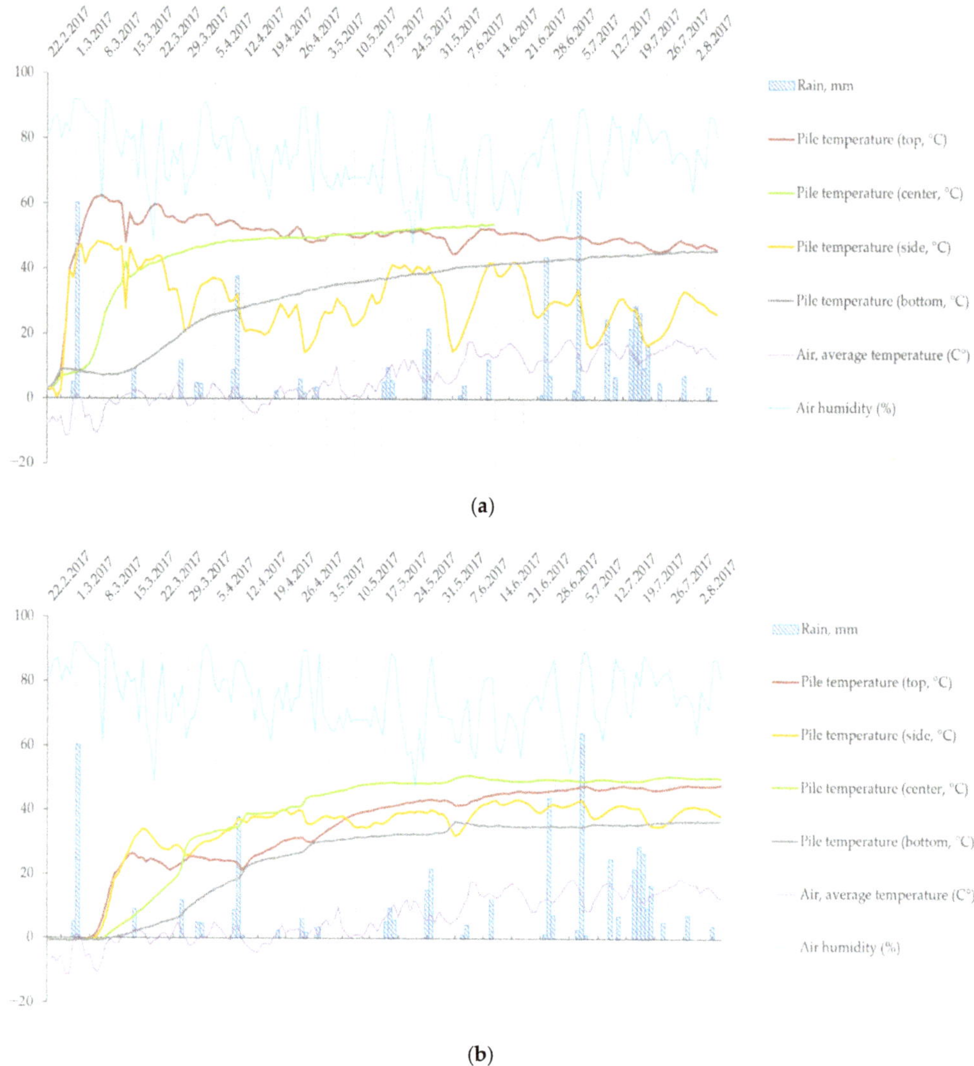

(a)

(b)

Figure 2. Temperature development inside the non-covered (**a**) and covered (**b**) bark piles according to the data gathered by thermocouples. The data shown are from sector one after 1 month of storage.

2.1.2. Changes in Carbohydrates and Lignin

Of the two studied bark piles, holocellulose was only determined from the zero samples and 24-week samples. In both piles, the holocellulose content of bark was equal at the beginning of storage (ca. 35%), and its relative proportion increased slightly towards the end of storage (because of the quicker loss of extractives). In addition, the relative total amount of lignin in bark more than doubled during storage, and the highest lignin concentrations (ca. 45%) were found at these sampling points, at which the extractive fractions were the lowest.

If no degradation occurs for hemicelluloses and cellulose, their relative proportion will increase (as in lignin). Nevertheless, the relative amounts of hemicelluloses and cellulose remained nearly the same throughout storage, indicating their slight degradation. Only on the side and top of the pile did the relative proportion between hemicelluloses and cellulose change, resulting in an overall 4% decrease in cellulose and an increase in hemicelluloses.

Similar findings of increased lignin and carbohydrate content during storage have been reported by Čabalova et al. recently [25]. However, contrary to the results presented here, the relative amount of hemicellulose was reported to have decreased while the amount of cellulose increased. This difference may be explained by the used solvent and extraction method. Compared to the unpressurised Soxhlet extraction used by Čabalova et al. [25], our hot-water extraction at 120 °C is quite harsh and may have resulted in carbohydrates that would otherwise have been included in hemicellulose and cellulose fractions to be included in the extractives fraction.

2.2. Biofuel Properties of Stored Bark

2.2.1. Temperature Development Inside Bark Piles

The data logged from the thermocouples together with the climate conditions from a transportable weather station (air temperature, humidity and amount of rain) are displayed in Figure 2a,b. The thermocouple data revealed that the thermal activity inside the pile started almost immediately after piling the material. In general, both the centre and top of the piles experienced the highest temperatures (with a maximum at around 60 °C), whereas the side and bottom of the piles were cooler. It is also noticeable that the insides of the pile (centre and bottom) experienced a constant increase in temperature, whereas the outermost layers (top and side) experienced heavy fluctuations and correlation with rain and ambient temperature, especially on the side of the pile. Similar dependence of temperature on sampling location was also observed by Routa et al. and Krigstin et al. [23,31]. The occurrence and amount of rain was clearly most significant in June and July, towards the end of the storage period. The top of the pile was also affected by the rising steam from inside the pile. Comparing the two piles (Figure 2a,b) revealed that the snow-covered pile was initially frozen for 10 days and that the overall temperature of the pile during storage was slightly lower.

2.2.2. Heating Values of Stored Bark

The heating values of the studied bark samples, their moisture and their ash, carbon, hydrogen and nitrogen contents are presented in Table 2. The results show that the average moisture content of all bark samples was approximately 57%. The sampling location also affected the moisture content of the bark. For example, in the non-covered bark pile, the moisture content was elevated to 61% at the top of the pile, remained at its original value in the middle and decreased to 41% on the side of the pile. This increased moisture on the top samples may be explained by the steam rising from inside the pile, as microbiological and chemical reactions lead to self-heating of the pile. In the snow-covered pile, presumably because of the melting of the snow cover, the 24-week samples had a high moisture content (62–70%) at all sampling locations, especially on the side and top.

The ash content of the samples underwent a gradual increase from the zero-sample level of 3.2%, especially on the side and top of the bark piles, after storage for 24 weeks, reaching peaks of 4.2% and 8.5% on the top of the non-covered and covered piles, respectively. Similar initial ash content of *Picea abies* industrial bark has been reported previously [32]. The unusually high ash content on the top of the snow-covered pile after 24 weeks of storage is most probably explained by the inorganic impurities (e.g., sand) that were mixed in with the snow that was used for covering. After the snow melted, the inorganic material accumulated on top. Moreover, the carbon content of the dry bark samples increased slightly from an initial level of 51.4% at all sampling locations, except on the top of the snow-covered pile, reaching a maximum of 52.8% at the top of the non-covered pile. The hydrogen content of the dry bark samples decreased from an initial level of 5.8% to an average of 5.6% at all sampling points, especially on the side and top of the piles and particularly in the snow-covered pile. The nitrogen content of the dry bark samples increased from an initial level of 0.47% to an average of 0.55% at all sampling points. This increase was most pronounced, especially on the side and top of the piles. However, the effective heating value remained very stable at approximately 19.3 MJ/kg at all sampling points. These heating values are slightly higher than those reported by Routa et al. for *Picea abies* bark at around 18.9 MJ/kg [24]. After storage for 24 weeks, the heating values decreased to 18.1 MJ/kg only on the top of the snow-covered pile due to increased ash content.

Table 2. Moisture, ash, carbon, hydrogen and nitrogen content of the studied bark samples and their effective heating values.

Storage Time, Weeks	Sampling Location	Moisture Content, %	Ash Content, %	Carbon Content [1], %	Hydrogen Content [2], %	Nitrogen Content [3], %	Effective Heating Value, MJ/kg
				Normal Pile			
0		57.38 ± 0.68	3.21 ± 0.02	51.4	5.82	0.47	19.14 ± 0.02
4	Middle	59.89 ± 1.05	3.30 ± 0.01	51.3	5.80	0.53	19.10 ± 0.01
4	Side	52.20 ± 1.22	3.53 ± 0.01	52.2	5.74	0.52	19.40 ± 0.01
4	Top	56.92 ± 0.64	3.46 ± 0.02	52.1	5.78	0.53	19.56 ± 0.03
12	Middle	61.40 ± 0.86	3.45 ± 0.01	51.1	5.73	0.53	18.78 ± 0.00
12	Side	53.09 ± 0.81	3.75 ± 0.02	51.7	5.63	0.55	19.37 ± 0.01
12	Top	51.65 ± 0.32	3.74 ± 0.01	52.2	5.59	0.54	19.40 ± 0.02
24	Middle	57.83 ± 0.40	3.53 ± 0.05	52.5	5.71	0.52	19.48 ± 0.01
24	Side	40.79 ± 0.82	3.85 ± 0.00	52.5	5.50	0.56	19.47 ± 0.02
24	Top	61.01 ± 0.71	4.17 ± 0.04	52.8	5.45	0.60	19.52 ± 0.01
				Snow-Covered Pile			
0		56.01 ± 0.89	3.12 ± 0.01	51.3	5.77	0.47	19.11 ± 0.01
24	Middle	62.05 ± 0.73	3.77 ± 0.12	51.8	5.65	0.50	19.36 ± 0.01
24	Side	64.33 ± 0.44	4.92 ± 0.08	51.5	5.34	0.61	19.09 ± 0.02
24	Top	69.50 ± 0.45	8.47 ± 0.35	49.9	5.27	0.56	18.13 ± 0.02

[1] Measurement uncertainty ±2%. [2] Measurement uncertainty ±4%. [3] Measurement uncertainty for values <0.3 is ±30%, and for values >0.3 is ± 15%.

2.3. Qualitative and Quantitative Results for Bark Extracts Obtained by Gas Chromatography with a Flame Ionisation Detector/Mass Selective Detector (GC-FID/MS)

2.3.1. Lipophilic and Hydrophilic Extractive Groups

The quantified lipophilic and hydrophilic extractive groups determined using GC-FID/MS methods are presented in Figures 3 and 4, respectively. The lipophilic extractives totalled 11% of all bark extractives, and their main extractive groups were resin, fatty acids, diterpenoids, sterols, steryl esters and triglycerides. In contrast, the hydrophilic extractives totalled 89% of the extractives. Their main groups were sugars, sugar alcohols, organic acids, stilbenes, sesquistilbenes and distilbenes, with the minor groups being flavonoids and other alcohols. The group defined as 'others' contained extractives that, despite being visible on the GC chromatograms, could not be identified or whose concentrations were very small. The 'unidentified' group referred to extractives that could not be detected by GC because of their low volatility or high molar weight. The relative amount of unidentified compounds increased during storage, suggesting an increase in polymerisation reactions.

As shown in Figure 3, overall, there was only a slight decrease in the total amount of lipophilic extractives over a storage period of 24 weeks. The most notable changes in the chemical composition of the lipophilic extract were as follows: a decrease in resin acids from 33% to 23%, a decrease in fatty acids from 22% to 12%, a decrease in triglycerides from 14% to 2% and an increase in unidentified compounds from 6% to 44%. Thus, the results suggest that the storage of bark increases the polymerisation reactions of lipophilic compounds. The results indicate that the rate of degradation gradually slowed as the storage progressed. The overall increase in new unidentified compounds was 2.5 mg/g/storage week after

4 weeks of storage and slowed down to 0.2 mg/g/storage week after 12 and 24 weeks of storage. The concentration of lipophilic extractives decreased on the top and side of the bark pile and increased in the middle of the pile. This finding was confirmed by comparing the results obtained on week 12 and week 24 for the zero sample of the non-covered pile and the 24-week sample of the covered pile. For a more detailed analysis of the degradation pattern of individual lipophilic compounds, see Figures 5–8. The results from our previous sawlog bark study indicate that there is much variation between individual sawlog barks, particularly in the amount of lipophilic extractives, sometimes reaching even above 70 mg/g of dry matter [10].

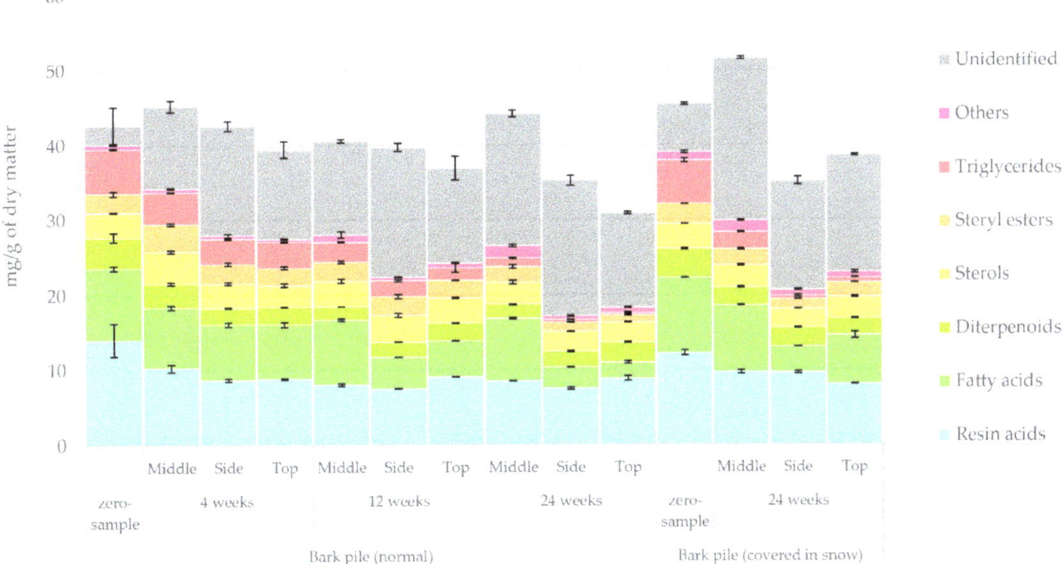

Figure 3. Lipophilic extractive groups in bark samples during pile storage.

The results outlined in Figure 4 show a clear and gradual change in the total amount of hydrophilic extractives and a dramatic decrease in the concentration of many hydrophilic extractive groups in bark resulting from pile storage. The unidentified bark extractives composed of polymeric compounds, such as condensed tannins (CTs) and oligo- and polymeric sugars, represented almost half of all hydrophilic extractives. Mono- and disaccharides represented the second-largest extractive group. The most significant changes in the relative proportion of extractives in the hydrophilic water extracts (zero sample vs. 24-week sample) were as follows: a decrease in sugars from 28% to 17% and an increase in unidentified compounds from 42% to 61%. Stilbenes, sesquistilbenes, distilbenes, flavonoids and other phenolics also experienced a major decrease in concentration, but this did not affect the total extract amount as much. Unlike with the lipophilic extractives, the relative increase in unidentified compounds seemed to result from the decrease in other compounds and not from the increase in polymerisation. For a more in-depth analysis of the hydrophilic extractive groups, see Figures 9–13. A major difference is seen here to sawlog bark, where the concentration of hydrophilics remained at the level of 300 mg/g of dry bark for up to 12 weeks of winter storage [10]. This amounted to approximately 59% less hydrophilic extractives in pile-stored bark at week 12, most likely due to microbial degradation.

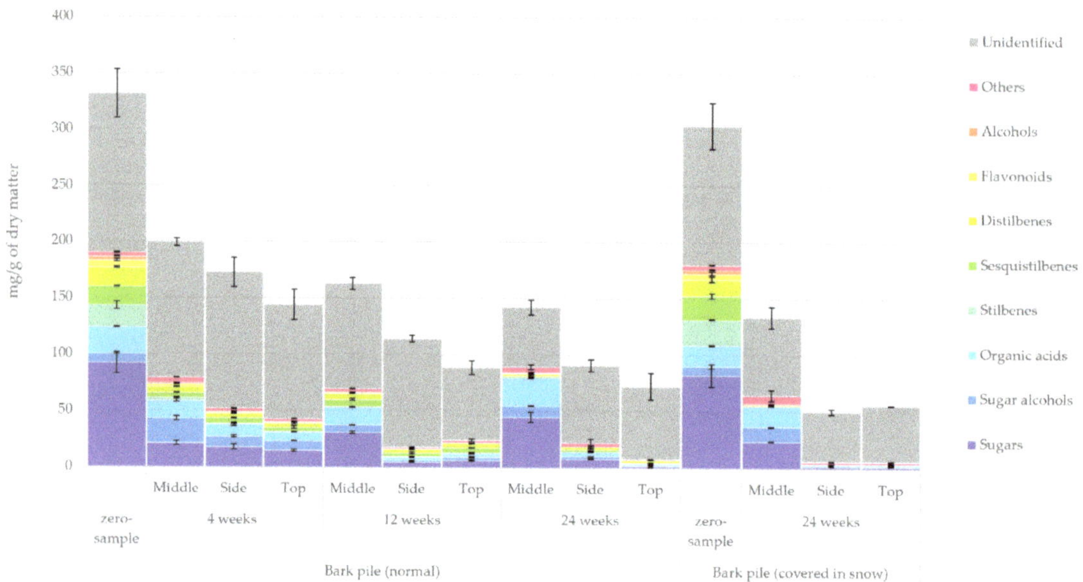

Figure 4. Hydrophilic extractive groups in bark samples during pile storage.

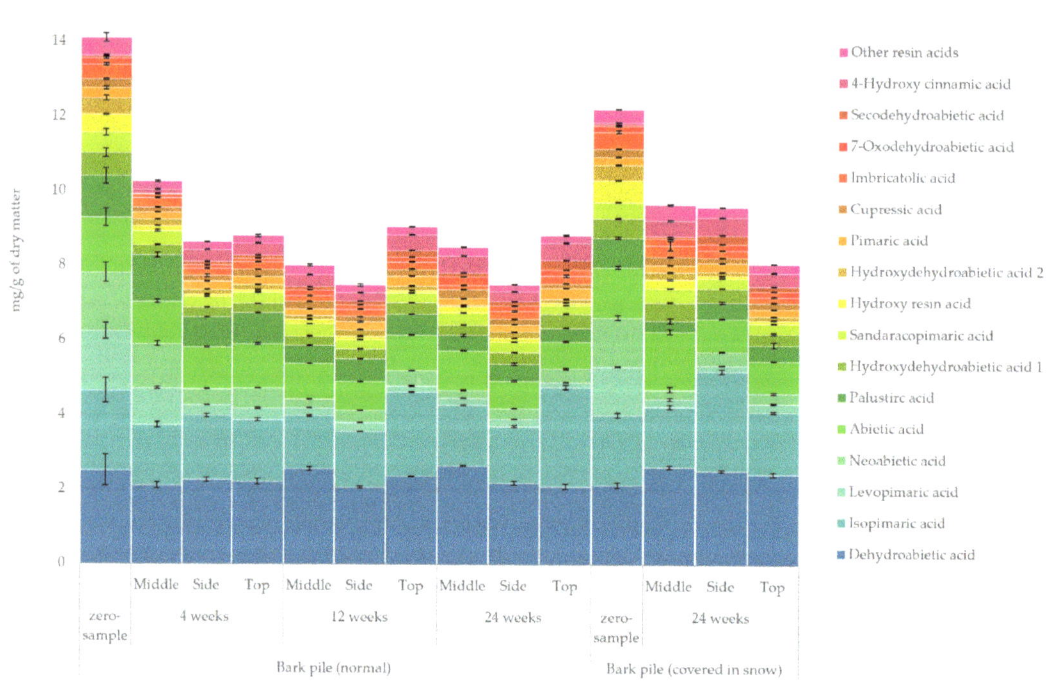

Figure 5. Quantified amounts of individual resin acids in the lipophilic extracts of stored bark.

2.3.2. Resin Acids

The quantified amount of resin acids in the lipophilic bark extracts determined using GC-FID/MS is presented in Figure 5. The results demonstrate a considerable overall decrease in the amount of resin acids during pile storage over the first 4 weeks of storage. After this initial decrease, the total amount of resin acids did not change much, and there was no apparent trend with sampling location. The general stability of resin acids has also been reported previously [10,33]. The most remarkable changes in the relative proportion of resin acids (zero sample vs. 24-week sample) were the increase in dehydroabietic acid from 18% to 28% and in isopimaric acid from 15% to 23% and the decrease in levopimaric acid from 11% to 2% and in neoabietic acid from 11% to 3%. The absolute values of the most prominent resin acids, namely dehydroabietic and isopimaric acid, remained more or less constant throughout storage. Although some reports indicate that certain fungi can reduce the amount of resin acids markedly, the way in which the degradation of resin acids halted after 4 weeks suggests that the initial drop correlated instead with the increased pile temperature [34]. This is also supported by the decrease in neoabietic and levopimaric acids, which are the most prone to thermal oxidation, Diels–Alder reaction, isomerisation and radical reactions because of their conjugated double-bond structure.

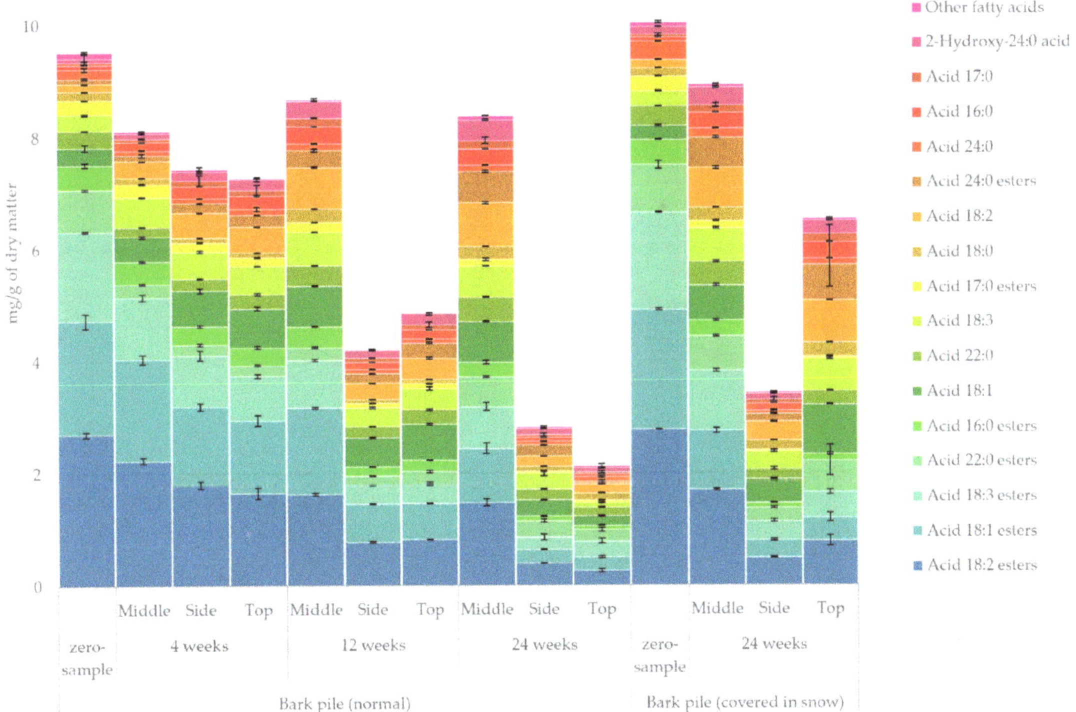

Figure 6. Quantified amounts of individual fatty acids in the lipophilic extracts of stored bark.

2.3.3. Fatty Acids

The quantified amount of fatty acids in the lipophilic bark extracts determined using GC-FID/MS is presented in Figure 6. The changes in triglycerides and fatty acids during storage in many raw materials have been known for a long time. Fatty acids can react either by their conjugated double bonds or carboxylic acid group, leading to various different derivatives [35]. The hydrolysis of triglycerides and consequent polymerisation of the released fatty acids was reported by Ekman among the major chemical changes in wood material during storage [36]. Similarly, Nielsen et al. attributed the decrease in fatty acids during the storage of softwood chips and sawdust to polymerisation and oxidation reactions [37]. It is noteworthy that the total amount of fatty acids dropped considerably during storage, especially on the top and side of the pile, whereas the fatty acids in the middle of the pile on the other hand appeared to be remarkably well-shielded from degradation (although a change in chemical composition was observed). This clearly indicates that the degradation is connected with hydrolysation and oxidation reactions caused by external influences. Esterified fatty acids constituted the vast majority (83%) of total fatty acids at the beginning of storage. The most significant changes (zero sample vs. 24-week sample) in the relative amount of fatty acids were a decrease in fatty acid esters 18:1, 18:2 and 18:3 from 21% to 11%, from 28% to 16% and from 17% to 9%, respectively, and an increase in acids 18:1 and 18:2 and esters of acid 24:0 from 3% to 9%, from 1% to 8% and from 1% to 6%, respectively. From this, the conversion of esterified fatty acids into non-esterified fatty acids seems evident. It should be considered that the degradation of triglycerides during storage (shown in Figure 3) also releases free fatty acids. Routa et al. reported fast degradation of triglycerides during the storage of Scots pine bark, which seemingly led to an increase in the total amount of fatty acids during storage [23].

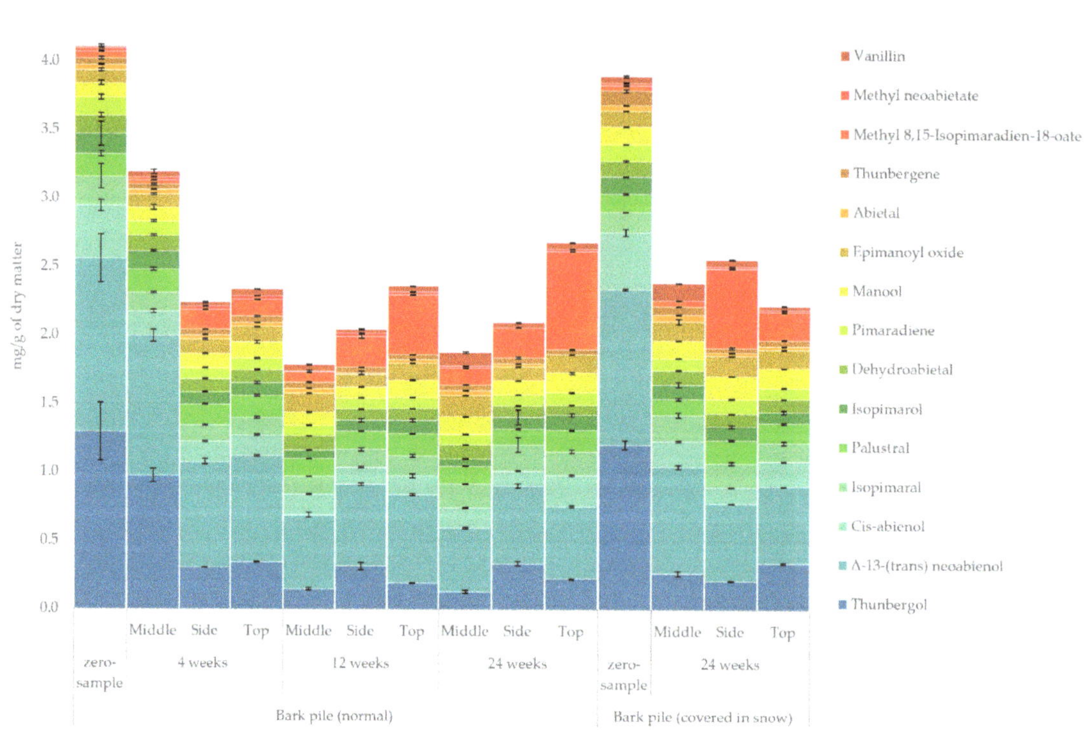

Figure 7. Quantified amounts of individual diterpenoids in the lipophilic extracts of the stored bark.

2.3.4. Diterpenoids

The quantified amount of diterpenoids in the lipophilic bark extracts determined using GC-FID/MS is presented in Figure 7. The amount of diterpenoids at the beginning of storage was slightly above the levels reported by Krogell et al. (0.7 mg/g and 3.2 mg/g in inner and outer bark, respectively) [21]. A considerable overall decrease in diterpenoids was observed during the 24-week storage. Thunbergol, which is associated with anti-fungal, anti-oxidative and anti-tumour activities, was the primary diterpenoid with Δ^{13}-(*trans*)neoabienol [38,39]. The most prominent changes (zero sample vs. 24-week sample) in the relative amount of diterpenoids were an increase in methyl 8,15-isopimaradien-18-oate from 1% to 16% and a decrease in thunbergol and Δ^{13}-(*trans*)neoabienol from 32% to 10% and from 31% to 24%, respectively. That methyl 8,15-isopimaradien-18-oate was formed primarily on the side and at the top of the piles indicates a formation through oxidation reaction. Nielsen et al. also reported that diterpenoid degradation is affected by oxidation and polymerisation reactions [37]. Thunbergol loss was expected because it is also entirely lost during tall oil distillation [40]. Our previous study regarding sawlog bark also indicated a loss of thunbergol with the increase in ambient temperature [10].

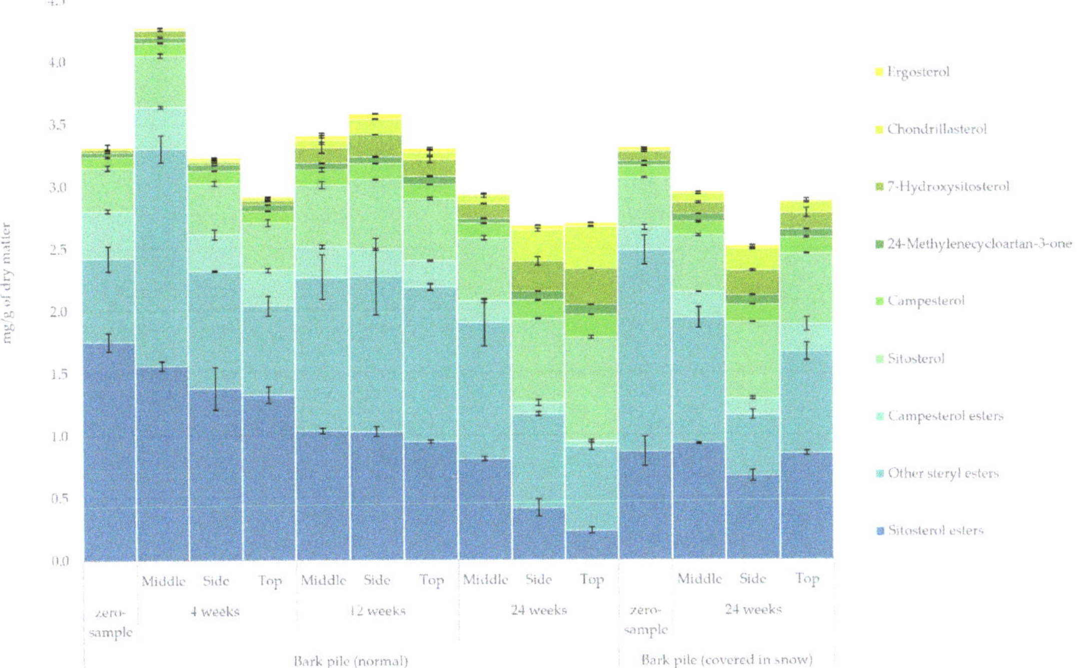

Figure 8. Quantified amounts of sterols and steryl esters in the lipophilic extracts of stored bark.

2.3.5. Sterols

The quantified amount of sterols and steryl esters in the lipophilic bark extracts determined using GC-FID/MS is presented in Figure 8. The major sterol in *Picea abies* is β-sitosterol, a prominent antibacterial and antioxidant agent [41]. The total amount of sterols ranged between 3.2–4.8 mg/g of dry matter and only a slight overall decrease was observed. Routa et al. reported similar sterol levels and only slight degradation during 8 weeks of Scots pine storage [23]. Assarson had reported similar resistance to degradation in unsaponifiable compounds (including sterols) in *Picea abies* chip pile storage [42]. The most prominent changes in the relative amount of sterols (zero sample vs. 24-week sample) were a decrease in the esters of sitosterol and campesterol from 53% to 17% and from 12% to 4%, respectively, and an increase in sitosterol, chondrillasterol and 7-hydroxysitosterol from 10% to 24%, from 0% to 8% and from 1% to 8%, respectively. Given these results, it seems that esterified sterols underwent gradual conversion into free sterols during storage. In addition, ergosterol, chondrillasterol and 7-hydroxysterol were formed as a result of storage, especially on the side and at the top of the pile, again indicating a formation through oxidation reactions [43].

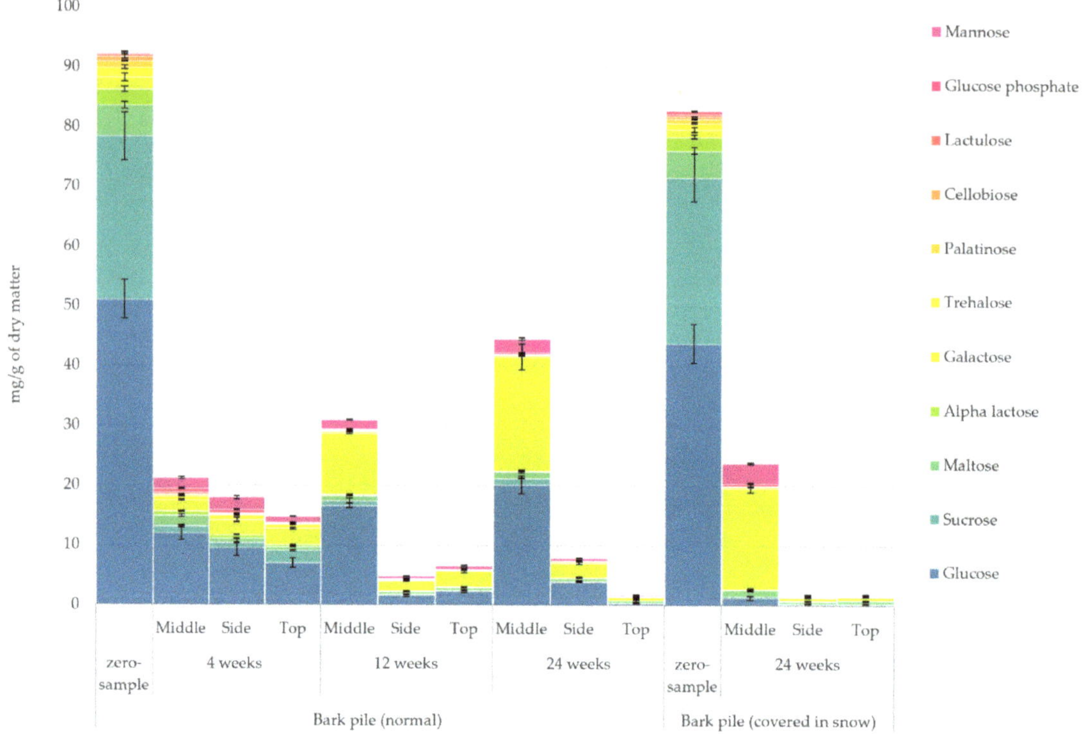

Figure 9. Quantified amounts of mono- and disaccharides in the hydrophilic extracts of stored bark.

2.3.6. Sugars

The quantified amount of simple sugars in the hydrophilic bark extracts determined using GC-FID/MS is presented in Figure 9. Mono- and disaccharides underwent major degradation during pile storage, with only approximately 20% of the sugars remaining after storage for 24 weeks. The sampling location resulted in an increasingly greater difference in the concentration of sugars. At the end of storage, the concentration of sugars at the top of the pile decreased to vanishingly low levels, with the concentration at the side of the pile being only slightly higher. The middle of the pile, on the other hand, exhibited an increased concentration after the initial decrease at week 4. The most significant changes in the relative proportion of sugars were an increase in galactose from 2% to 41% and a decrease in sucrose and glucose from 30% to 2% and from 55% to 45%, respectively. It is generally understood that the rapid loss of saccharides happens due to them being among the first to be consumed by micro-organisms [44,45]. Leaching should, however, be considered as a possibility, especially as a consequence of the steam released during pile storage [5,28,46]. Concentrations of galactose and mannose in the middle of the pile by leaching might have been observed here. In his dissertation, Sauro

Bianchi noted the prevalence of hemicellulose-derived saccharides in water extracts above 80 °C [46]. Noting that the extraction temperature that was used in this study was 120 °C, the presence of saccharides from hemicellulose should be expected. The presence of mannose after storage for 4 weeks and the increased amount of galactose may be, at least partly, explained by the degradation of galactoglucomannan, the main water-soluble hemicellulose in Norway spruce [47]. As a polymeric carbohydrate, galactoglucomannan would be included in the 'unidentified' hydrophilic extractive group (Figure 4). It is also worth noting that the degradation of lactose (4-*O*-β-D-galactopyranosyl-D-glucopyranose) released galactose units.

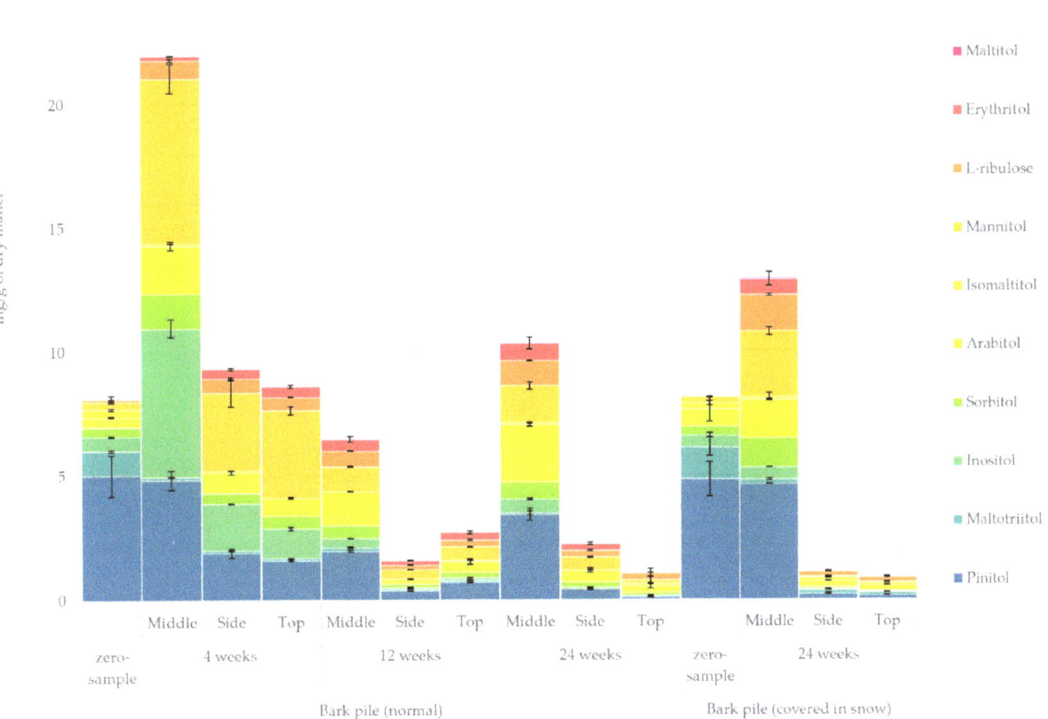

Figure 10. Quantified amounts of individual sugar alcohols in the hydrophilic extracts of stored bark.

2.3.7. Sugar Alcohols

The quantified amount of sugar alcohols in the hydrophilic bark extracts determined using GC-FID/MS is presented in Figure 10. A significant overall variation was observed in the amount of sugar alcohols during storage. After storage for 4 weeks, a sharp increase was detected in the sugar alcohol concentration in the middle of the pile, whereas on the side and at the top of the pile, the total amount remained the same. After 4 weeks, maltotriitol and isomaltitol almost disappeared, whereas inositol and maltitol dramatically increased. Moreover, L-ribulose and erythritol were produced. At the end of the 24-week storage, the amount of sugar alcohols significantly decreased, with only the middle of the pile having a slightly elevated amount of total sugar alcohols. The most significant changes in the relative amount of individual sugar alcohols in the samples (zero sample vs. 24-week storage) were an increase in arabitol, mannitol and L-ribulose from 5% to 23%, from 4% to 16% and from 1% to 11%, respectively, and a decrease in pinitol and maltotriitol from 62% to 29% and from 12% to 0%, respectively. The literature regarding the storage of wood and forestry sidestreams does not discuss the fate of sugar alcohols much. Our previous study regarding the storage of sawlog bark found the sugar alcohol levels to remain constant (c.a. 10 mg/g level) during winter storage until week 12 and then drop to 3 mg/g at 24 weeks of storage [10]. The increase in sugar alcohols observed here, at week 4, should probably be attributed to the hydrogenation reactions of sugars—a process that has also been utilised in the production of value-added chemicals and food ingredients [48]. It is possible that the initial conversion of some sugars to sugar alcohols happened followed by their rapid leaching towards the middle of the pile. This would include the conversion of maltose to maltitol. Production of L-ribulose would, however, suggest a microbial and enzymatic conversion [49]. Similarly the formation of inositol happens through enzymatic phosphorylation of glucose to glucose phosphate (see the residues in Figure 9) followed by isomerisation of glucose phosphate to inositol-phosphate and finally dephosphorylation to inositol [50].

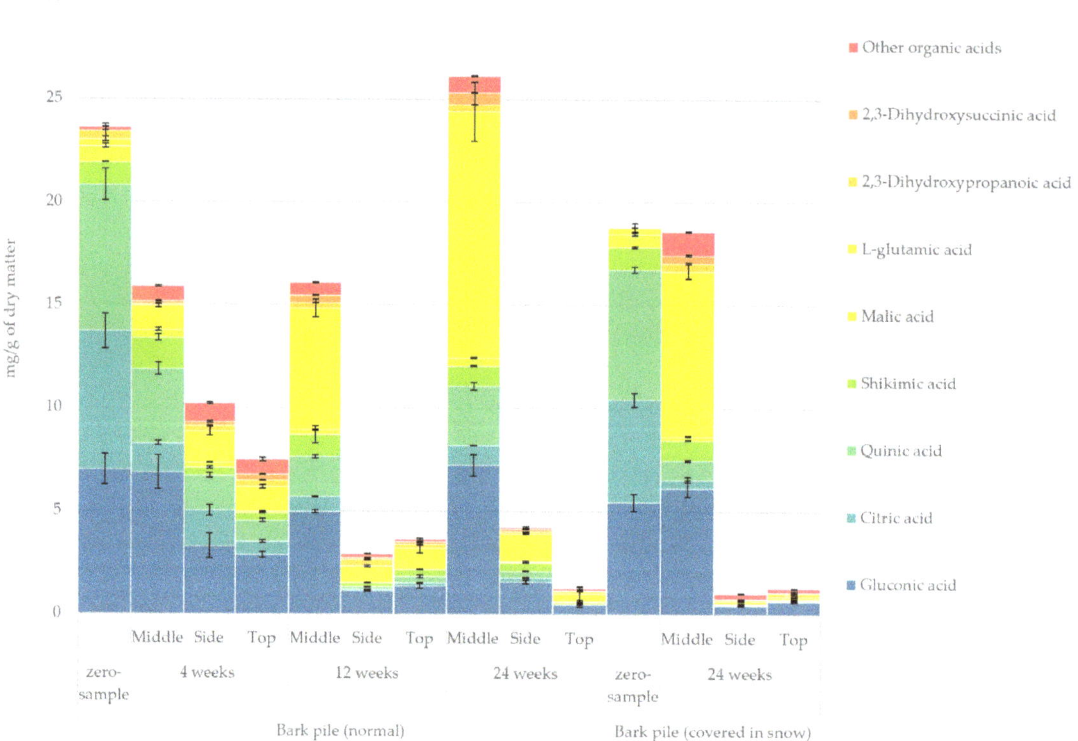

Figure 11. Quantified amounts of individual organic acids in the hydrophilic extracts of stored bark.

2.3.8. Organic Acids

The quantified amount of organic acids in the hydrophilic bark extracts determined using GC-FID/MS is presented in Figure 11. A considerable overall decrease was observed in the amount of organic acids during storage. At the beginning of storage, gluconic acid, citric acid and quinic acid constituted the vast majority of all organic acids. The presence and leaching of organic acids during wood storage has been noted several times before [28,51]. According to Fuller, the presence of even mild acetic acid in pile storage can lead to the shortening of the cellulose fragments in wood [5]. The most significant changes in the relative proportion of organic acids in the samples (zero sample vs. 24-week sample) were an increase in L-glutamic acid from 1% to 43% and a decrease in citric acid and quinic acid from 28% to 4% and from 30% to 10%, respectively. Notably, the concentration of organic acids on the side and at the top of the pile decreased rapidly, whereas in the middle of the pile, an increase was observed from week 12 to week 24. Contrary to these results, the production of new organic acids was not observed in our previous study regarding sawlog storage of bark [10]. Generally, L-glutamic acid is an amino acid by-product of microbiological fermentation of plant proteins (e.g., gluten) with, for instance, glucose as the carbon source [52]. Thus, the significant increase observed in L-glutamic acid also indicated an increase in microbial degradation during storage. Among other degradation products, 2,3-dihydroxysuccinic acid (tartaric acid) was also formed as a fermentation product—a common degradation product in aged fruits and wines [53].

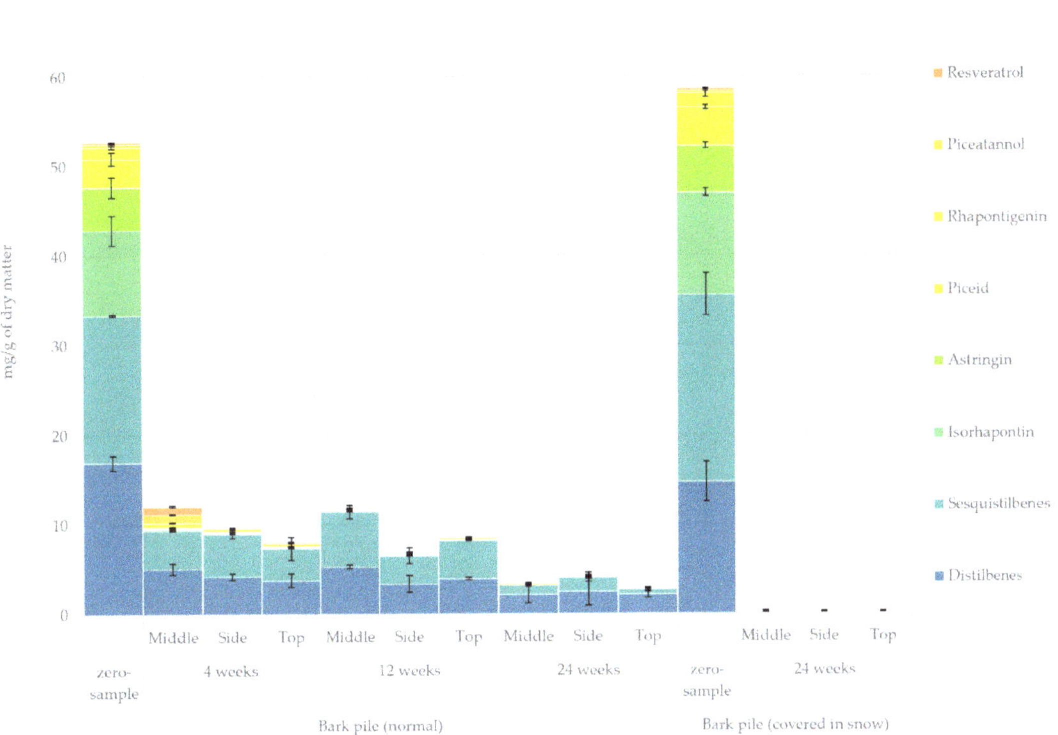

Figure 12. Quantified amounts of stilbenoids in the hydrophilic extracts of stored bark.

2.3.9. Stilbenes

Stilbenes are among the most attractive organic compounds and potential platform chemicals obtained from spruce bark. However, stilbenes are usually lost at a particularly fast rate, not only because they are hydrophilic and may be leached by rainwater but also because of their high anti-oxidative capacity and reactivity under UV light to form phenanthrene derivatives via photo-oxidative cyclisation [18].

The quantified amount of stilbenoids in the hydrophilic bark extracts determined using GC-FID/MS is presented in Figure 12. During storage, a radical overall loss of stilbenes was observed in the study samples, especially during the first few weeks of storage. After storage for 4 weeks, only 23% of the original stilbenes remained, and the stilbene monoglucosides isorhapontin, astringin and piceid, totalling 90% of the original monoglucosides, were almost completely removed. However, the concentrations of the aglycones resveratrol, piceatannol and rhapontigenin increased by 23% at week 4 as a result of the hydrolysis reactions of the glucosides. Moreover, distilbenes and sesquistilbenes constituted 63% of the total stilbenes at the beginning of storage, but only 13% and 6% of the original distilbenes and sesquistilbenes, respectively, remained at the end of storage.

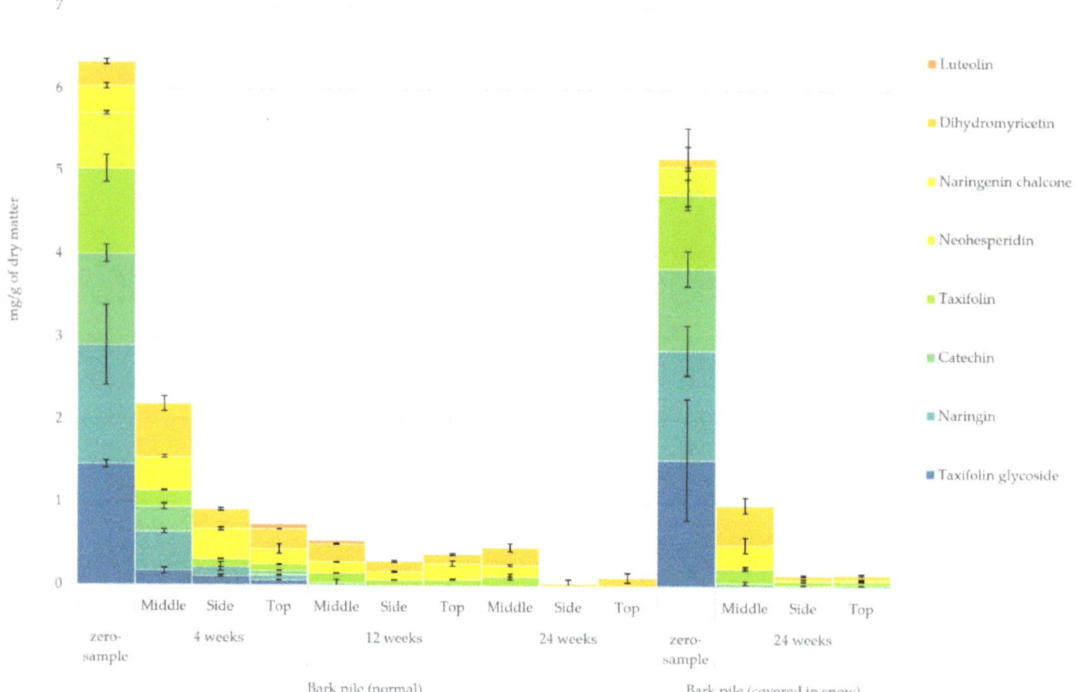

Figure 13. Quantified amounts of individual flavonoids in the hydrophilic extracts of stored bark.

The average concentrations of stilbene monoglucosides, sesquistilbenes and distilbenes in piled bark (from both covered and non-covered piles) were found to be 21.2, 18.7 and 15.8 mg/g of dry matter, respectively. On the other hand, as reported in a previous study, the average amounts of stilbene monoglucosides, sesquistilbenes and distilbenes in the bark of freshly felled (winter-stored) saw logs were found to be 23.5, 10.8 and 10.1 mg/g of dry matter, respectively [10]. Thus, it seems that while the initial amount of stilbene monoglucosides in bark pile and sawlog bark is closely paralleled, the amount of sesqui- and distilbenoids is greater in chipped and piled bark. This may be coincidental, given that the stilbene levels of individual saw logs may considerably vary. However, while the initial amount of stilbenoids was slightly greater in the piled bark, after just 4 weeks of storage, the winter-stored saw logs retained 79% more stilbenoids than those retained by the piled bark. This finding highlights the impact that the storage method can have on individual extractives. To effectively utilise piled bark for its stilbene content, either protective measures need to be taken to ensure their preservation, or the bark needs to be further processed rapidly (within days of the initial piling).

Stilbene concentrations presented here were markedly higher than those reported by Krogell et al. [21]; however, Jyske et al. have reported at least twice as high concentrations of stilbene glucosides in the bark of 18–37 year old *Picea abies* trees [54]. It should, however, be noted that while Jyske et al. [54] looked at stilbene concentration at different bark zones and heights (inner bark having highest stilbene concentrations), our results reflect more the average stilbene concentration in sidestream *Picea abies* bark from sawmills without further distinctions. Stilbene levels similar to those presented by Jyske et al. [54] have also been reported in the root bark of Norway spruce [55].

2.3.10. Flavonoids

The quantified amount of simple flavonoids in the hydrophilic bark extracts determined using GC-FID/MS is presented in Figure 13. The initial amount of flavonoids was approximately twice as high as that reported by Krogell et al. [21].The loss of flavonoids seemed to follow a path similar to that of stilbenes, with a dramatic concentration decrease after just 4 weeks of storage. Slower flavonoid degradation was observed in our previous study regarding *Picea abies* sawlog bark [10]. The most prominent flavonoids were taxifolin glycoside, naringin, catechin, taxifolin and neohesperidin. Notably, dihydromyricetin, which has potent anti-oxidative properties, was found to be the most resilient among flavonoids [56]. Its amount was even found to be somewhat increased during storage (e.g., through the bio-conversion of other flavonoids). The most significant changes in the relative proportion of extractives in the samples (zero sample vs. 24-week sample) were an increase in dihydromyricetin and naringenin chalcone from 5% to 59% and from 5% to 23%, respectively, and a decrease in taxifolin glycoside, naringin, catechin and neohesperidin from 23%, 23%, 17% and 11% to 0%, respectively. Flavonoids (similarly to stilbenes) are lost particularly rapidly to photo-degradation because of their tendency as phenolic compounds to form unstable phenoxy radicals [12,13,18].

2.4. High-Performance Liquid Chromatography (HPLC) Analysis of Proanthocyanidins

Overall, the thiolytic degradation of spruce bark CTs (procyanidins) produced (epi)catechins and (epi)catechin thioethers as major reaction products and (epi)gallocatechins and (epi)gallocatechin thioethers as minor products, indicating that spruce bark CTs are a mixture of procyanidins and prodelphinidins, as observed in previous studies [11,57–59]. As shown in Figure 14, the initial CT content was 3.0–3.2 g/100 g, but it decreased rapidly during storage. After 4 weeks, the total content of CTs was found to exhibit a great variation (0.556–1.451 g/100 g) between the different samples, but this variation always remained below 50% of the original amount. After 12 weeks, the concentration reached 0.384–0.472 g/ 100 g, and only minor changes were observed for the rest of the storage duration. The final CT content in the normally stored bark pile was found to be 0.251–0.365 g/100 g after 24 weeks, which is equal to approximately 10% of the original content. A recent study on Scots pine reported a similar drastic and rapid loss in the CT content during pile storage of bark [23].

The average CT content in the snow-covered piles was somewhat higher after storage for 24 weeks, and notable differences were observed between the samples. These samples were obtained from different pile locations, which might partly explain the variations observed in the CT content. In this study, the highest CT content was determined twice in the samples taken from the middle of the pile (after storage for 4 weeks and 24 weeks for normally stored and snow-covered piles, respectively). It is possible that the bark in the middle of the pile was better protected from environmental stress than that on the side or at the top of the pile. This also means that the CTs were less exposed to detrimental reactions. Similarly, a recent study has shown that the outer bark is expected to protect the inner bark, with the CTs in the outer bark degrading much faster than in the inner bark during the summertime fresh-air storage of spruce logs [11]. However, further research is still needed to confirm the significance of location in a pile for the recovery of CTs and other constituents in spruce bark.

The average DP in spruce bark CTs was found to be the highest at the beginning of the experiment, but it decreased during storage, indicating that the polymerisation of CTs is the first step in the degradation process. However, the oxidation of CTs during storage might result in degradation and the formation of new covalent bonds between CTs and other macromolecules, producing new polymers partially resistant to thiolysis [59,60]. As a result, both the content and the DP of CTs are somewhat under-estimated with the current determination method. Furthermore, the relative proportion of prodelphinidins in CTs was found to slightly increase during storage. The same finding was observed in the CTs of spruce logs stored in the open air [11]. This may indicate that prodelphinidins in spruce bark CTs are more resistant to environmental stress than procyanidins.

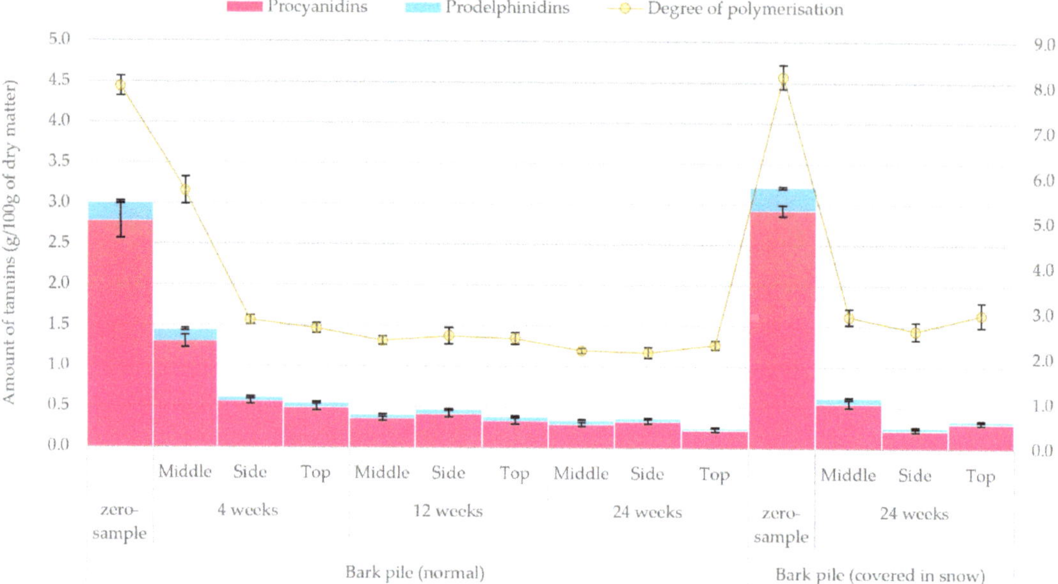

Figure 14. Quantified amounts of tannins (procyanidins, prodelphinidins) and the degree of polymerisation (DP) in freeze-dried bark samples under pile storage.

2.5. Carbohydrate Analysis

2.5.1. Acid Hydrolysis and High-Performance Anion-Exchange Chromatography (HPAEC) Analysis of Monosaccharides

The results obtained from the HPAEC analysis of extractive-free bark monosaccharides (i.e., holocellulose) are presented in Figure 15. The initial amount of holocellulosic monosaccharides in the bark samples was found to be 58% in extractives-free bark in the normal bark pile and 54% in the snow-covered pile. After a storage period of 24 weeks, the amount in both piles decreased to approximately 42% of extractives-free bark. These values, however, correlate to approximately 35.8% of the initial amount of holocellulose in dry bark (according to Figure 1) and 37.5% in dry bark after 24 weeks of storage. Thus, the total holocellulose content (as % of dry bark) increased 1.7%. In our previous study regarding *Picea abies* sawlog bark storage, the amount of holocellulose was initially 33.9% of dry bark and increased to 37.7% in 24 weeks (a 3.8% increase) [10]. Čabalová et al. also reported relatively increased cellulose content during storage for 8 months [25]. Generally, glucose was by far the most prominent monosaccharide. The notable changes in the relative proportion of monosaccharides (zero sample vs. 24-week sample) were an increase in glucose from 66% to 74% of dry matter and a decrease in arabinose from 13% to 5% of dry matter. Moreover, mannose decreased slightly more in the samples from the snow-covered pile.

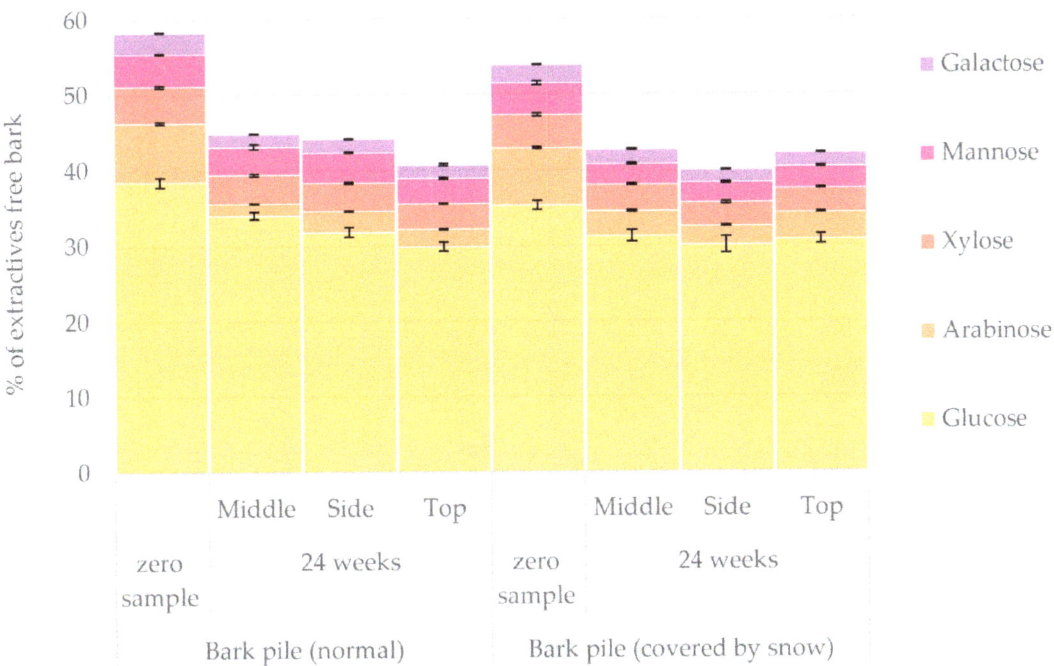

Figure 15. Quantified amounts of holocellulosic monosaccharides in bark samples at the beginning and end of normal and snow-covered pile storage.

2.5.2. Acidic Methanolysis

The results obtained from the acidic methanolysis of extractive-free bark monosaccharides (i.e., hemicelluloses) are presented in Figure 16. These results indicate that the overall amount of hemicelluloses decreased by 21%. Such a decrease occurred during the first 4 weeks of storage, and the total amount of hemicellulosic monosaccharides remained constant throughout the storage period, although changes in the composition occurred. The most notable changes in the relative proportion of hemicellulosic groups in the samples (zero sample vs. 24-week sample) were an increase in glucose and xylose from 10% to 22% and from 13% to 21%, respectively, and a decrease in galacturonic acid and arabinose from 31% to 18% and from 23% to 12%, respectively. Conversion of galacturonic acid to galactose was probably also observed. A similar trend was observed with regard to the sampling location in each pile, and the concentration of extractives was probably also observed at weeks 4 and 12. The highest concentration was found in the middle of the piles, whereas the top and side of the piles showed greater signs of degradation. Notably, the hemicellulosic monosaccharides presented here are basically a subset of the results presented in Figure 15. By comparing the results for holocellulosic and hemicellulosic monosaccharides (in the normal bark pile), we were able to observe that the cellulosic monosaccharides were primarily composed of glucose and mannose. The apparent increase in some hemicelluloses, such as glucose and xylose, could be explained (similarly to the increase in lignin in bark (see Section 2.1.2)) as a relative increase caused by the faster degradation of extractives and other carbohydrates. Relative increases in hemicelluloses were also observed in our previous study regarding single stem *Picea abies* bark storage [10]. It should also be noted that while the total amount of carbohydrates as mg/g of extractives-free bark (in Figure 15) decreased, the relative amount of carbohydrates as % of dry bark (i.e., bark containing extractives; see Figure 1) slightly increased during storage. A similar relative increase in cellulose and lignin due to short term storage of *Picea abies* bark has also been recently reported by

Čabalova et al. [25]. This effect could be likened to the concentration of carbohydrates by weight observed in dried fruits.

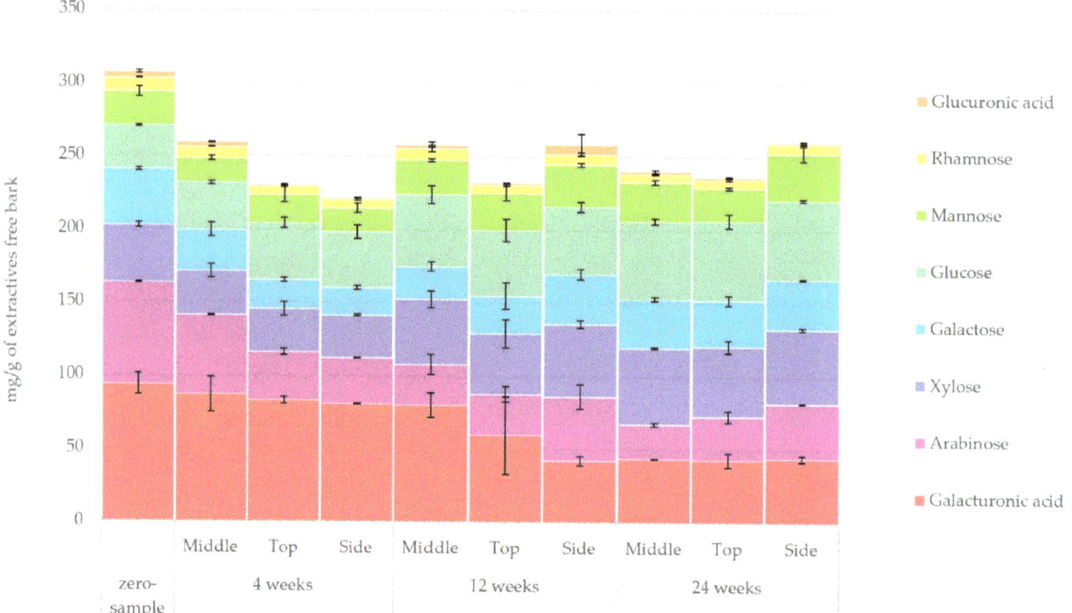

Figure 16. Quantified amounts of hemicellulosic carbohydrates in the extractive-free bark samples under normal pile storage.

3. Materials and Methods

3.1. Experimental Setup of Storage Studies and Sampling

All the bark used in this study was provided by the UPM-Kymmene Oyj sawmill in Ostrobothnia, and all the bark pile setups were located outside in the factory yard in Pietarsaari. The two 450 m^3 bark piles used in this storage study were constructed on 20 and 21 February 2017. The piles consisted of *Picea abies* bark that was debarked a maximum of 48 h before the construction of the pile. However, most of the material was even fresher. It should be noted that since the bark originated as a sidestream from a standard operating sawmill, no exact measurements of individual trees from which the bark was obtained (their height, width, age, etc.) were available. It is known that the used trees were gathered within a 200 km range from Pietarsaari, mostly from private forest owners in Ostrobothnia. The sampling points and dimensions of the bark in the non-covered pile are outlined in Figure 17. The sampling locations were chosen from areas of piles expected to have significant variations in temperature and moisture content, according to earlier storage studies [29]. The length of the constructed pile was 17.6 m, and it was divided into three sectors. Sector one was opened for sampling after 4 weeks, sector two after 12 weeks and finally sector three at the end of the storage study, after 24 weeks. Thermocouples were placed inside the pile in the locations indicated in Figure 17a, and the temperature was measured in each sector until the sector was opened for sampling. At each sampling time, bark samples were taken from the exact locations of the thermocouples, except for the bottom of the pile.

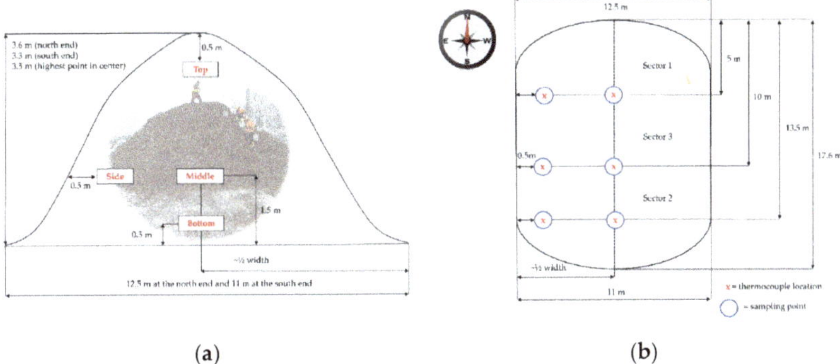

(a)   (b)

Figure 17. Schematic representations of the measures of the bark pile and (**a**) the thermocouple locations inside the bark pile and (**b**) the sampling points and sectors.

3.2. Sample Pre-Treatment and Basic Characterisation

First, the bark was ground to a finer particle size with a Jens Algol System woodchipper (Jenz GmbH, Petershagen, Germany). Then, a standard method (CEN/TS 14774-2:2004) was used to determine the fresh bark samples' moisture content [61]. Next, the samples were dried at 105 °C until a constant mass was achieved. All measurements were performed in duplicate.

The bark was then lyophilised (for at least 3 days) and ground with a Retsch SM 100 cutting laboratory mill (Retsch GmbH, Haan, Germany) equipped with a bottom sieve with trapezoidal holes (perforation size < 1.0 mm) for chemical analysis. Samples were stored in a frozen state (below −20 °C). Then, the dry matter content of the lyophilised bark samples was determined by drying 1 g of bark powder at 105 °C in an oven overnight in tared crucibles.

3.3. Calorific Values and Carbon, Hydrogen and Nitrogen (CHN) Measurements of Bark Samples

First, the moisture content (on a wet basis) of the bark samples was analysed according to the same method as referred to in Section 3.2, and the ash content was determined according to the standard method SFS-EN 14775 [62]. A bomb calorimeter (IKA C 5000; IKA-Werke GmbH & Co., Staufen, Germany) was used to determine the calorific heating value (qP_{gross}) of the bark dry matter. Samples were dried, milled (Retsch SM-1 mill; Retsch GmbH, Haan, Germany) and pelletised before analysis with the bomb calorimeter. Next, the calorimetric heating values were determined and the gross calorific values were calculated using the standard method CEN/TS 14918:2005 [63]. Then, the carbon, hydrogen and nitrogen concentrations were analysed using the standard method SFS-EN ISO 16948:2015 at the laboratory of Ahma Environment Ltd. [64]. The following equation was used to calculate the effective heating value (qP_{net}):

$$qP_{net} = qP_{gross} - 2.45 \times 0.09 H_2 \tag{1}$$

where qP_{net} is the effective heating value (kJ·kg^{-1}), qP_{gross} is the calorific heating value (kJ·kg^{-1}), 2.45 MJ kg^{-1} is the latent heat of vaporisation of water at 20 °C, 0.09 is a factor expressing that one part of hydrogen and eight parts of oxygen form nine parts of water, and H_2 is the hydrogen content of the oven-dried biomass.

3.4. Chemicals

The solvents used in the sample preparation of extractives were analytical-grade acetone (BDH), HPLC-grade *n*-hexane (VWR), methyl *tert*-butyl ether (Lab-Scan), pyridine (BDH), 95% ethanol (EtOH, >94%, ETAX A; Altia Corporation) and *n*-butanol (Merck).

Bis(trimethylsilyl)trifluoroacetamide (BSTFA) and trimethylchlorosilane (TMCS) were obtained from Regis Technologies (Morton Grove, IL, USA) for silylation.

The compounds used as internal standards in the GC analysis of extractives were heneicosanoic acid (99%; Sigma-Aldrich Finland, Espoo, Finland) and betulinol ($\geq$98%; Sigma-Aldrich Finland, Espoo, Finland), cholesteryl margarate ($\geq$97%; TCI, Portland, OR, USA) and 1,3-dipalmitoyl-2-oleoylglycerol ($\geq$99%; Sigma). NaOH (>98%; VWR), HCl (37%; VWR), Na_2CO_3 ($\geq$99.8; Sigma-Aldrich Finland, Espoo, Finland), sulphuric acid (95–97%; Sigma-Aldrich Finland, Espoo, Finland), and bromocresol green (>95%; Sigma-Aldrich Finland, Espoo, Finland) were also used in the analysis.

Cysteamine ($\geq$98%; Sigma-Aldrich Finland, Espoo, Finland), 37% aqueous hydrochloride (Thermo Fisher Scientific, Waltham, MA, USA) and HPLC-grade methanol ($\geq$99.8%; VWR International, Helsinki, Finland) were used for the thiolysis of CTs. Then, CT degradation products, that is, free flavan-3-ols (terminal units) and their cysteaminyl derivatives (extension units), were quantified using external standards of catechin, epicatechin, gallocatechin and epigallocatechin (Sigma-Aldrich Finland) and thiolysed procyanidin B2 (Extrasynthese, Lyon, France). HPLC-grade acetonitrile (VWR International) and formic acid ($\geq$98%; Sigma-Aldrich, Espoo, Finland) were used for HPLC determination of thiolysed CTs.

3.5. ASE Extraction

Bark samples were extracted with a Dionex Accelerated Solvent Extractor (Dionex, ASE 100, Sunnyvale, CA, USA) using *n*-hexane and water as solvents to extract lipophilic and hydrophilic extractives, respectively. The extraction temperature was set to 120 °C, with a static extraction time of 10 min, flush of extraction cell of 60%, nitrogen purge for 70 s and extraction pressure of 1500 psi. For each extraction procedure, 2 g of bark powder was loaded to a 34 mL extraction cell plugged with a cellulose filter. Each sample was first extracted with *n*-hexane and then with water, and the extractive-free bark was consequently lyophilised and stored for carbohydrate analysis. All extraction procedures were performed in duplicate for each sample.

3.6. Gravimetric Analysis of Total Dissolved Solids and Preparation of Stock Solutions

Overall, the TDSs of bark extracts were determined gravimetrically. The *n*-hexane extracts were evaporated to near dryness in a rotary evaporator and subsequently transferred to tared Kimax test tubes in acetone and finally evaporated to dryness under nitrogen flow. The weight of the dried extract was the TDS of the *n*-hexane extracts. A stock solution (100 mL) was then prepared by dissolving the extract in acetone.

Stock solutions of hydrophilic extracts were prepared by diluting the raw extract to 100 mL with ultra-high-quality (UHQ) water. Some of the stock solutions (10 mL) were lyophilised, and the TDS of the hydrophilic extracts was determined according to the weight of the lyophilised sample.

3.7. Analysis of Bark Extractives with Chromatographic Methods

3.7.1. Qualitative Analysis of Bark Extracts by Gas Chromatography with Mass Selective Detection (GC-MS)

To perform a qualitative analysis, 3 mg of extracts (based on dry weight) was dried (either by nitrogen flow or lyophilisation) and dissolved in 500 µL of pyridine and 300 µL of a silylation reagent (TMCS). The silylation process was accelerated by keeping the sample in a 70 °C oven for 1 h. The samples were then analysed using a Hewlett Packard 5973 GC-MS instrument equipped with an HP-5 column (30 m $\times$ 0.32 mm, with a 0.25 µm film). Next, the samples were injected at 290 °C and detected with a mass selective detector at 300 °C. Notably, the method used for the analysis was the same as in our previous study [10].

3.7.2. Quantitative Analysis of Bark Extracts by GC-FID

To perform a quantitative analysis, approximately 3 mg samples of bark extracts were dried with internal standards. The mixtures were then dissolved in 500 μL of pyridine and 300 μL of a silylation reagent and kept in an oven for 1 h.

To analyse the extractive groups, 100 μg of four internal standards was used: heneicosanoic acid, betulin, cholesteryl margarate and 1,3-dipalmitoyl-2-oleoylglycerol. An Agilent 6850 GC-FID instrument equipped with a short HP-1/simulated distillation column (7.5 m × 0.53 mm, with a 0.15 μm film) was used for the analysis. The samples were injected on-column at 90 °C and detected using FID at 320 °C. The temperature program used was the same as in our previous study [10].

To perform an individual extractive analysis, 100 μg of heneicosanoic acid and the same amount of betulin were added as internal standards. An Agilent 6850 GC-FID instrument equipped with a long HP-5 column (30 m × 0.32 mm, with a 0.25 μm film) was used for the analysis. The samples were then injected at 290 °C and detected at 300 °C. The temperature program used was the same as in our previous study [10].

To analyse the esterified lipophilic extractives, the samples were hydrolysed and derivatised for analysis as described by Halmemies et al. (2021) [10].

3.7.3. Analysis of Proanthocyanidins by High-Performance Liquid Chromatography (HPLC)

A thiolytic degradation method as described by Korkalo et al. was applied to determine CTs (proanthocyanidins) in the lyophilised bark samples [65]. First, a ground sample (10–20 mg) was mixed with 1 mL of a depolymerisation reagent (3 g of cysteamine dissolved in 56 mL of methanol acidified with 4 mL of 13 M HCl) and incubated for 60 min at 65 °C. During incubation, the samples were vortexed for a few seconds every 15 min. Thiolysis was stopped by transferring the samples into an ice bath. The cooled samples were then filtrated into HPLC vials and analysed on an Agilent 1290 Infinity UHPLC instrument equipped with a Zorbax Eclipse Plus C_{18} column (50 × 2.1 mm i.d., 1.8 m; Agilent Technologies, Santa Clara, CA, USA). The binary mobile phase consisted of 0.5% formic acid (aq.) and acetonitrile. Elution was started with 2% acetonitrile isocratically for 2 min, followed by a linear gradient to 5% in 3 min, to 15% in 7 min, to 20% in 3 min, to 35% in 5 min, to 90% in 1 min and back to the initial condition in 2 min. The post-time was 2 min before the next injection. The flow rate was 0.5 mL/min, and the injection volume was 2 μL. Elution was monitored using diode array detection (DAD; λ_1 = 270 nm, λ_2 = 280 nm) and fluorescence detection (FLD; λ_{ex} = 275 nm, λ_{em} = 324 nm).

3.8. Carbohydrate Analyses

Acid hydrolysis and acidic methanolysis were used to analyse the carbohydrate (holocellulose, cellulose and hemicelluloses) and acid-soluble and acid-insoluble lignin (see below) content of extractive-free bark. Holocellulose is defined as the sum of cellulosic and hemicellulosic carbohydrates. The holocellulose and lignin content was first determined using acid hydrolysis, and then the hemicellulose content was determined using acidic methanolysis. Then, the cellulose content of the samples was determined as the difference between holocellulose and hemicelluloses.

3.8.1. Acid Hydrolysis

Separation of holocellulose, acid-insoluble lignin and acid-soluble lignin from the extractive-free bark samples was performed according to the TAPPI standard T 222 [66]. For the acid hydrolysis samples, 200 mg of lyophilised extractive-free bark was weighed in a test tube. Then, around 4 mL of 72% cold sulphuric acid was added, and the test tubes were kept in a water bath at 30 °C for 1 h. Every 5 min, the mixtures were stirred with a glass rod. Next, the samples were transferred to 250 mL autoclave bottles, washed with 112 mL of UHQ water and then placed in an autoclave (MELAG Autoklav 23, Berlin, Germany) at a pressure of 1 bar (~121 °C) for 1 h.

Solid acid-insoluble lignin was then separated from the mixtures by filtration with a tared borosilicate glass filter (Munktell MGA 413004, Falun, Sweden) in a vacuum funnel. Insoluble lignin was gravimetrically determined by drying the residues together with the used filter papers (of known weight) in an oven at 105 °C to a constant weight. The filtrates were then diluted to 500 mL with UHQ water and consequently analysed with HPAEC for their holocellulose-derived monosaccharide content and with UV–Vis spectroscopy for their soluble lignin content.

3.8.2. High-Performance Anion-Exchange Chromatography (HPAEC) Analysis of Holocellulose-Derived Monosaccharides

First, HPAEC was used to analyse the monosaccharides formed during the acid hydrolysis from the 500 mL dilutions. Standard solutions for HPAEC were prepared using a sulphuric acid concentration corresponding to the samples' background: cold 72% sulphuric acid (3 mL) was diluted to 500 mL with UHQ water. Fucose (500 ppm) was used as an internal standard. The preparation of the standard solutions is described in detail in our previous study [10].

Bark samples (500 mL UHQ water dilution) from acid hydrolysis were analysed with HPAEC (Dionex) using 1 M sodium acetate, 0.5 M sodium acetate plus 0.1 M NaOH and 0.3 M NaOH solutions as eluents. The analytes were then separated in CarboPac PA1 + Quard PA1 columns and detected with an ED50 detector using carbohydrate pulsing. The post-column elute was pumped by an IC25 isocratic pump.

Samples for HPAEC analysis were prepared by pipetting 2 mL of an internal standard solution to a 20 mL volumetric flask and filling the flask with the diluted sample (500 mL) from the acid hydrolysis. This solution (1.0–1.5 mL) was then transferred into an HPLC vial by filtrating it through a syringe filter (Phenex-RC, 0.2 μm).

3.8.3. UV–Vis Measurement of Acid-Soluble Lignin

The amount of acid-soluble lignin was determined from the 500 mL dilution following acid hydrolysis via UV–Vis spectroscopy at 205 nm according to the TAPPI standard UM 250 using an extinction coefficient of 120 L/(g·cm) (for softwood) [67].

3.8.4. Acidic Methanolysis

The amount of hemicellulose in spruce bark samples was analysed from extractive-free lyophilised bark using acidic methanolysis. An internal standard solution was prepared by dissolving 10 mg of sorbitol into 100 mL of methanol. To prepare an external standard solution, a 10 mg mixture of arabinose, galactose, glucose, xylose, mannose, galacturonic acid and glucuronic acid was dissolved into 100 mL of UHQ water. Then, a methanolysis reagent was prepared by cooling 100 mL of methanol in an ice bath and carefully adding and mixing 16 mL of acetyl chloride into the cold methanol. Next, the reagent was stored at −20 °C.

For methanolysis, 2 mL of the methanolysis reagent was added to 2–3 mg of extractive-free bark samples and to a dried monosaccharide standard sample (1 mL). The samples were then sonicated in an ultrasound bath and kept at 100 °C in an oven for 3 h. Pyridine (80 μL) and an internal standard (1 mL) were next added to the samples, and the solvent was evaporated. Then, 80 μL of pyridine and 250 μL of a silylation reagent were added, and the samples were sonicated in an ultrasound bath and kept in a shaker at room temperature for 40 min. Next, the samples were filtrated with glass wool for GC analysis with an Agilent 6850 gas chromatograph equipped with an HP-5 column (30 m × 0.32 mm, with a 0.25 μm film). Finally, the samples were injected at 260 °C and detected by FID at 290 °C. The method used was the same as in our previous study [10].

3.9. Statistical Analysis

One-way analysis of variance (ANOVA) was used to assess the effect of storage time (0, 4, 12 or 24 weeks) and sampling location (side, middle or top of the pile) on the concentrations of bark components, DP values of CTs, ash content and effective heating value. Logarithmic transformation was used for the variables that were sufficiently non-normal to cause concern about the validity of the normality assumption. In addition, the Kruskal–Wallis test, the non-parametric equivalent of one-way ANOVA, was used for the variables that were not normally distributed even after logarithmic transformation.

In turn, an independent-samples *t*-test was used to test the statistical differences in the concentrations, DP, ash content and net calorific value between the snow-covered and non-covered bark piles. Again, logarithmic transformation was used for the non-normal variables.

4. Conclusions

According to our study of bark storage in piles (both non-covered and covered with snow), spruce bark is a valuable raw material that is rich in hydrophilic extractives. However, it is worth noting that the material losses experienced as a result of pile storage (even during the winter) are dramatic, even after only a few weeks of storage. The loss of hydrophilic, phenolic extractive groups, such as stilbenes and tannins, was particularly notable. Significant proportions of the losses of extractives are to be attributed to the microbiological degradation and increase in pile temperature, which initiates and facilitates further degrading chemical reactions. In addition, exposure to UV light and the leaching of extractives from piles also cause losses of these compounds.

A clear trend was observed with regard to the sampling location in storage piles and the concentrations of the studied chemical compounds. In particular, both extractives and carbohydrates were found to have high concentrations in the middle of the pile with prolonged storage durations, indicating in some instances a leaching of compounds from elsewhere in the piles. Despite the covering of the other pile, no significant difference was observed between the degradation results of the non-covered and snow-covered bark piles. Other covering options and their impact on bark extractives should be further investigated.

From the results, it was evident that if piled bark material is to be valorised for its extractive content, storage periods should be as short as possible. Even four weeks of piled bark storage seems too long a time for stilbenes and tannin products. It was demonstrated that a simple hydrophilic extraction (e.g., with hot water) can effectively remove many potential platform chemicals for further purification steps. Methods, such as these, ought to be considered especially by bio-refinery plants that handle sidestream bark material. In addition, the logistics of bark material delivery to refineries needs to be planned in order to eliminate unnecessary exposure to weathering and moisture.

Supplementary Materials: The following supporting information can be downloaded online, Table S1: Values for Figure 1, Table S2: Values for Figure 3, Table S3: Values for Figure 4, Table S4: Values for Figure 5, Table S5: Values for Figure 6, Table S6: Values for Figure 7, Table S7: Values for Figure 8, Table S8: Values for Figure 9, Table S9: Values for Figure 10, Table S10: Values for Figure 11, Table S11: Values for Figure 12, Table S12: Values for Figure 13, Table S13: Values for Figure 14, Table S14: Values for Figure 15, Table S15. Values for Figure 16.

Author Contributions: Conseptualisation, J.N., H.E.B., E.S.H. and O.L.; methodology, E.S.H.; software, E.S.H., J.H. and M.H.; validation, E.S.H. and J.H.; formal analysis, E.S.H. and J.H.; investigation, J.N., H.E.B. and O.L.; resources, J.N.; data curation, E.S.H., J.H. and M.H.; writing–original draft preparation, E.S.H.; writing–review and editing, J.N., H.E.B., E.S.H. J.H., O.L., M.H. and R.A.; visualisation, E.S.H.; supervision, H.E.B. and R.A.; project administration, J.N.; funding acquisition, J.N. All authors have read and agreed to the published version of the manuscript.

Funding: The authors would like to express their gratitude for the funding received from the European Regional Development Fund, Interreg Botnia Atlantica, for the project BioHub (20200866), which made all the experimental work possible.

Institutional Review Board Statement: Not applicable.

Informed Consent Statement: Not applicable.

Data Availability Statement: Detailed data regarding the presented figures are available within the Supplementary Materials. Other data are available upon request from the corresponding author.

Acknowledgments: Thanks are due to Jaakko Miettinen, Tero Takalo, Reetta Kolppanen and Anna Claydon for their skilful technical assistance in the laboratory and field, and also to our partners at SLU who took part in the planning and assembly of the storage piles.

Conflicts of Interest: The authors declare no conflict of interest.

Sample Availability: Samples discussed in this article are not available for further study.

References

1. Hassan, M.K.; Villa, A.; Kuittinen, S.; Jänis, J.; Pappinen, A. An Assessment of Side-Stream Generation from Finnish Forest Industry. *J. Mater. Cycles Waste Manag.* **2019**, *21*, 265–280. [CrossRef]
2. Berman, A.Y.; Motechin, R.A.; Wiesenfeld, M.Y.; Holz, M.K. The Therapeutic Potential of Resveratrol: A Review of Clinical Trials. *NPJ Precis. Oncol.* **2017**, *1*, 1–9. [CrossRef] [PubMed]
3. Singh, A.P.; Kumar, S. Applications of Tannins in Industry. In *Tannins-Structural Properties, Biological Properties and Current Knowledge*, 1st ed.; Aires, A., Ed.; Intech Open: London, UK, 2020; pp. 117–136.
4. Jirjis, R. Storage and Drying of Wood Fuel. *Biomass Bioenergy* **1995**, *9*, 181–190. [CrossRef]
5. Fuller, W.S. Chip Pile Storage—A Review of Practices to Avoid Deterioration and Economic Losses. *Tappi J.* **1985**, *68*, 48–52.
6. Brand, M.A.; de Muñiz, G.I.B.; Quirino, W.F.; Brito, J.O. Storage as a Tool to Improve Wood Fuel Quality. *Biomass Bioenergy* **2011**, *35*, 2581–2588. [CrossRef]
7. Anerud, E.; Routa, J.; Bergström, D.; Eliasson, L. Fuel Quality of Stored Spruce Bark–Influence of Semi-Permeable Covering Material. *Fuel* **2020**, *279*, 118467. [CrossRef]
8. Krigstin, S.; Wetzel, S. A Review of Mechanisms Responsible for Changes to Stored Woody Biomass Fuels. *Fuel* **2016**, *175*, 75–86. [CrossRef]
9. Anerud, E.; Jirjis, R.; Larsson, G.; Eliasson, L. Fuel Quality of Stored Wood chips—Influence of Semi-Permeable Covering Material. *Appl. Energy* **2018**, *231*, 628–634. [CrossRef]
10. Halmemies, E.S.; Brännström, H.E.; Nurmi, J.; Läspä, O.; Alén, R. Effect of Seasonal Storage on Single-Stem Bark Extractives of Norway Spruce (*Picea abies*). *Forests* **2021**, *12*, 736. [CrossRef]
11. Jyske, T.; Brännström, H.; Sarjala, T.; Hellström, J.; Halmemies, E.; Raitanen, J.; Kaseva, J.; Lagerquist, L.; Eklund, P.; Nurmi, J. Fate of Antioxidative Compounds within Bark during Storage: A Case of Norway Spruce Logs. *Molecules* **2020**, *25*, 4228. [CrossRef]
12. Zahri, S.; Belloncle, C.; Charrier, F.; Pardon, P.; Quideau, S.; Charrier, B. UV Light Impact on Ellagitannins and Wood Surface Colour of European Oak (*Quercus petraea* and *Quercus robur*). *Appl. Surf. Sci.* **2007**, *253*, 4985–4989. [CrossRef]
13. George, B.; Suttie, E.; Merlin, A.; Deglise, X. Photodegradation and Photostabilisation of Wood—The State of the Art. *Polym. Degrad. Stab.* **2005**, *88*, 268–274. [CrossRef]
14. Malan, F.S. Some Notes on the Effect of Wet-Storage on Timber. *S. Afr. For. J.* **2004**, *202*, 77–82. [CrossRef]
15. Nozomi, M.; Takashi, F.; Takafumi, M.; Miyao, I.; Kenji, T. Effect of Wood Biomass Components on Self-Heating. *Bioresour. Bioprocess* **2021**, *8*, 1.
16. Bhat, T.K.; Singh, B.; Sharma, O.P. Microbial Degradation of Tannins—A Current Perspective. *Biodegradation* **1998**, *9*, 343–357. [CrossRef] [PubMed]
17. Dorado, J.; Van Beek, T.A.; Claassen, F.W.; Sierra-Alvarez, R. Degradation of Lipophilic Wood Extractive Constituents in *Pinus sylvestris* by the White-Rot Fungi *Bjerkandera* sp. and *Trametes versicolor*. *Wood Sci. Technol.* **2001**, *35*, 117–125. [CrossRef]
18. Mallory, F.B.; Mallory, C.W. Photocyclization of Stilbenes and Related Molecules. *Org. React.* **2004**, *30*, 1–456.
19. Olsson, V. Wet Storage of Timber: Problems and Solutions. Master's Thesis, KTH Royal Institute of Technology, Stockholm, Sweden, 2005.
20. Bianchi, S.; Koch, G.; Janzon, R.; Mayer, I.; Saake, B.; Pichelin, F. Hot Water Extraction of Norway Spruce (*Picea abies* [Karst.]) Bark: Analyses of the Influence of Bark Aging and Process Parameters on the Extract Composition. *Holzforschung* **2016**, *70*, 619–631. [CrossRef]
21. Krogell, J.; Holmbom, B.; Pranovich, A.; Hemming, J.; Willför, S. Extraction and Chemical Characterization of Norway Spruce Inner and Outer Bark. *Nord. Pulp Pap. Res. J.* **2012**, *27*, 6–17. [CrossRef]
22. Lappi, H.; Läspä, O.; Nurmi, J. Decrease in Extractives of Chain-Flail Residue. *For. Refine Info Sheet, WP3* **2014**, *13*, 1–3.
23. Routa, J.; Brännström, H.; Hellström, J.; Laitila, J. Influence of Storage on the Physical and Chemical Properties of Scots Pine Bark. *Bioenergy Res.* **2021**, *14*, 575–587. [CrossRef]
24. Routa, J.; Brännström, H.; Laitila, J. Effects of Storage on Dry Matter, Energy Content and Amount of Extractives in Norway Spruce Bark. *Biomass Bioenergy* **2020**, *143*, 105821. [CrossRef]
25. Čabalová, I.; Bélik, M.; Kučerová, V.; Jurczyková, T. Chemical and Morphological Composition of Norway Spruce Wood (*Picea abies*, L.) in the Dependence of its Storage. *Polymers* **2021**, *13*, 1619. [CrossRef] [PubMed]
26. Schuller, W.H.; Moore, R.N.; Lawrence, R.V. Air Oxidation of Resin Acids. II. the Structure of Palustric Acid and its Photosensitized Oxidation[2]. *J. Am. Chem. Soc.* **1960**, *82*, 1734–1738. [CrossRef]

27. Hemingway, R.W.; Nelson, P.J.; Hillis, W.E. Rapid Oxidation of the Fats and Resins in *Pinus Radiata* Chips for Pitch Control. *Tappi* **1971**, *54*, 95–98.
28. Hedmark, Å.; Scholz, M. Review of Environmental Effects and Treatment of Runoff from Storage and Handling of Wood. *Bioresour. Technol.* **2008**, *99*, 5997–6009. [CrossRef] [PubMed]
29. Nurmi, J. Longterm Storage of Fuel Chips in Large Piles. *Folia For.* **1990**, *767*, 1–18.
30. Jylhä, P.; Halmemies, E.; Hellström, J.; Hujala, M.; Kilpeläinen, P.; Brännström, H. The Effect of Thermal Drying on the Contents of Condensed Tannins and Stilbenes in Norway Spruce (*Picea abies* [L.] *Karst.*) Sawmill Bark. *Ind. Crops. Prod.* **2021**, *173*, 114090. [CrossRef]
31. Krigstin, S.; Helmeste, C.; Jia, H.; Johnson, K.E.; Wetzel, S.; Volpe, S.; Faizal, W.; Ferrero, F. Comparative Analysis of Bark and Woodchip Biomass Piles for Enhancing Predictability of Self-Heating. *Fuel* **2019**, *242*, 699–709. [CrossRef]
32. Neiva, D.M.; Araújo, S.; Gominho, J.; Carneiro, A.d.C.; Pereira, H. An Integrated Characterization of *Picea abies* Industrial Bark regarding Chemical Composition, Thermal Properties and Polar Extracts Activity. *PLoS ONE* **2018**, *13*, e0208270. [CrossRef]
33. Josefsson, P.; Nilsson, F.; Sundström, L.; Norberg, C.; Lie, E.; Jansson, M.B.; Henriksson, G. Controlled Seasoning of Scots Pine Chips using an Albino Strain of Ophiostoma. *Ind. Eng. Chem. Res.* **2006**, *45*, 2374–2380. [CrossRef]
34. DiGuistini, S.; Wang, Y.; Liao, N.Y.; Taylor, G.; Tanguay, P.; Feau, N.; Henrissat, B.; Chan, S.K.; Hesse-Orce, U.; Alamouti, S.M. Genome and Transcriptome Analyses of the Mountain Pine Beetle-Fungal Symbiont *Grosmannia Clavigera*, a Lodgepole Pine Pathogen. *Proc. Natl. Acad. Sci. USA* **2011**, *108*, 2504–2509. [CrossRef]
35. Gunstone, F.D. Chemical Reactions of Fatty Acids with Special Reference to the Carboxyl Group. *Eur. J. Lipid Sci. Technol.* **2001**, *103*, 307–314. [CrossRef]
36. Ekman, R. Resin during Storage and in Biological Treatment. In *Pitch Control, Wood Resin and Deresination*; Back, E.L., Allen, L.H., Eds.; TAPPI Press: Atlanta, GA, USA, 2000; pp. 185–195.
37. Nielsen, N.P.K.; Nørgaard, L.; Strobel, B.W.; Felby, C. Effect of Storage on Extractives from Particle Surfaces of Softwood and Hardwood Raw Materials for Wood Pellets. *Eur. J. Wood Wood Prod.* **2009**, *67*, 19. [CrossRef]
38. de Lima, E.J.; Alves, R.G.; Anunciação, T.A.D.; Silva, V.R.; Santos, L.D.S.; Soares, M.B.; Cardozo, N.; Costa, E.V.; Silva, F.; Koolen, H.H. Antitumor Effect of the Essential Oil from the Leaves of *Croton matourensis aubl.* (*Euphorbiaceae*). *Molecules* **2018**, *23*, 2974. [CrossRef]
39. Axelsson, K.; Zendegi-Shiraz, A.; Swedjemark, G.; Borg-Karlson, A.; Zhao, T. Chemical Defence Responses of Norway Spruce to Two Fungal Pathogens. *For. Pathol.* **2020**, *50*, 12640. [CrossRef]
40. Holmbom, B.; Avela, E. *Studies on Tall Oil from Pine and Birch*; Åbo Akademi: Åbo, Finland, 1971.
41. Burčová, Z.; Kreps, F.; Greifová, M.; Jablonský, M.; Ház, A.; Schmidt, Š.; Šurina, I. Antibacterial and Antifungal Activity of Phytosterols and Methyl Dehydroabietate of Norway Spruce Bark Extracts. *J. Biotechnol.* **2018**, *282*, 18–24. [CrossRef]
42. Assarsson, A.; Croon, I. Studies on Wood Resin, especially the Change in Chemical Composition during Seasoning of the Wood, Part 1. Changes in the Composition of the Ethyl Ether Soluble Part of the Extractives from Birch Wood during Log Seasoning. *Sven. Papp.* **1963**, *21*, 876–883.
43. Xu, G.; Guan, L.; Sun, J.; Chen, Z. Oxidation of Cholesterol and B-Sitosterol and Prevention by Natural Antioxidants. *J. Agric. Food Chem.* **2009**, *57*, 9284–9292. [CrossRef]
44. Anerud, E.; Krigstin, S.; Routa, J.; Brännström, H.; Arshadi, M.; Helmeste, C.; Bergström, D.; Egnell, G. Dry Matter Losses during Biomass Storage-Measures to Minimize Feedstock Degradation. *IEA Bioenergy: Task 43.* 2019, pp. 1–45. Available online: https://task43.ieabioenergy.com/wp-content/uploads/sites/11/2020/01/EIA-Dry-Matter-Loss_Final.pdf (accessed on 6 January 2022).
45. Schwarze, F.W. Wood Decay Under the Microscope. *Fungal Biol. Rev.* **2007**, *21*, 133–170. [CrossRef]
46. Bianchi, S. Extraction and Characterization of Bark Tannins from Domestic Softwood Species. Ph.D. Thesis, Faculty of Mathematics, Informatics and Natural Sciences, Department of Biology, University of Hamburg, Hamburg, Germany, 2016.
47. Timell, T.E. Isolation of Polysaccharides from the Bark of Gymnosperms. *Sven. Papp.* **1961**, *64*, 651–660.
48. Herrera, V.A.S.; Mendoza, D.E.R.; Leino, A.; Mikkola, J.; Zolotukhin, A.; Eränen, K.; Salmi, T. Sugar Hydrogenation in Continuous Reactors: From Catalyst Particles Towards Structured Catalysts. *Chem. Eng. Process Process Intensif.* **2016**, *109*, 1–10. [CrossRef]
49. Guo, Z.; Long, L.; Ding, S. Characterization of an L-Arabinose Isomerase from *Bacillus Velezensis* and its Application for L-Ribulose and L-Ribose Biosynthesis. *Appl. Biochem. Biotechnol.* **2020**, *192*, 935–951. [CrossRef] [PubMed]
50. Eisenberg Jr, F.; Parthasarathy, R. Measurement of biosynthesis of myo-inositol from glucose 6-phosphate. In *Methods in Enzymology*; Elsevier: Amsterdam, The Netherlands, 1987; Volume 141, pp. 127–143.
51. Alakoski, E.; Jämsén, M.; Agar, D.; Tampio, E.; Wihersaari, M. From Wood Pellets to Wood Chips, Risks of Degradation and Emissions from the Storage of Woody biomass–A Short Review. *Renew. Sust. Energ. Rev.* **2016**, *54*, 376–383. [CrossRef]
52. Li, T. The Production of Glutamic Acid by Fermentation. Master's Thesis, Univerity of Missouri, Columbia, SC, USA, 1965.
53. Danilewicz, J.C. Role of Tartaric and Malic Acids in Wine Oxidation. *J. Agric. Food Chem.* **2014**, *62*, 5149–5155. [CrossRef]
54. Jyske, T.; Laakso, T.; Latva-Mäenpää, H.; Tapanila, T.; Saranpää, P. Yield of Stilbene Glucosides from the Bark of Young and Old Norway Spruce Stems. *Biomass Bioenergy* **2014**, *71*, 216–227. [CrossRef]
55. Mulat, D.G.; Latva-Mäenpää, H.; Koskela, H.; Saranpää, P.; Wähälä, K. Rapid Chemical Characterisation of Stilbenes in the Root Bark of Norway Spruce by Off-line HPLC/DAD–NMR. *Phytochem. Anal.* **2014**, *25*, 529–536. [CrossRef] [PubMed]
56. Zhang, Y.S.; Ning, Z.X.; Yang, S.Z.; Wu, H. Antioxidation Properties and Mechanism of Action of Dihydromyricetin from *Ampelopsis Grossedentata*. *Yao Xue Xue Bao Acta Pharm. Sin.* **2003**, *38*, 241–244.

57. Bianchi, S.; Gloess, A.N.; Kroslakova, I.; Mayer, I.; Pichelin, F. Analysis of the Structure of Condensed Tannins in Water Extracts from Bark Tissues of Norway Spruce (*Picea abies* [Karst.]) and Silver Fir (*Abies alba* [Mill.]) using MALDI-TOF Mass Spectrometry. *Ind. Crop. Prod.* **2014**, *61*, 430–437. [CrossRef]
58. Hammerbacher, A.; Paetz, C.; Wright, L.P.; Fischer, T.C.; Bohlmann, J.; Davis, A.J.; Fenning, T.M.; Gershenzon, J.; Schmidt, A. Flavan-3-ols in Norway Spruce: Biosynthesis, Accumulation, and Function in Response to Attack by the Bark Beetle-Associated Fungus *Ceratocystis polonica*. *Plant Physiol.* **2014**, *164*, 2107–2122. [CrossRef]
59. Matthews, S.; Mila, I.; Scalbert, A.; Donnelly, D.M. Extractable and Non-Extractable Proanthocyanidins in Barks. *Phytochemistry* **1997**, *45*, 405–410. [CrossRef]
60. Kraus, T.E.; Dahlgren, R.A.; Zasoski, R.J. Tannins in Nutrient Dynamics of Forest Ecosystems-a Review. *Plant Soil* **2003**, *256*, 41–66. [CrossRef]
61. *CEN/TS 14774-2: 2004*; Solid Biofuels—Methods for the Determination of Moisture Content-Oven Dry Method-Part 2: Total Moisture-Simplified Method. British Standards Institute: London, UK, 2004.
62. *BS EN 14775: 2009*; Solid Biofuels—Determination of Ash Content. British Standards Institution: London, UK, 2009.
63. *CEN/TS 14918: 2005*; Solid Biofuels—Method for the Determination of Calorific Value. British Standards Institution: London, UK, 2006.
64. *ISO, E.N. 16948: 2015*; Solid Biofuels—Determination of Total Content of Carbon, Hydrogen and Nitrogen. European Committee for Standardization (CEN): Brussels, Belgium, 2015.
65. Korkalo, P.; Korpinen, R.; Beuker, E.; Sarjala, T.; Hellström, J.; Kaseva, J.; Lassi, U.; Jyske, T. Clonal Variation in the Bark Chemical Properties of Hybrid Aspen: Potential for Added Value Chemicals. *Molecules* **2020**, *25*, 4403. [CrossRef] [PubMed]
66. TAPPI Test Methods 222 om-02. Acid-Insoluble Lignin in Wood and Pulp. In *2002–2003 TAPPI Test Methods*; TAPPI: Tokyo, Japan, 2002.
67. TAPPI Useful Methods 250. Acid-Soluble Lignin in Wood and Pulp. In *1991 TAPPI Useful Methods*; TAPPI: Tokyo, Japan, 1991.

Article

Physicochemical Characteristics of Biofuel Briquettes Made from Pecan (*Carya illinoensis*) Pericarp Wastes of Different Particle Sizes

Maginot Ngangyo Heya [1], Ana Leticia Romo Hernández [1], Rahim Foroughbakhch Pournavab [2], Luis Fernando Ibarra Pintor [3], Lourdes Díaz-Jiménez [4], Michel Stéphane Heya [5,*], Lidia Rosaura Salas Cruz [1,*] and Artemio Carrillo Parra [6,*]

1 Agronomy School (FA), Universidad Autónoma de Nuevo León (UANL), Francisco Villa S/N, Col. Ex-Hacienda "El Canadá", Escobedo 66050, Nuevo Leon, Mexico; nheyamaginot@yahoo.fr (M.N.H.); ana.romohrnd@uanl.edu.mx (A.L.R.H.)

2 Department of Botany, Biological Science School (FCB), Universidad Autónoma de Nuevo León (UANL), Ave. Pedro de Alba S/N Cruz & Ave. Manuel L. Barragán, San Nicolas de los Garza 66451, Nuevo Leon, Mexico; rahim.forough@gmail.com

3 Facultad de Ingeniería en Tecnología de la Madera, Universidad Michoacana de San Nicolás de Hidalgo, Gral. Francisco J. Múgica S/N Ciudad Universitaria, Morelia 58040, Michoacan, Mexico; luis.pintor@umich.mx

4 Laboratorio de Revaloración de Residuos, Sustentabilidad de los Recursos Naturales y Energía, Cinvestav Saltillo, Ramos Arizpe 25900, Coahuila, Mexico; lourdes.diaz@cinvestav.edu.mx

5 Biotechnology Institute, Biological Science School (FCB), Universidad Autónoma de Nuevo León (UANL), Ave. Pedro de Alba S/N Cruz & Ave. Manuel L. Barragán, San Nicolas de los Garza 66451, Nuevo Leon, Mexico

6 Institute of Silviculture and Wood Industry (ISIMA), Universidad Juárez del Estado de Durango (UJED), Boulevard del Guadiana 501, Ciudad Universitaria, Torre de Investigación, Durango 34120, Durango, Mexico

* Correspondence: heyamichelstephane@yahoo.fr (M.S.H.); biolidiasalas@yahoo.com.mx (L.R.S.C.); acarrilloparra@ujed.mx (A.C.P.)

Citation: Ngangyo Heya, M.; Romo Hernández, A.L.; Foroughbakhch Pournavab, R.; Ibarra Pintor, L.F.; Díaz-Jiménez, L.; Heya, M.S.; Salas Cruz, L.R.; Carrillo Parra, A. Physicochemical Characteristics of Biofuel Briquettes Made from Pecan (*Carya illinoensis*) Pericarp Wastes of Different Particle Sizes. *Molecules* **2022**, *27*, 1035. https://doi.org/10.3390/molecules27031035

Academic Editor: Mohamad Nasir Mohammad Ibrahim

Received: 30 December 2021
Accepted: 21 January 2022
Published: 3 February 2022

Publisher's Note: MDPI stays neutral with regard to jurisdictional claims in published maps and institutional affiliations.

Abstract: Pecan nut (*Carya illinoensis*) pericarp is usually considered as a waste, with no or low value applications. Its potential as a densified solid biofuel has been evaluated, searching for alternatives to generating quality renewable energy and reducing polluting emissions in the atmosphere, based on particle size, that is an important feedstock property. Therefore, agro-industrial residues from the pecan nut harvest were collected, milled and sieved to four different granulometry: 1.6 mm (N° 12), 0.84 mm (N° 20), 0.42 mm (N° 40), and 0.25 mm (N° 60), used as raw material for biofuel briquette production. The carbon and oxygen functional groups in the base material were investigated by Fourier transform infrared spectroscopy (FTIR) and proximate analyses were performed following international standards, for determining the moisture content, volatile materials, fixed carbon, ash content, and calorific value. For the biofuel briquettes made from base material of different particle sizes, the physical characteristics (density, hardness, swelling, and impact resistance index) and energy potential (calorific value) were determined to define their quality as a biofuel. The physical transformation of the pecan pericarp wastes into briquettes improved its quality as a solid biofuel, with calorific values from around 17.00 MJ/kg for the base material to around 18.00 MJ/kg for briquettes, regardless of particle size. Briquettes from sieve number 40 had the highest density (1.25 g/cm^3). Briquettes from sieve number 60 (finest particles) presented the greater hardness (99.85). The greatest susceptibility to swelling (0.31) was registered for briquettes with the largest particle size (sieve number 20). The IRI was 200 for all treatments.

Keywords: solid waste-to-biofuel; biomass densification; granulometric distribution; proximate analysis; functional groups for energy storage

1. Introduction

Pecan (*Carya illinoensis*) nuts are dry fruits native to California in the US, with the biggest production coming from Mexico (around 124,000 tons/year) [1]. Pecan nut processing generates important amounts of waste in the form of shells and pericarp (40–50%) [2,3] that are generally dumped into the environment or burned, representing an important loss of biomass and, at the same time, constituting a cause of some environmental problems. For these wastes to be better used it is necessary to know their composition and some characteristics [4,5].

It is well known that agro-industrial wastes, as well as forest residues, municipal solid wastes, and refused-derived fuels represent different types of biomass energy resources [6]. Nevertheless, their direct combustion has negative aspects owing to their intrinsic properties, such as low density, low calorific value per unit volume, and high moisture content. Moreover, the direct burning of agricultural residues is inefficient due to associated transportation, storage, and handling problems [7,8]. Therefore, it is necessary to develop strategies for converting biomass to secondary fuels with better characteristics than the base material [9]. Currently, briquetting is one of the suitable evolving technologies of waste materials conversion to solid biofuels for energy purposes. Thus, biofuel briquettes provide a sustainable approach for the improvement and efficient utilization of agricultural and/or other biomass residues [10].

However, biomass briquetting depends on feed parameters that influence the extrusion process. The two most important feedstock properties are particle size and moisture content [11]. For moisture content, the optimal level is defined by the mandatory technical standard EN 18134-2, while for particle size, which is also considered as great influencer of final briquette quality [12], there is no mandatory technical standard to define optimal range of particle size. Furthermore, briquette hardness and quality can be checked by water test, where a quality briquette should fall to the bottom in a moment due to its higher specific density than water [13].

According to Jenkins et al. [14], these physical properties are closely linked to molecular structure, affecting the combustion characteristics of a particular fuel or fuel blend. In this way, the functional groups prevalent in biomass derived molecules may be related to the fuel properties such as combustion characteristics and emissions, being carbon and oxygen functional groups the bio-derived molecules with a real application as fuels [14]. The retention of oxygen atoms, abundant in biomass, provides certain advantages, including complete fuel combustion and less harmful exhaust emissions [15,16]. For carbon atoms that are the molecular building blocks of cellulosic biomass, the number is limited to 5- and 6-carbon containing species. The effect of chain extension must also be considered, especially for more specialized applications [14].

Pecan waste products are rich carbon sources [17], with high concentrations of tannins and phenols [18,19]. The lignocellulosic composition of these pecan wastes, in terms of their content in cellulose, hemicellulose, lignin, and ash from the shells, are 5.6, 3.8, 70, and 5.85%, respectively [17]. In the case of pericarp they are 30, 26, 41, and 1.7% [20], and from the pecan branches they are 38.7, 30.2, 23.3, and 0.4%, respectively [21].

Although pecan wastes have been studied as a feedstock for bioenergy purposes [22], and their main components have been determined [23,24], specific information on their properties for biofuel briquettes elaboration is scarce. Therefore, the present study aimed to determine the most appropriate particle size from pecan pericarp feedstock materials for biofuel briquettes production, to optimize the briquetting process, for the prevention of material lose during production, transportation, and storage; and to investigate how the energy content and fuel quality of pecan briquettes differ among the functional groups.

2. Results

2.1. Proximate Analysis of the Pecan Pericarp Base Materials

The results of the proximate analyses of the base materials from pecan pericarp residues are presented in Figure 1. The moisture content of the four analyzed granulometry

1.6 mm (sieve N° 12), 0.84 mm (sieve N° 20), 0.42 mm (sieve N° 40), and 0.25 mm (sieve N° 60), did not present significant differences, as the probability (p value) were 0.993, with values ranging between 7.63 and 7.92% (Figure 1a), values that are <12% as indicated by the EN-14774-1 standard [25]. The highest moisture content was recorded in granulometry 1.6 mm (N° 12) and 0.84 mm (N° 20), corresponding to the finest materials.

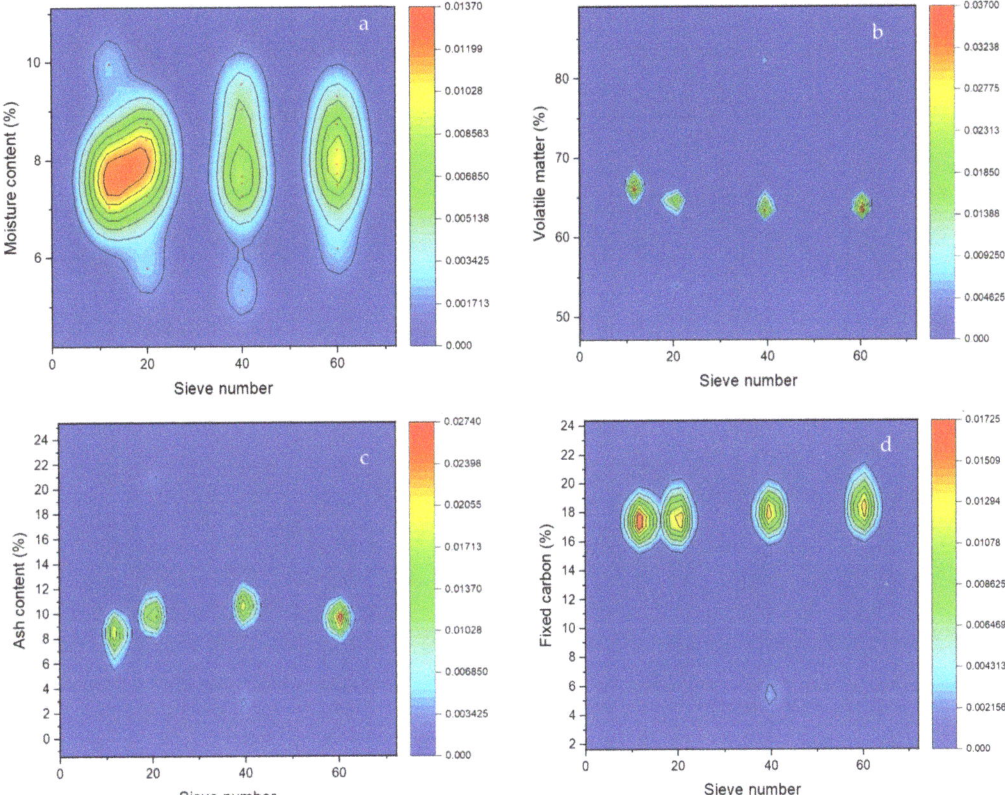

Figure 1. Results of the proximate analyses of pecan pericarp wastes of different particle sizes: 1.6 mm (N° 12), 0.84 mm (N° 20), 0.42 mm (N° 40) and 0.25 mm (N° 60). (**a**)—Moisture content, (**b**)—Volatile matter, (**c**)—Ash content, (**d**)—Fixed carbon.

Regarding the volatiles, there were no significant differences (p value = 0.595) among the granulometric numbers analyzed, with values ranged between 63.01 and 66.68% (Figure 1b), being below that established by the UNE-EN-15148 standard [26]. The finest material (particle size N° 12) presented the highest content of volatiles.

For the ash content, there were no significant differences between the treatments (p value = 0.156). The highest ash content was presented with granulometry 0.42 mm (N° 40), and the lowest ash content was presented in granulometry 1.6 mm (N° 12), with values of 11.55% and 8.20%, respectively (Figure 1c). These values are outside the UNE-EN-14775 standard [27].

The fixed carbon did not show significant differences between the granulometry (p value = 0.727). The fixed carbon content recorded the range from 16.09 to 18.59%, values that correspond to particle sizes N° 40 (0.42 mm) and N° 60 (0.25 mm), respectively (Figure 1d). The finest materials presented very similar values (17.49% and 17.73 for materials of particle size 12 and 20, respectively).

2.2. Energy Content of the Base Materials and the Resulting Briquettes, at Different Particle Sizes, from Pecan Pericarp Residues

The calorific value analysis shows no significant differences in particle size between the base material (p value = 0.50) and the briquettes (p value = 0.66). The calorific value ranged from 17.04 to 17.54 MJ/kg for the base material (Figure 2A), and from 18.19 to 18.45 MJ/kg for the briquettes (Figure 2B), which suggests that the physical transformation of the pecan pericarp raw material into biofuel briquettes considerably improves its calorific potential. However, the highest calorific value was registered for the smallest particle sizes with the raw material (0.42 mm and 0.25 mm, corresponding to sieves number 40 and 60, respectively). At the same time, the briquettes presented a similar trend of calorific value for the materials from the different particle sizes analyzed, with greater energy power. This means that the size of particles is decisive for the energy value of a biofuel, since it affects the contact among particles, smaller particles has more contact surface and thus heat transmission is easy.

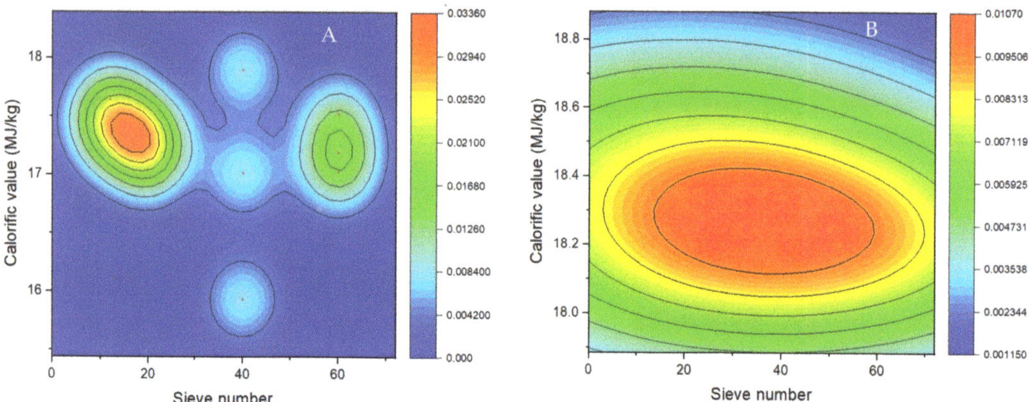

Figure 2. Calorific values of pecan pericarp residues at the different particle sizes 1.6 mm (N° 12), 0.84 mm (N° 20), 0.42 mm (N° 40) and 0.25 mm (N° 60). (**A**). Raw material; (**B**). Briquettes.

These calorific values are within the range established by the EN-14918 standard [28], both for the base material and for the resulting bio-briquettes. This means that the base material from pecan pericarp residues already constitutes an interesting source of biofuel, which would be more efficient if transformed into briquettes.

2.3. Morphological Characteristics of the Different Particle Sizes from the Pecan Pericarp Residues

Images from the scanning electron microscopy (SEM) analysis of pecan pericarp particle sizes are presented in Figure 3, showing that the different particles are ovoid in shape and of homogeneous size as a result of a correct reduction process. This homogeneity in the shapes and sizes of the particles is important, since there is greater cohesion at a homogeneous size.

2.4. Fourier Transform Infrared (FTIR) Analysis of Functional Groups in the Pecan Pericarp Residues

The main functional groups present in the pecan pericarp residues are shown in Figure 4, where the highest stretching can be observed in the particles size 0.25 mm (N° 60), and the lowest stretching, in the particle sizes 1.6 mm (N° 12). However, the trend of the signals by the functional groups is the same in the material from all the particle sizes, being the band between 3000 and 3600 cm^{-1} attributed to the stretching of hydroxyl (OH) groups, and the band registered between 2750 and 3000 cm^{-1}, corresponding to methyl (CH$_3$) groups. The peak at 1600 cm^{-1} corresponds to the carbonyl group (C=O), with the corresponding stretching vibration from 1490–1615 cm^{-1}. The stretching signals at 1200–1490 cm^{-1} are attributed to the methylene groups (CH$_2$), which show a peak at 1450 cm^{-1}. A higher absorbance characterized the carboxyl group (COOH), reaching its peak at 1000 cm^{-1}, within the stretching vibration from 700–1200 cm^{-1}. However, as far as the CH functional group is concerned, the absorbance differs greatly between particles, there were two well-defined peaks in the particles of N° 12. At the same time, in 20 and 40 particle size, that peaks practically disappeared, and in particle size N° 60, it reappeared but one alone, in lower intensity.

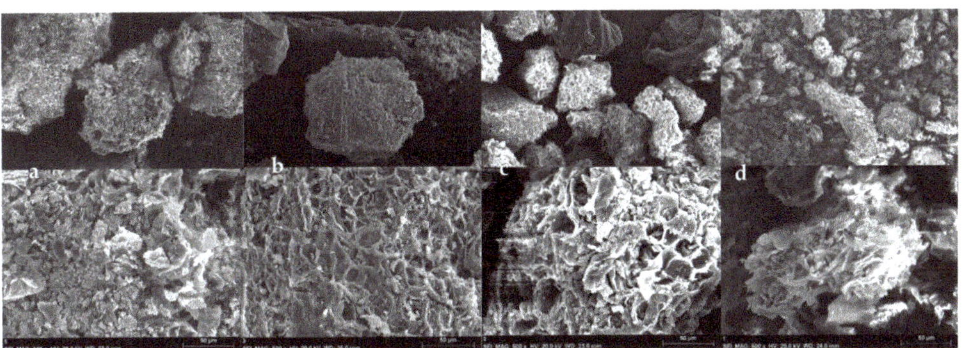

Figure 3. Typical morphology of the particle sizes in samples from pecan pericarp residues. (**a**)—Sieve number 12 (1.6 mm); (**b**)—N° 20 (0.84 mm); (**c**)—N° 40 (0.42 mm); (**d**)—N° 60 (0.25 mm).

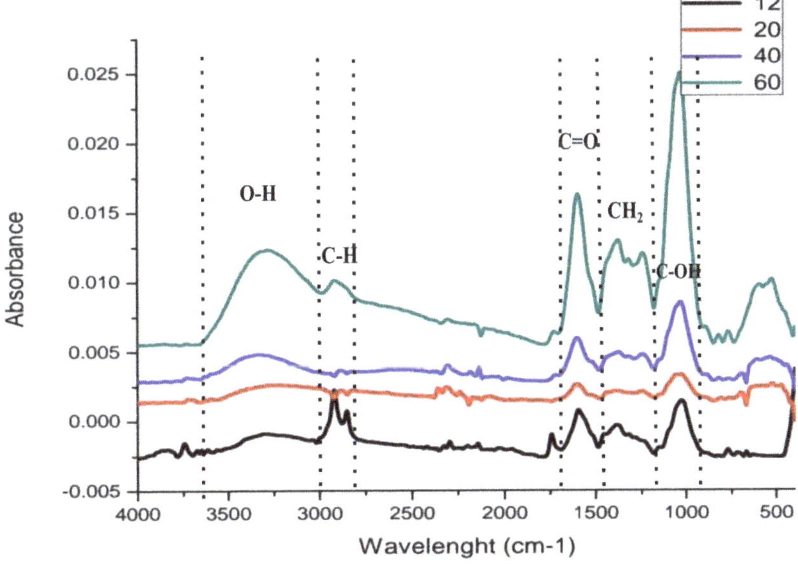

Figure 4. FT-IR spectrograms of different particle sizes from pecan pericarp residues, indicating the stretching of the signals corresponding to the most representative functional groups.

The intensity of the infrared spectrogram bands was materialized numerically by obtaining the highest point through the OriginLab program. As Figure 5 shows, the intensity of the spectrometric stretching was particle size-dependent, being the highest values obtained with the particles size 60 and the lowest values with the particle size 12, for all the functional groups (O-H, C=O, CH_2, C-OH).

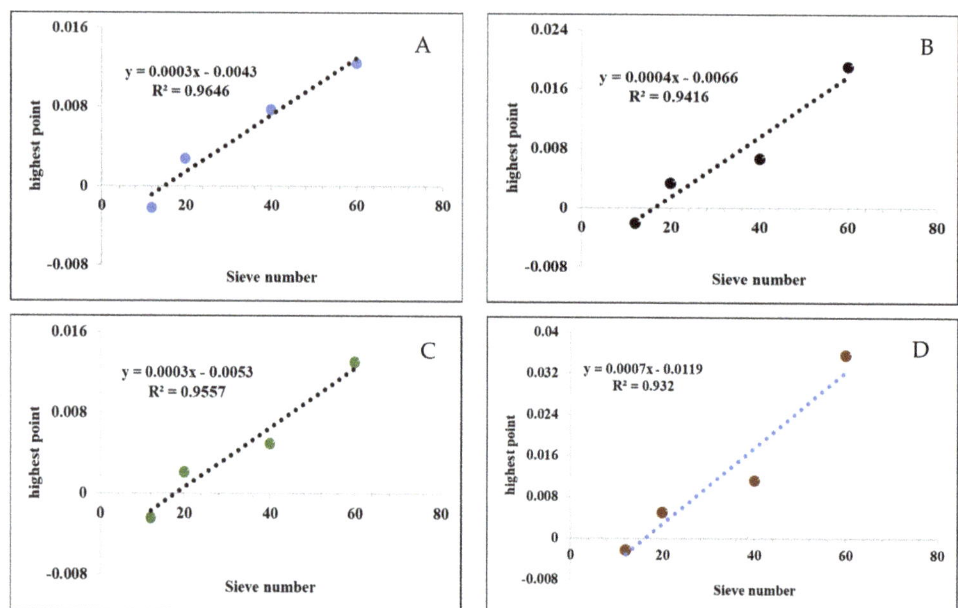

Figure 5. Numerical materialization of the intensity of the most representative signals of the FT-IR spectrograms as a function of particle size. (**A**) hydroxyl groups (O-H); (**B**) carbonyl group (C=O); (**C**) methylene group (CH_2); (**D**) carboxyl group (C-OH).

2.5. Physical Properties of the Biofuel briquettes from Different Particle Sizes of Pecan Pericarp Residues

At the 0.05 level, the density means of biofuel briquettes from pecan pericarp are significantly different (*p* value = 0.02), with values from 1.22 to 1.25 g/cm^3. Briquettes from sieve number 40 had the highest density, while sieve numbers 12 and 20 had the lowest density (Figure 6A), suggesting that thicker materials produce lower-density briquettes than finer materials. However, the appropriate particle size for briquettes of good density is 0.42 mm material from the sieve number 40, since with the smallest particle size 0.25 mm (sieve number 60, the density dropped to 1.23 g/cm^3. This trend is the same with hardness, where the finest particles presented greater resistance to impact, with a hardness of 99.85 for biofuel briquettes from particle size of sieve number 60 (Figure 6B).

Regarding the swelling index, the briquettes with the largest particle size presented the greatest susceptibility to swelling, with a value of 0.31 for material from sieve number 20. In contrast, briquettes from sieve number 60 presented a value of 0.15 (Figure 6C).

After dropping briquettes three times from 1.85 m height, they did not lose weight so there were no broken pieces. Therefore, the IRI was 200 for all treatments, representing the maximum value that can be presented.

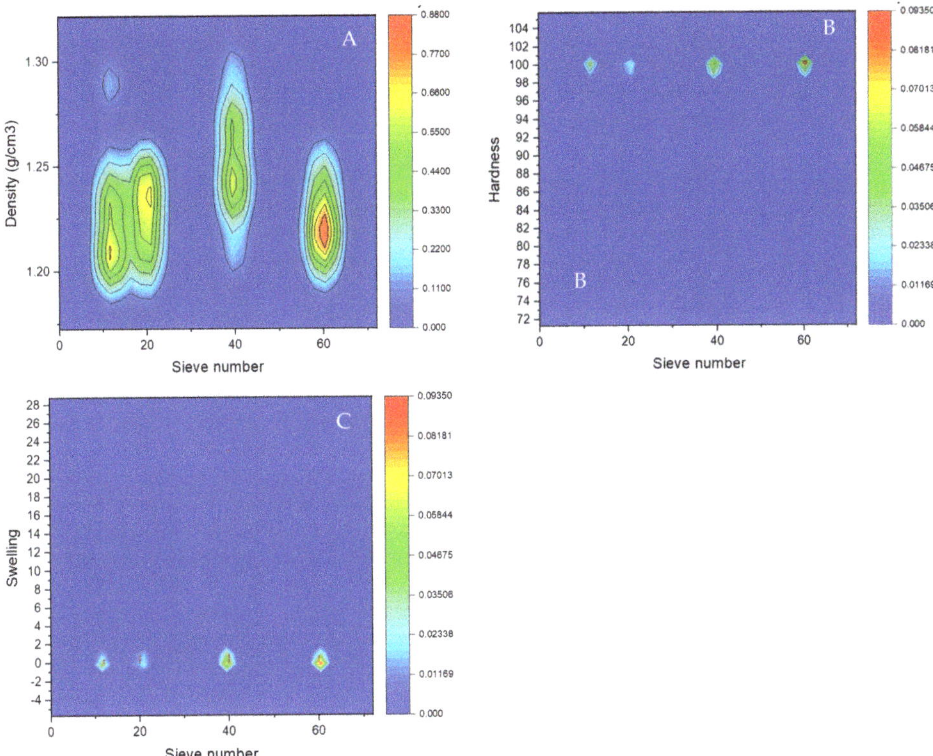

Figure 6. Physical characteristic of biofuel briquettes from pecan pericarp wastes of different particle sizes. (**A**) Density; (**B**) Hardness; (**C**) Swelling.

3. Discussion

3.1. Characteristics of the Pecan Pericarp Raw Material at Different Particle Sizes for Energy Use

The characterization of the base material from pecan pericarp residues was done through proximate analysis, determining the moisture, volatile, ash contents, and fixed carbon. The moisture was around 7.8%, meeting the requirements of the EN-14774-1 standard [25], which established the moisture content in the feed biomass as a very critical factor, and that may be around 10–12%. The moisture present in biomaterials forms steam under high-pressure conditions, which then hydrolyses the hemicellulose and lignin into lower molecular carbohydrates, lignin products, sugar polymers, and other derivatives [9], hence the need to keep the moisture content at a specific value. For five analyzed biomaterials (rapeseed oilcake, pine sawdust, Virginia mallow chips, Rape straw, and Willow chips), Stolarski et al. [29] reported a moisture content range from 9.8 to 18.3%. For palm oil residues (shell and fiber), Husain et al. [30] pointed out a moisture content of 12%. Chen et al. [31] suggested that the moisture content should be range between 10 and 15% for biomaterials of energy purposes. According to Kaur et al. [9], more than 15% moisture content produces poor and weak briquettes. Felfli et al. [32] advocate producing briquettes from agro-industrial materials (e.g., rice husk, coffee husk, bagasse, soybean husk, or sawdust) for their moisture content of 15%.

As for the volatile, the values oscillated around 65%, being below that established in the UNE-EN-15148 standard [26]. The volatile compounds come from both the organic part of the biomass and the inorganic part; and a high content of these compounds indicates the presence of an organic load when they are in a range of 74–89%, which provides a susceptibility to thermal degradation in processes such as combustion and pyrolysis.

However, a low material volatile content is advantageous since it reduces gases emissions such as condensable or not condensable containing CO, CO_2, CH_4 during combustion. Barroso [33] found a lower percentage of volatile matter that volatilizes during combustion, concluding that a high percentage does not volatilize. This may justify the percentage obtained in this study, lower than what is established by the standard.

The percentage of volatiles that were not released during combustion can explain the high value obtained for the ashes of the pecan pericarp, which were 8–12%, above the maximum permissible limits of international standards. However, fuels with levels of more than 20% ash are not enabling for heat generation due to the residuality of chemical compounds that interfere in the combustion process.

Low fixed carbon content was found in the pecan pericarp (16.09–18.59%). According to Demirbas [34] and Ngangyo-Heya [35], the low content of fixed carbon increases friability and brittleness, decreasing compressive strength, cohesion, and energy.

However, the calorific value of the walnut pericarp (around 17.00 MJ/kg) resulted acceptable for the production of the second-generation solid biofuel, thus constituting an alternative to protect forests from clearing for the production of charcoal.

Although not all the properties were fully met for the generation of quality biofuels, the ash content of pecan pericarp was slightly high. On the other side, the fixed carbon and the volatile matter were little lower than the normal. The values obtained, especially moisture content and the calorific value, meet the standard ranges established by the standardized norms, constituting pecan pericarp materials as suitable for their use as biofuels. However, the improvement of its characteristics by densification may allow the development of eco-friendly materials, since they are renewable and abundant resources, according to Shekar and Ramachandra [36] and Spiridon et al. [37].

FTIR is one of the most used techniques for characterization of lignocellulosic materials, since it is able to identify the main functional groups that are present in its constitution, to analyze the relationship of some bands through lateral order index and total crystalline index, which allow understanding the crystalline cellulose structure and the influence of crystalline and amorphous domains on the properties of those materials. It is worth mentioning that in the last decade, this method has been widely used for the identification and quantification of compounds in solid mixtures (e.g., soil, plant extracts, etc.) as well as liquids. The strong absorption registered in the materials of particle size 60 suggests a high bioavailability of the functional groups in the finer lignocellulosic materials than in the thick ones. Biomass produces pure linear hydrocarbons, where hexane and pentane were reported to be the main components of cellulose [38]. It is reported that an increase in the number of carbon atoms in a fuel molecule has a profound effect on a range of physical properties, owing to the increase in intermolecular forces related to the surface area of the molecule. This could be seen in particle size change of the pecan pericarp which, although it was a physical modification of the material, presented a greater amount of the functional groups OH, CH, C=O, C-OH in the smaller particles (sieve N° 60), indicating the effect of the mill (raising the temperature), which could have caused the rupture of certain structural elements that facilitated the presence of greater bonds as mentioned. The carbon functional groups are reported to have a profound effect on melting, flash, and boiling points while keeping the carbon number the same. As intermolecular forces increase with increasing carbon number, the density, boiling point, melting point, and flash point.

3.2. Physical and Energy Properties of Biofuel Briquettes from Pecan Pericarp Residues at Different Particle Sizes

The briquettes for energy use correspond to a solid fuel of renewable origin, obtained by the densification of biomass. They are compressed at high pressure without the presence of additives, obtaining a homogeneous product with very low humidity [33]. The main characteristic of briquettes is their high density and homogeneity [33]. Density is an important parameter to characterize the briquetting process. The standards define the interval of briquette density values from 1 to 1.4 g/cm^3 [39], which means that the materials of all particle sizes used in this study are suitable to produce good quality briquettes. However, it was found that the briquettes made from materials with larger particle sizes presented the lowest density, compared to briquettes produced with fine materials, due to the spaces left by larger particles, while in briquettes with smaller particles, the spaces are reduced. This agrees with the results of Tirado [40], who found that the compaction process, which consists of the application of pressure and temperature, causes a decrease in the volume of the walnut shell particles, increasing the density of the briquettes. The suitable material, as the present study suggests, was particles of size N° 40 (0.42 mm) to reach the highest density. Higher density leads to a higher energy/volume ratio desirable in terms of transportation, storage, and handling. Briquettes with higher density have a longer burning time [9]. The density of bio-waste briquettes depends on the density of the original bio-waste, the briquetting pressure, and, to a certain extent, on the briquetting temperature and time.

Regarding the homogeneity, Tirado [40] found that the smaller the particle size, the briquettes have greater homogeneity, with which he established an optimal particle size of 2 mm. This further shows that for the densification process, the particle sizes are of great importance, since it has been shown that as the particle size increases, there is a decrease in the bonding force between particles, and the speed of heat transmission is reduced due to the void spaces. Jha and Yadav [41] proposed a mixture of particle sizes. They found that different sizes of particles improve the packing dynamics and contribute to high static strength.

The hardness and impact resistance are other characteristics that determine the physical quality of briquettes, since they simulate the compression force due to the weight of the other briquettes that can be supported during storage or transport. According to previous studies, finer grinding of feedstock material results in solid biofuels of higher quality [42,43]. MacBain [44] asserted that larger particles accept less moisture which causes fractures in solid biofuels in contrast with finer particles. Thus, briquette quality increases with decreasing of feedstock particle size, as indicated in the present study, that the finest particles had greater resistance. However, this trend is limited, according to Kaliyan and Morey [42]. Extremely small particles exhibit a negative influence and decreasing mechanical hardness, which is the main indicator of mechanical quality of briquettes and disproportionately large particles.

On the other hand, the moisture content has a remarkable effect on the briquettes hardness. Previous studies showed that more than 15% moisture content produces poor and weak briquettes [42]. Mani et al. [45] have demonstrated the neediness to establish the initial moisture content of the biomass feed to maintain equilibrium and thus avoid swelling of briquette during storage, transportation, and disintegration when exposed to humid atmospheric conditions. Brunerová and Brožek [46] proved that the moisture of over a long period of stored briquettes changes, namely in dependence on the storage conditions. Feed moisture of 10–12% produces briquettes of 8–10% moisture, and such briquettes are strong and free of cracks [47]. From the above mentioned information, the majority of authors have experiences about briquetting of materials of moisture between 9 and 18%. However, it emerges from the present work that, the moisture content of the base material can drop to around 7%, and still produce briquettes of good quality, with a high capacity to resist shock during transport and storage, but, an excessive drop in the moisture content of the base material can lead to a decrease in the mechanical properties of the briquettes.

The moisture content can vary depending on the particle size of the materials. In this way, Mani et al. [45] and Gilbert et al. [48] proposed a size reduction of raw biomass to obtain better properties of the product by drying, mixing, and briquetting. Biomass material of 6–8 mm size with 10–20% powdery component gives the best results [42]. Mitchual et al. [11] analyzed briquettes produced from tropical hard-wood sawdust, and found that best particle size ranges between 1–2 mm. However, other authors exhibited that suitable particle size ranges between 10–15 mm for briquettes from municipal solid waste [49] or ranges between 6–8 mm size for briquettes made from combination of three hard-wood species [50]. For their part, Tumuluru et al. [51] studied briquettes produced from wheat, oat, canola, and barley straw, indicating the best particle size between 25–32 mm. In the present work on the pecan nut pericarp, the appropriate particle size for the production of good quality briquettes is around 1 mm. Choice of optimal particle size apparently partially depends on concrete feedstock material, but in general, it is not disputed that overall optimal particle size is not defined yet.

4. Materials and Methods

4.1. Base Materials from Pecan Pericarp Residues and the Biofuel Briquettes Elaboration

The pecan pericarp residues were collected from orchards in southern Nuevo Leon, Mexico, immediately after the harvesting, and were left to dry at room temperature for a minimum time of 10 days. Then, the material was ground through a knife mill and since particles produced by knife milling are not uniform in their size, the particle size distribution was determined according to the UNE-EN 17827-2 standard [52], with a series of metal mesh sieves of standard specifications and sizes, placed in sequential order (from largest to smallest) and arranged from the top to the bottom, to allow a complete reduction of the particle size (12, 20, 40 and 60 corresponding to 1.60 mm, 0.84 mm, 0.42 mm, and 0.25 mm openings, respectively). The representative fraction of each size was collected, and used as raw material for the production of biofuel briquettes. A vertical orientation laboratory briquetting machine (LIPPEL®, Agrolândia, Santa Catarina, Brazil) was used, which presents a solid base with a cylinder of 20 mm in diameter and an integrated thermostat that allows the temperature to be modified, two pistons that apply pressure and facilitate the briquette extraction. Each one was made with 200 cm^3 of the base material, and the operating conditions were 15 MPa of pressure, 90 °C of temperatures and five minutes pressure, similar to that developed by Zepeda-Cepeda [53] and Ramírez-Ramírez [54], for briquettes of 3.3 cm in diameter and 8.0 cm in length. Eight briquettes were made per treatment (granulometry) for subsequent analyzes.

4.2. Proximate Analysis and Energy Content of the Pecan pericarp Base Materials

The term "proximate analysis" is used to indicate the ASTM standardized test to define the quality of a fuel [36,55], based on determining the moisture content, volatile matter, ash content, and fixed carbon. Regarding the energy content, the calorific value was determined from the content of fixed carbon (FC) and volatile matter (VM), according to the formula described by Cordero et al. [56]. Although all these properties are somewhat interrelated, they are measured and valued separately, as indicated in Table 1.

Table 1. Parameters analyzed according to the ASTM standards.

	Relation	Observation
Moisture content	$MC = \frac{(Wi - Wd)}{Wi} * 100$	Wi = initial weight of the raw material; Wd = dry weight of the raw material after 3 h in an oven at 105 °C.
Volatile matter	$VM = \frac{(Wd - Wv)}{Wd} * 100$	Wv = weight of the materials after placing them in a muffle furnace at 950 °C.
Ash content	$Ash = \frac{Wa}{Wv} * 100$	Wa = ash weight after muffling at 750 °C during 6 h.
Fixed carbon content	$FC = 100 - (MC + VM + Ash)$	The fixed carbon content was obtained, subtracting the moisture, volatiles, and ash content, from 100%.
Calorific value	$CV = 354.3\ FC + 170.8\ VM$	The calorific value is a function of the fixed carbon and volatiles.

4.3. Fourier Transform Infrared (FTIR) Spectroscopy for Analysis of Functional Groups from Pecan Pericarp Residues

The phytochemical characterization of the particles was carried out by infrared analysis, using a IFS 66 FT-IR spectrophotometer (Perkin-Elmer Frontier, Waltham, USA) with digitization of spectra that allows to obtain electronic files of the analyzes. All samples were analyzed under the same conditions, using 64 scans, in a range from 4000–400 cm^{-1}, at a resolution of 4 cm^{-1}. The analysis was carried out with 1 mg of each particle size sample, and the majority functional groups were identified according to the Spectrometric Identification of Organic Compounds manual, and the absorbance of their bands was obtained from a local baseline corrected and normalized at 2900 cm^{-1}, between adjacent valleys [57]. Subsequently, the intensities of the spectrometric signals were materialized by calculating the highest point using the OriginLab 8.0 software (OriginLab Corporation, Northampton, MA, USA), with the aim of comparing the availability of possible biofuel compounds in the particles of interest.

4.4. Morphological Analysis of Different Particle Sizes from the Pecan Pericarp Residues

The different particle sizes of the pecan pericarp residues materials were analyzed through scanning electron microscopy (SEM). The samples were dehydrated to constant weight, at a temperature of 100 ± 3 °C and to each one, a coating with copper was applied so as not to decompose organic matter, in a Sputter Coater equipment, during a time of 15 min at 10 mAh, to obtain SEM images in a model JSM6400 scanning electron microscope (JEOL, JSM-6400 Tokyo, Japan), coupled to an X-ray spectrometer, with an electron beam acceleration potential of 20 kV at 9.

4.5. Physical Properties Analysis of the Biofuel Briquettes from Different Particle Sizes of Pecan Pericarp Residues

4.5.1. Density

The density was determined according to the UNE-EN-16127 standard [58]. Equation (1) was used:

$$\varphi = \frac{m}{V} \tag{1}$$

where φ is the density of the briquettes, m the mass of the sample, and v the volume of the sample.

The volume of the samples was determined by Equation (2):

$$V = \pi r^2 h \tag{2}$$

Measurements were made after seven days of conditioning at 20 °C and 65% relative humidity, so that the briquette stabilized and that the dilation effect did not affect the results [59].

4.5.2. Hardness

The hardness consisted of estimating and analyzing the briquette ability to resist during the storage, transport and/or compression processes. It is based on weighing each sample, dropping it three times from a height of 1.85 m, and weighing again the piece that remained larger, the calculation of this property was made as described by Kaliyan and Morey [42] with Equation (3):

$$\text{Hardness} = 100 - \frac{W_s - W_p}{W_s} \times 100 \tag{3}$$

where W_s is the weight of the briquette sample, W_p is the weight of the piece that remained larger.

4.5.3. Swelling

The swelling index was used to see the ability of agglomeration between the particles during the briquetting process. It was determined by Equation (4):

$$\text{Swelling} = \frac{W_s - W_p}{W_s} \times 100 \tag{4}$$

where W_s is the weight of the briquette sample, W_p is the weight of the piece that remained larger.

4.5.4. Impact Resistance Index (IRI)

The impact resistance test simulates the forces encountered during emptying of densified products from trucks onto ground, or from chutes into bins. The IRI was calculated as described by Richards [60] with Equation (5):

$$IRI = (100 * N)/n \tag{5}$$

where N is the number of drops, and n is the total number of pieces after N drops. Small pieces weighing less than 5% of the original weight of the logs were not included in the IRI calculation.

4.6. Statistical Analysis

Since the data resulting from the proximate analysis are percentage values, they were transformed with the square root function of the p arcsine, where p = the proportion of the dependent variable [61]. Subsequently, data normality tests were performed for each variable, using the Kolmogorov–Smirnov test. For all the results, a completely randomized experimental design was applied, evaluating 4 treatments with 7 repetitions: T1. N° 12 (1.60 mm), T2. N° 20 (0.841 mm), T3. N° 40 (0.420 mm), T4. N° 60 (0.250 mm), and the statistical package used for the analysis of the data obtained was Origin 2021b. An analysis of variance was performed to verify the significant differences between the variables evaluated, with a 95% confidence interval.

5. Conclusions

The proximate analyses carried out on the materials of different particle sizes obtained from pecan (*Carya illinoinensis*) pericarp allowed us to determine the moisture content, volatile matter, ash content, and fixed carbon, which turned out to be adequate to define these materials as good quality biofuels. Its physical transformation through briquetting significantly increased its bioenergetic potential, with calorific values going from 17.00 MJ/kg for the base material to 18.00 MJ/kg for briquettes, due to the high density reached, its hardness and strong impact resistance, giving a solid biofuel that is easy to transport, store, and handle. In this way, the particle size was illustrated as a determining factor in the quality of the briquettes, since the finer materials presented higher density (1.25 g/cm^3 for granulometry 0.42 mm from a number 40 sieve), and higher hardness (99.85 for biofuel briquettes made from number 60 sieve particle size materials) than larger materials (obtained using number 12 and 20 sieves). Likewise, a greater bioavailability of the main functional groups was registered in the finest materials, which increases the briquettes' energy characteristics. Based on the results obtained, the studied biomass can produce suitable densified biofuels.

Author Contributions: Conceptualization, L.R.S.C., L.D.-J., M.N.H. and A.L.R.H.; methodology, M.N.H., A.L.R.H., L.R.S.C., A.C.P., M.S.H. and L.F.I.P.; software, M.N.H., M.S.H., A.C.P. and L.F.I.P.; validation, L.R.S.C., L.D.-J., M.N.H., A.C.P. and R.F.P.; formal analysis, M.N.H., A.C.P., R.F.P., M.S.H. and L.F.I.P.; investigation, M.N.H., A.C.P. and M.S.H.; resources, A.L.R.H., L.R.S.C., A.C.P., M.S.H. and L.F.I.P.; data curation, M.N.H., A.L.R.H., A.C.P., L.F.I.P. and M.S.H.; writing—original draft preparation, M.N.H., A.C.P., M.S.H. and L.F.I.P.; writing—review and editing, all the authors; visualization, all the authors; supervision, L.R.S.C., L.D.-J. and M.N.H.; project administration, L.R.S.C. and M.N.H.; funding acquisition, L.R.S.C. and A.L.R.H. All authors have read and agreed to the published version of the manuscript.

Funding: This research received no external funding.

Institutional Review Board Statement: Not applicable.

Informed Consent Statement: Not applicable.

Data Availability Statement: Not applicable.

Acknowledgments: We thank Sara Catalina Garza Rodriguez, a student in the Agricultural Engineering program at the Faculty of Agronomy UANL, for her valuable contribution in the preparation of this work, within the framework of the Scientific Summer 2021 (PROVERICYT) program. We also thank Guillermo Cristian G. Martínez Ávila for his observations for the improvement of the manuscript.

Conflicts of Interest: The authors declare no conflict of interest.

References

1. Nuts, I.N.C. *Dried Fruits, Statistical Yearbook*; International Nut and Dried Fruits Press: Reus, Spain, 2018; p. 80.
2. Pinheiro do Prado, A.C.; Monalise Aragão, A.; Fett, R.; Block, J.M. Antioxidant properties of pecan nut [Carya illinoinensis (Wangenh.) C. Koch] shell infusion. *Grasas Aceites* **2009**, *60*, 330–335. [CrossRef]
3. Stafne, E.T.; Rohla, C.T.; Carroll, B.L. Pecan shell mulch impact on "loring" peach tree establishment and first harvest. *Horttechnology* **2009**, *19*, 775–780. [CrossRef]
4. Do Prado, A.C.P.; Manion, B.A.; Seetharaman, K.; Deschamps, F.C.; Barrera, A.D.; Block, J.M. Relationship between antioxidant properties and chemical composition of the oil and the shell of pecan nuts [Carya illinoinensis (Wangenh) C. Koch]. *Ind. Crop. Prod.* **2013**, *45*, 64–73. [CrossRef]
5. Agustin-Salazar, S.; Cerruti, P.; Medina-Juárez, L.Á.; Scarinzi, G.; Malinconico, M.; Soto-Valdez, H.; Gamez-Meza, N. Lignin and holocellulose from pecan nutshell as reinforcing fillers in poly (lactic acid) biocomposites. *Int. J. Biol. Macromol.* **2018**, *115*, 727–736. [CrossRef] [PubMed]
6. Paist, A.; Kask, Ü.; Kask, L.; Vrager, A.; Muiste, P.; Padari, A.; Pärn, L. Potential of biomass fuels to substitute for oil shale in energy balance in Estonian energy sector. *Oil Shale* **2005**, *22*, 369–380.
7. Pallavi, H.V.; Srikantaswamy, S.; Kiran, B.M.; Vyshnavi, D.R.; Ashwin, C.A. Briquetting Agricultural Waste as an Energy Source. *J. Environ. Sci. Comput. Sci. Eng. Technol.* **2013**, *2*, 160–172.
8. Proto, A.R.; Palma, A.; Paris, E.; Papandrea, S.F.; Vincenti, B.; Carnevale, M.; Guerriero, E.; Bonofiglio, R.; Gallucci, F. Assessment of wood chip combustion and emission behavior of different agricultural biomasses. *Fuel* **2021**, *289*, 119758. [CrossRef]
9. Kaur, A.; Roy, M.; Kundu, K. Densification of biomass by briquetting: A review. *Int. J. Recent Sci. Res.* **2017**, *8*, 20561–20568.
10. Li, Y.; Liu, H. High-pressure Densification of Wood Residues to Form an Upgraded Fuel. *Biomass Bioenergy* **2000**, *19*, 177–186. [CrossRef]
11. Mitchual, S.J.; Frimpong-Mensah, K.; Darkwa, N.A. Effect of species, particle size and compacting pressure on relaxed density and compressive strength of fuel briquettes. *Int. J. Energy. Environ. Eng.* **2013**, *4*, 30. [CrossRef]
12. Saptoadi, H. The Best Biobriquette Dimension and its Particle Size. *Asian J. Energy Environ.* **2008**, *9*, 161–175.
13. Križan, P. Research of factors influence on quality of wood briquettes. *J. Acta Montan. Slovaca* **2007**, *12*, 223–230.
14. Jenkins, R.W.; Moore, C.M.; Semelsberger, T.A.; Chuck, C.J.; Gordon, J.C.; Sutton, A.D. The Effect of Functional Groups in Bio-Derived Fuel Candidates. *Chem. Sus. Chem.* **2016**, *9*, 922–931. [CrossRef] [PubMed]
15. Hoppe, F.; Benedikt, H.; Thewes, M.; Kremer, F.; Pischinger, S.; Dahmen, M.; Hechinger, M.; Marquardt, W. Tailor-made fuels for future engine concepts. *Int. J. Engine Res.* **2016**, *17*, 16–27. [CrossRef]
16. Sendzikiene, E.; Makareviciene, V.; Janulis, P. Influence of fuel oxygen content on diesel engine exhaust emissions. *Renew. Energ.* **2006**, *31*, 2505–2512. [CrossRef]
17. Antal, M.; Allen, S.; Dai, X.; Shimizu, B.; Tam, M.; Gronli, M. Attainment of the theoretical yield of carbon from biomass. *Ind. Eng. Chem. Res.* **2000**, *39*, 4024–4031. [CrossRef]
18. De-La-Rosa, L.A.; Alvarez-Parrilla, E.; Shahidi, F. Phenolic compounds and antioxidant activity of kernels and shells of Mexican pecan (*Carya illinoinensis*). *J. Agric. Food Chem.* **2011**, *59*, 152–162. [CrossRef]
19. Mian, I.H.; Rodríguez-Kábana, R. Organic amendments with high tannin and phenolic contents for control of Meloidogyne arenaria in infested soil. *Nematropica* **1982**, *12*, 221–234.
20. Hernández-Montoya, V.; Mendoza-Castillo, D.; Bonilla-Petriciolet, A.; Montes-Morán, M.; Pérez-Cruz, M. Role of the pericarp of *Carya illinoinensis* as biosorbent and as precursor of activated carbon for the removal of lead and acid blue 25 in aqueous solutions. *J. Anal. Appl. Pyrolysis* **2011**, *92*, 143–151. [CrossRef]
21. Petterson, R.C. The chemical composition of wood. Advances in Chemistry Series 207. *Chem. Solid Wood* **1984**, *207*, 57–126. [CrossRef]
22. Mumbach, G.D.; Alves, J.L.F.; da Silva, J.C.G.; Di Domenico, M.; Arias, S.; Pacheco, J.G.A.; Marangoni, C.; Machado, R.A.F.; Bolzan, A. Prospecting pecan nutshell pyrolysis as a source of bioenergy and bio-based chemicals using multicomponent kinetic modeling, thermodynamic parameters estimation, and Py-GC/MS analysis. *Renew. Sust. Energ. Rev.* **2022**, *153*, 111753. [CrossRef]
23. Loredo-Medrano, J.A.; Bustos-Martínez, D.; Rivera-De-la-Rosa, J.; Carrillo-Pedraza, E.S.; Flores-Escamilla, G.A.; Ciuta, S. Particle pyrolysis modeling and thermal characterization of pecan nutshell. *J. Therm. Anal. Calorim.* **2016**, *126*, 969–979. [CrossRef]
24. Aldana, H.; Lozano, F.J.; Acevedo, J.; Mendoza, A. Thermogravimetric characterization and gasification of pecan nut shells. *Bioresour. Technol.* **2015**, *198*, 634–641. [CrossRef] [PubMed]
25. *EN 14774-1*; Solid Biofuels—Determination of Moisture Content—Oven Dry Method—Part 1: Total Moisture—Reference method. EN: Stockholm, Sweden, 2010.

26. *UNE-EN 15148*; Solid Biofuels—Determination of the Content of Volatile Matter. Asociación Española de Normalización y Certificación (AENOR): Madrid, Spain, 2010.

27. *UNE-EN 14775*; Solid Biofuels—Determination of Ash Content. Asociación Española de Normalización y Certificación (AENOR): Madrid, Spain, 2010.

28. *EN 14918*; Solid Biofuels—Determination of Calorific Value. EN: London, UK, 2009; 1–52.

29. Stolarski, M.J.; Szczukowski, S.; Tworkowski, J.; Krzyżaniak, M.; Gulczyński, P.; Mleczek, M. Comparison of quality and production cost of briquettes made from agricultural and forest origin biomass. *Renew. Energy* **2013**, *57*, 20–26. [CrossRef]

30. Husain, Z.; Zainac, Z.; Abdullah, Z. Briquetting of palm fibre and shell from the processing of palm nuts to palm oil. *Biomass Bioenergy* **2002**, *22*, 505–509. [CrossRef]

31. Chen, L.; Xing, L.; Han, L. Renewable energy from agro-residues in China: Solid biofuels and biomass briquetting technology. *Renew. Sust. Energ. Rev.* **2009**, *13*, 2689–2695. [CrossRef]

32. Felfli, F.F.; Rocha, J.D.; Filippetto, D.; Luengo, C.A.; Pippo, W.A. Biomass briquetting and its perspectives in Brazil. *Biomass Bioenergy* **2011**, *35*, 236–242. [CrossRef]

33. Barroso, T.S.; Elaboración de Pellets a Partir de Cáscara de Pecana Como Combustible Bioenergético. Thesis to Obtain the Professional Title of Environmental Engineer. Universidad César Vallejo, Peru. 2018. Available online: https://hdl.handle.net/20.500.12692/24713 (accessed on 25 September 2021).

34. Demirbaş, A. Sustainable cofiring of biomass with coal. *Energy Convers. Manag.* **2003**, *44*, 1465–1479. [CrossRef]

35. Ngangyo-Heya, M.; Rahim, F.P.; Carrillo-Parra, A.; Maiti, R.; Salas-Cruz, L.R. Timber-yielding plants of the Tamaulipan thorn scrub: Forest, fodder, and bioenergy potential. In *Biology, Productivity and Bioenergy of Timber-Yielding Plants*; Springer: Cham, Switzerland, 2017; pp. 1–119. [CrossRef]

36. Shekar, H.S.S.; Ramachandra, M. Green composites: A review. *Mater. Today Proc.* **2018**, *5*, 2518–2526. [CrossRef]

37. Spiridon, I.; Darie-Nita, R.N.; Hitruc, G.E.; Ludwiczak, J.; Cianga-Spiridon, I.A.; Niculaua, M. New opportunities to valorize biomass wastes into green materials. *J. Clean Prod.* **2016**, *133*, 235–242. [CrossRef]

38. De Beeck, B.O.; Dusselier, M.; Geboers, J.; Holsbeek, J.; Morr, E.; Oswald, S.; Giebeler, L.; Sels, B.F. Direct catalytic conversion of cellulose to liquid straight-chain alkanes. *Energy Environ. Sci.* **2015**, *8*, 230–240. [CrossRef]

39. Lehtikangas, P. Quality properties of pelletized sawdust, logging residues and bark. *Biomass Bioenergy* **2001**, *20*, 351–360. [CrossRef]

40. Tirado-Jijón, P.A. Estudio de Compactación de la Cáscara de Nuez Para Mejorar la Calidad de Briquetas de Biomasa. Bachelor's Thesis, Universidad Técnica de Ambato, Ambato, Ecuador, 2015. Available online: http://repositorio.uta.edu.ec/jspui/handle/123456789/10366 (accessed on 11 October 2021).

41. Jha, P.; Yadav, P. Briquetting of Saw Dust. *Appl. Mech. Mater.* **2012**, *110*, 1758–1761. [CrossRef]

42. Kaliyan, N.; Morey, R.V. Factors affecting strength and durability of densified biomass products. *Biomass Bioenergy* **2009**, *33*, 337–359. [CrossRef]

43. Karunanithy, C.; Wang, Y.; Muthukumarappan, K.; Pugalendhi, S. Physiochemical characterization of briquettes made from different feedstocks. *Biotechnol. Res. Int.* **2012**, *2012*, 165202. [CrossRef]

44. MacBain, R. *Pelleting Animal Feed*; American Feed Manufacturing Association: Chicago, IL, USA, 1966.

45. Mani, S.; Tabil, L.G.; Sokhansanj, S. Effects of Compressive Force, Particle Size and Moisture Content on Mechanical Properties of Biomass Pellets from Grasses. *Biomass Bioenergy* **2006**, *30*, 648–654. [CrossRef]

46. Brunerová, A.; Brožek, M. Optimal feedstock particle size and its influence on final briquette quality. In Proceedings of the 6th International Conference on Trends in Agricultural Engineering, Prague, Czech Republic, 7–9 September 2016; pp. 7–9.

47. Grover, P.D.; Mishra, S.K. *Biomass Briquetting: Technology and practices. Regional Wood Energy Development Programme in Asia GCP*; RAS/154/NET; Food and Agriculture Organization of the United Nations: Bangkok, Thailand, 1996.

48. Gilbert, P.; Ryu, C.; Sharif, V.; Switchenbank, J. Effect of processing parameters on pelletisation of herbaceous crops. *Fuel* **2009**, *88*, 1491–1497. [CrossRef]

49. Young, P.; Kennas, S. Feasibility and Impact Assessment of a Proposed Project to Briquette Municipal Solid Waste for Use as a Cooking Fuel in Rwanda. *Intermed. Technol. Consult.* **2003**, *1*, 1–59.

50. Emerhi, E.A. Physical and combustion properties of briquettes produced from sawdust of three hardwood species and different organic binders. *Adv. Appl. Sci. Res.* **2011**, *2*, 236–246.

51. Tumuluru, J.S.; Tabil, L.G.; Song, Y.; Iroba, K.L.; Meda, V. Impact of process conditions on the density and durability of wheat, oat, canola, and barley straw briquettes. *Bioenerg. Res.* **2015**, *8*, 388–401. [CrossRef]

52. *UNE-EN 17827-2*; Biocombustibles Sólidos. Determinación de la Distribución de Tamaño de Partícula para Combustibles sin Comprimir. Parte:2 Método del Tamiz Vibratorio con Abertura de Malla Inferior o Igual a 3.15 mm. Asociación Española de Normalización y Certificación (AENOR): Madrid, Spain, 2016.

53. Zepeda-Cepeda, C.O.; Goche-Télles, J.R.; Palacios-Mendoza, C.; Moreno-Anguiano, O.; Núñez-Retana, V.D.; Heya, M.N.; Carrillo-Parra, A. Effect of sawdust particle size on physical, mechanical, and energetic properties of pinus durangensis briquettes. *Appl. Sci.* **2021**, *11*, 3805. [CrossRef]

54. Ramírez-Ramírez, A.; Carrillo-Parra, A.; Ruíz-Aquino, F.; Hernández-Solís, J.J.; Pintor-Ibarra, L.F.; González-Ortega, N.; Orihuela-Equihua, R.; Carrillo-Ávila, N.; Rutiaga-Quiñones, J.G. Evaluation of Selected Physical and Thermal Properties of Briquette Hardwood Biomass Biofuel. *Bioenerg. Res.* **2022**, 1–8. [CrossRef]

55. Ndudi, A.; Gbabo, A. The Physical, Proximate and Ultimate Analysis of Rice Husk Briquettes Produced from a Vibratory Block Mould Briquetting Machine. *Int. J. Innov. Sci. Eng. Technol.* **2015**, *2*, 814–822. [CrossRef]
56. Cordedo, T.; Marquez, F.; Rodriguez-Mirasol, J.; Rodriguez, J.J. Predicting heating values of lignocellulosics and carbonaceous materials from proximate analysis. *Fuel* **2001**, *80*, 1567–1571. [CrossRef]
57. Carrillo, I.; Mendonça, R.T.; Ago, M.; Rojas, O.J. Comparative study of cellulosic components isolated from different Eucalyptus species. *Cellulose* **2018**, *25*, 1011.e29. [CrossRef]
58. *UNE-EN-16127*; Solid Biofuels—Determination of Length and Diameter of Pellets. European Committee for Standardization: Brussels, Belgium, 2012.
59. Križan, P.; Matú, M.; Šooš, Ľ.; Beniak, J. Behavior of beech sawdust during densification into a solid biofuel. *Energies* **2015**, *8*, 6382–6398. [CrossRef]
60. Richards, S.R. Physical testing of fuel briquettes. *Fuel* **1990**, *25*, 89–100. [CrossRef]
61. Schefler, W.C. *Bioestadística*; No. 310.9/S416; Fondo Educativo Interamericano: Mexico City, Mexico, 1981.

Article

Use of Onion Waste as Fuel for the Generation of Bioelectricity

Rojas-Flores Segundo [1],*, Magaly De La Cruz-Noriega [1], Nélida Milly Otiniano [1], Santiago M. Benites [2], Mario Esparza [3] and Renny Nazario-Naveda [4]

[1] Instituto de Investigación en Ciencias y Tecnología de la Universidad Cesar Vallejo, Trujillo 13001, Peru; mdelacruzn@ucv.edu.pe (M.D.L.C.-N.); notiniano@ucv.edu.pe (N.M.O.)

[2] Vicerrectorado de Investigación, Universidad Autónoma del Perú, Lima 15842, Peru; santiago.benites@autonoma.pe

[3] Laboratorio Generbim (Genetica, Reproduccion y Biologia Molecular), Escuela de Medicina Humana, Facultad de Medicina Humana, Universidad Privada Antenor Orrego, Trujillo 13001, Peru; mrodrigount@yahoo.com

[4] Grupo de Investigación en Ciencias Aplicadas y Nuevas Tecnologías, Universidad Privada del Norte, Trujillo 13007, Peru; renny.nazario@upn.edu.pe

* Correspondence: segundo.rojas.89@gmail.com

Abstract: The enormous environmental problems that arise from organic waste have increased due to the significant population increase worldwide. Microbial fuel cells provide a novel solution for the use of waste as fuel for electricity generation. In this investigation, onion waste was used, and managed to generate maximum peaks of 4.459 ± 0.0608 mA and 0.991 ± 0.02 V of current and voltage, respectively. The conductivity values increased rapidly to $179{,}987 \pm 2859$ mS/cm, while the optimal pH in which the most significant current was generated was 6968 ± 0.286, and the $^\circ$ Brix values decreased rapidly due to the degradation of organic matter. The microbial fuel cells showed a low internal resistance ($154{,}389 \pm 5228\ \Omega$), with a power density of 595.69 ± 15.05 mW/cm^2 at a current density of 6.02 A/cm^2; these values are higher than those reported by other authors in the literature. The diffractogram spectra of the onion debris from FTIR show a decrease in the most intense peaks, compared to the initial ones with the final ones. It was possible to identify the species *Pseudomona eruginosa*, *Acinetobacter bereziniae*, *Stenotrophomonas maltophilia*, and *Yarrowia lipolytica* adhered to the anode electrode at the end of the monitoring using the molecular technique.

Keywords: organic waste; generation; electricity; onion; microbial fuel cells

Citation: Segundo, R.-F.; De La Cruz-Noriega, M.; Milly Otiniano, N.; Benites, S.M.; Esparza, M.; Nazario-Naveda, R. Use of Onion Waste as Fuel for the Generation of Bioelectricity. *Molecules* **2022**, *27*, 625. https://doi.org/10.3390/molecules27030625

Academic Editors: Mohamad Nasir Mohamad Ibrahim, Patricia Graciela Vázquez and Mohd Hazwan Hussin

Received: 24 November 2021
Accepted: 14 January 2022
Published: 19 January 2022

Publisher's Note: MDPI stays neutral with regard to jurisdictional claims in published maps and institutional affiliations.

1. Introduction

In recent decades, the rapid growth of human society has brought with it an increase in pollution, in which organic waste due to the lack of solid waste collection centers is dumped around supply centers [1,2]. This situation causes discomfort to homes near these markets due to bad smells, as well as birds and rodents originating from the extensive duration of accumulation [3]. According to the World Bank, people will produce 2.2 trillion tons of solid waste in 2025 [4], of which approximately 33% will be waste that will not be managed safely [5]. According to Szulc et al. (2021), each person produces an average of 0.74 kg per day, but it can vary from 0.11 to 4.54 kg due to people's diet [6]. Considering that some regions are developing rapidly, this will double or triple their waste production [7]. Due to the large amounts of waste, many researchers have looked for innovative ways to use it; for example, waste has been used in fertilizers [8], biopolymers [9], biogas generation [10], and bioelectricity [11]. In this sense, the use of organic waste to generate electricity through microbial fuel cells is being intensively investigated, because it can be used as a fuel to generate electricity and, once depleted, it can further be used for sowing, since the waste is still rich in minerals [12]. Microbial fuel cells (MFCs) are bioelectrochemical devices in which the chemical energy of the substrates is converted into electrical energy. There are many cell types, although in general the cell is connected to an external circuit; the composition of the cell consists of an anodic and cathodic chamber separated by a proton exchange membrane. The anodic chamber is customarily in anaerobiosis, and the cathodic chamber is in contact with oxygen [13–15]. In this sense, Cecconet et al. (2018) used agro-industrial waste in their cells, which were made with graphite electrodes, and managed to generate voltage peaks of 637 mV with an external resistance of 20.4 Ω on day 51. The excellent production of electricity obtained was attributed to colonization of the species of electroactive bacteria on the anode electrode [16]. In the same way, Asefi et al. (2019) used food waste as substrate and carbon felt as electrodes, and managed to generate voltage peaks of approximately 775 ± 2 mV and 422 mW/m^2, with the decline in values being attributed to the lack of nutrients in the last days of monitoring [17]. Although vegetable residues are difficult to use for the generation of electrical currents, Fogg et al. (2015) were one of the first to report the use of plants in MFCs for electricity generation, pointing out their great potential for use as a substrate in a cell due to their possessing many active redox mediators [18]. In the same sense, Shrestha et al. (2016) managed to generate peak voltages of approximately 0.78 V and current densities of 1.504 A/cm^2 in their MFCs [19]. These results were attributed to the high concentration of carbohydrates, amino acids, and species with redox activity [20]. Due to this, vegetable residues with high redox species and carbohydrate content are potentially excellent candidates for generating electrical current in MFCs when used as substrates. One of the most important agricultural products in the world is onion (*Allium cepa* L.), with 392,536 tons being produced in 2018, and its production continues to grow [21]. This product is a rich source of polyphenol dietary fibres and antioxidants, and onion skin is a waste that contains a high concentration of flavonoids compared to the edible part [22]. Flavanols have a higher concentration (280–400 mg/kg) in onions than in other vegetables; for example, broccoli has a concentration of 100 mg/kg and apple 50 mg/kg; thus, the onion contains active antimicrobial agents [23]. Due to the incredible popularity of this vegetable, large amounts of waste are being generated, which is used to flavour food products, especially dairy products such as flavoured cheese, sour cream, or meat products such as processed hams, cold cuts, and meat cans [24,25].

Worldwide, approximately 66 million tons of onions are produced each year. It is expected that in 2025, an increase of 88.6–92.8% of waste generated by this vegetable will occur, with the implementation of strategies to minimize contamination due to this product being of vital importance [26]. The accelerated increase in waste affects the environment, emphasising the need to design adequate strategies to minimize the social and environmental impacts on future generations [27]. In general, a 70% increase in the generation of urban solid waste is expected by 2050, which will occur if there are no changes in consumer habits in the process of disposing waste. The linear economy model follows a step-by-step purchase, manufacturing, and disposal scheme; this means raw materials are collected and used until they are finally discarded. This creates a domino effect where it results in a shortage of raw materials and increases in cost due to the handling and disposal of waste [28,29]. The circular economy and its acceptance have a direct effect on the fight against waste, since it uses waste as a product, contributing to the minimization of pollution [30]. To achieve positive changes with respect to the conservation of the environment, the use of innovative technologies for solid waste management must be counted on, taking the circular economy as the main focus, since it is framed around the valorisation of waste as a resource [31]. Countries and their citizens have the responsibility of producing structural changes in policies and plans related to the sustainable development of solid waste, achieving adequate treatment, management, utility, and final disposal through the circular economic model [32].

The main objective of this research is to generate electricity using onion waste (monitoring the physical-chemical parameters) and to identify the main microorganisms in anodic electrodes that generate such energy. In order to provide a clear picture, the voltage, current, pH, brix degrees, and conductivity were monitored for 35 days. The internal resistance, power density, current density, and the initial and final transmittance spectrum of the substrates were characterized by Fourier transform infrared spectroscopy (FTIR).

Molecular techniques identified electrogenic bacteria attached to the anode. This research provides a solution by which the waste originating from this vegetable can be used as fuel and the eco-friendly generation of electricity, to the benefit of farmers and export and import companies.

2. Materials and Methods

2.1. Construction of Single-Chamber Microbial Fuel Cells

Microbial fuel cells (three in total) were created using copper (Cu) at the anode and zinc (Zn) at the cathode in the absence of a proton exchange membrane, as shown in the prototype in Figure 1. A 600 mL polymethylmethacrylate tube was used as the MFC chamber, in which a 5 cm hole was drilled at one end so that the cathode had contact with the environment (O_2). The electrodes were 78.50 cm^2 in area; both electrodes were joined by an external resistance connected with copper wire (0.2 cm in diameter).

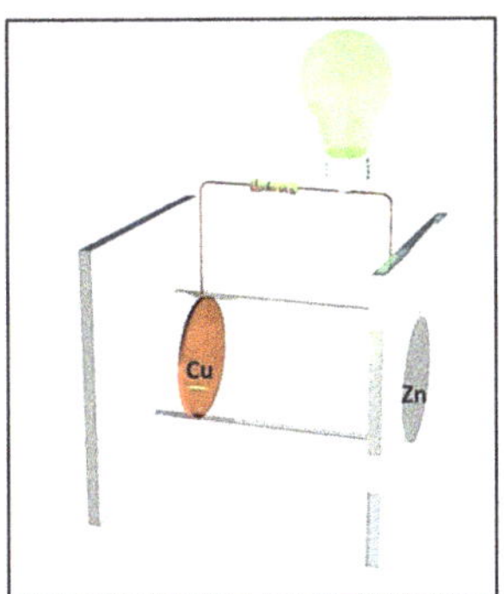

Figure 1. Scheme of the MFC prototype.

2.2. Onion Waste Collection and Preparation

An amount of 3 kg of decomposed onion was collected from the La Hermelinda Trujillo market, Peru, in sealed airtight bags and stored in a cooler. The debris was placed in plastic trays to remove any plastic, paper, or other foreign material. The samples were then washed with distilled water three times to remove dust, insects, or other impurities. Next, the samples were left to dry in an oven (Labtron, LDO-B10) for 24 h 30 ± 1.5 °C. Finally, the onion residues were placed in an extractor (Maqorito-400 rpm) to obtain 800 mL (150 mL for each MFC) of waste onion juice.

2.3. Characterization of Microbial Fuel Cells

The voltage and current values generated were monitored using a multimeter (Prasek Premium PR-85-USA) for a period of 35 days with an external resistance of 1000 Ω at 22 ± 2 °C. Current density (CD) and power density (PD) were calculated using the equations of CD = $V_{cell}^2 / R_{ext.} A$ and PD = $V_{cell} / R_{ext.} A$, where A (area) of the cathode has an approximate value of 78.50 cm^2 [33]. With external resistances (Rext.) of 0.3 (±0.1), 0.6 (±0.18), 1 (±0.3), 1.5 (±0.31), 3 (±0.6), 10 (±1.3), 20 (±6.5), 50 (±8.7), 60 (±8.2), 100 (±9.3), 120 (±9.8), 220 (±13), 240 (±15.6), 330 (±20.3), 390 (±24.5), 460 (±23.1), 531 (±26.8), 700 (±40.5) and 1000 (±50.6) Ω [16]. Changes in conductivity (CD-4301 conductivity meter), pH (110Series Oakton pH meter) and Brix degrees (RHB-32 Brix refractometer) were also measured. Transmittance values were measured by FTIR (Thermo Scientific IS50) and MFC resistance values were measured using an energy sensor (Vernier- ±30 V and ±1000 mA). The data points of the voltage, current, pH, conductivity, ° Brix, Current density (CD) and power density (PD) figures represent the average values from three replicates and the error bars represent the corresponding standard deviations.

2.4. Isolation of Electrogenic Microorganisms in Anodic Chamber

For isolation, a swab of the anode plate was made; then, it was seeded by the stria technique in culture media such as Brain Heart Infusion Agar, Nutritive Agar, Mac Conkey Agar, and Sabouraud Agar. They were incubated at 36 °C to isolate Gram-negative bacteria and 30 °C for fungi and yeasts. The isolation of microorganisms was carried out in duplicate.

2.5. Molecular Identification of Bacteria and Fungi

The Laboratory's Analysis and Research Center of "Biodes Laboratorios" carried out the molecular identification. From axenic cultures, they carried out DNA extraction using the CTAB technique. The MACROGEN Laboratory sequenced the PCR products, then analysed them by the MEGA X bioinformatics software (Molecular Evolutionary Genetics Analysis); finally, they were aligned and compared with the BLAST bioinformatics program (Basic Local Alignment Search Tool) to obtain the percentage of identity in identifying fungi and bacteria.

3. Results and Analysis

In Figure 2a, the voltage values observed during the 35 days of monitoring of the MFCs are shown, in which the successive increase of the values can be seen from the first (0.9033 ± 0.016 V) to the sixth day (0.991 ± 0.02), and then slowly decay until the last day (0.5463 ± 0.0345 V). The rapid generation of voltage in the first days is mainly due to the metallic electrode used. According to Hindatu et al. (2017), the use of an anodic electrode of this type has lower resistance and higher conductivity of the electrons generated to the cathode electrode [34]. The use of Zn and Cu as electrodes has already been studied; rapid generation of voltage on the first day was observed due to the current chemical present in the substrate, and because there was not enough time to adhere the microorganisms on the electrode [35,36]. However, as the days go by, the microorganisms exhibit increasing anodic behaviour due to the decomposition of the substrate [37]. At the same time, the voltage variations are generated by the substrate used, which contains components that affect the growth of microorganisms at the time of degradation [38]. Figure 2b shows the generated values of electric current during the 35 days of monitoring. It can be seen that, after the first day (3.006 ± 0.0837 mA), the current values increase to their maximum peak on the seventh day (4459 ± 0.0608 mA), following which a decrease is observed until the last day (10806 ± 0.0540 mA). The work carried out by Din et al. (2020), in which they used potato waste, mentions that the decrease in current values is due to the rapid hydrolysis of organic matter (substrate) due to the variation of the appropriate pH values [39].

On the other hand, glucose consumption by microorganisms under anaerobic conditions produces carbon dioxide, protons, and electrons in the anode chamber, which contributes to the increase in current values [40]. Likewise, the possible low resistance of the system (substrate, electrodes, and external circuit) contributes to making the flow of electrons more feasible, and the high current values during the monitoring period confirms the presence of an exoelectrogenic biofilm (community bacterially or electrochemically active) on the anode electrode [41,42]. These preliminary results on the use of onion waste for the generation of bioelectricity show the way for more research due to the excellent current and voltage results obtained, although as mentioned by Rahman et al. (2021), one of the essential points is the study of the availability of glucose by the substrates, which is what limits the release of electrons and leads to a decrease in electrical current [43].

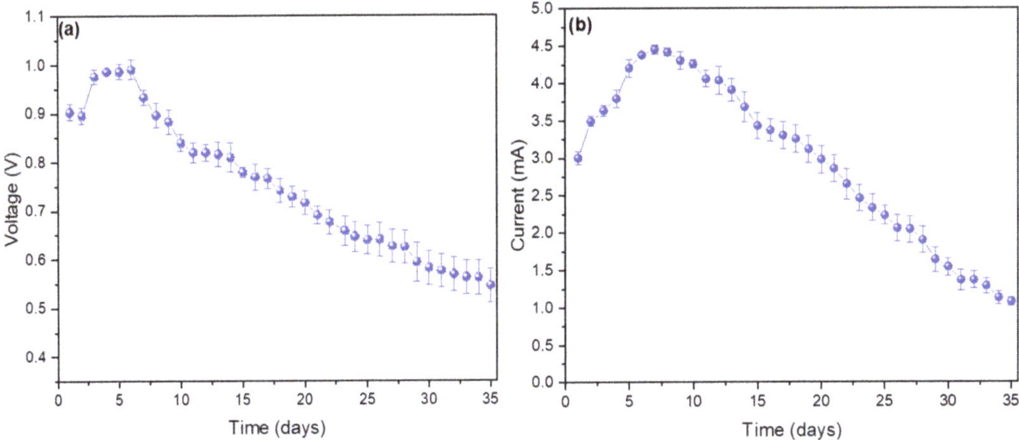

Figure 2. Monitoring of the values of (**a**) voltage and (**b**) current of the microbial fuel cells.

The maximum value of conductivity is obtained on the fifth day (179,987 ± 2859 mS/cm), as can be seen in Figure 3a, and conductivity then decreases continuously until the last day (28,667 ± 6110 mS/cm). The variations in conductivity are due to the creation and formation of sediments from the waste used during the electricity generation process [44]. While the increase in conductivity is mainly due to the reduction of the resistance of organic waste used as fuel, these values can be improved by adding inorganic salts to the substrate of the anode chamber [45,46]. In Figure 3b, it can be observed that the pH values vary from slightly acidic (3.77 ± 0.036) to neutral (6.968 ± 0.286), with the optimum pH for the generation of voltage and current being 4.35 ± 0.201. According to Geng et al. (2020), an MFC operating at neutral pH decreases the electrical parameters due to the competition between methanogens and electrigens contained in the substrate [47]. However, the optimal operation for each substrate and cell type varies due to the different operating parameters in electricity production [48]. For example, Ren et al. (2018) investigated the performance of their cells by adjusting the pH, managing to obtain the optimal pH for a value of 9, because the microorganisms present in their substrate obtained the appropriate conditions for their growth [49]. Figure 3c shows the ° Brix values exhibited by the cells during monitoring, which in the first days have a value of 4 and then drop to zero by the twelfth day, a value which is maintained until the last day of operation of the cells. Clark et al. 2018 [50] determined the presence of soluble solids (0.997, 0.1 ° Brix), pyruvate (0.825, 0.8 µmol g-1 FW), fructan (0.98, 1.9 mg g-1 FW), glucose (0.941, 1.1 mg g-1 FW), fructose (0.967, 1.0 mg g-1 FW) and sucrose (0.919, 1.7 mg g-1 FW) in onion crops by FTIR, as they are carbon sources for the growth of microorganisms.

Figure 4a shows the internal resistance values (Rint.) The resistance values of the microbial fuel cells during the 35 days (50,400 min) of monitoring did not show major fluctuations at 22 ± 2 °C, and the average value of Rint. Was 154,389 ± 5228 Ω. Previous studies show that low internal resistance is mainly due to good biofilm formation on the anode electrode due to electrogenic microorganisms present in the substrate [51,52]. In the same way, by containing a low Rint., electron transfer will occur more efficiently from anode to cathode; thus, microorganisms may prefer more direct ways of electron transfer due to their genetics [37]. Figure 4b shows the values of power density (PD) and voltage as a function of current density (CD), managing to generate a $PD_{max.}$ of 595.69 ± 15.05 mW/cm^2 in a CD 6.02 A/cm^2 and with a maximum voltage of 871.92 ± 7.9 mV. The PD and CD values obtained in this investigation were higher compared to those obtained with other substrates (*H. undatus*, *M. citrifolia* and *R. ulmifolius*) [53], using the same design and materials, which may be due to greater ease of degradation of the grape substrate [54]. In the same sense, in the work carried out by Yang et al. (2019), the values of CD and PD using sediments as substrate are lower than those obtained by us, thus demonstrating the great potential of fruit residues for the generation of electricity [55]. Although the CD and PD values can still be increased using a proton exchange membrane (MIP) between the anodic and cathode chamber, as demonstrated by Asensio et al. (2018) in his work, in which he used wastewater and different types of IPM as a substrate (Nafion-117 HRT3.16, Nafion-117 HRT 6.32d, Neosepta CMX, and Neosepta AMX), achieving a higher current density (~850 mA/m^2) using the NAFION-117 HRT 3.16 as MIP [56].

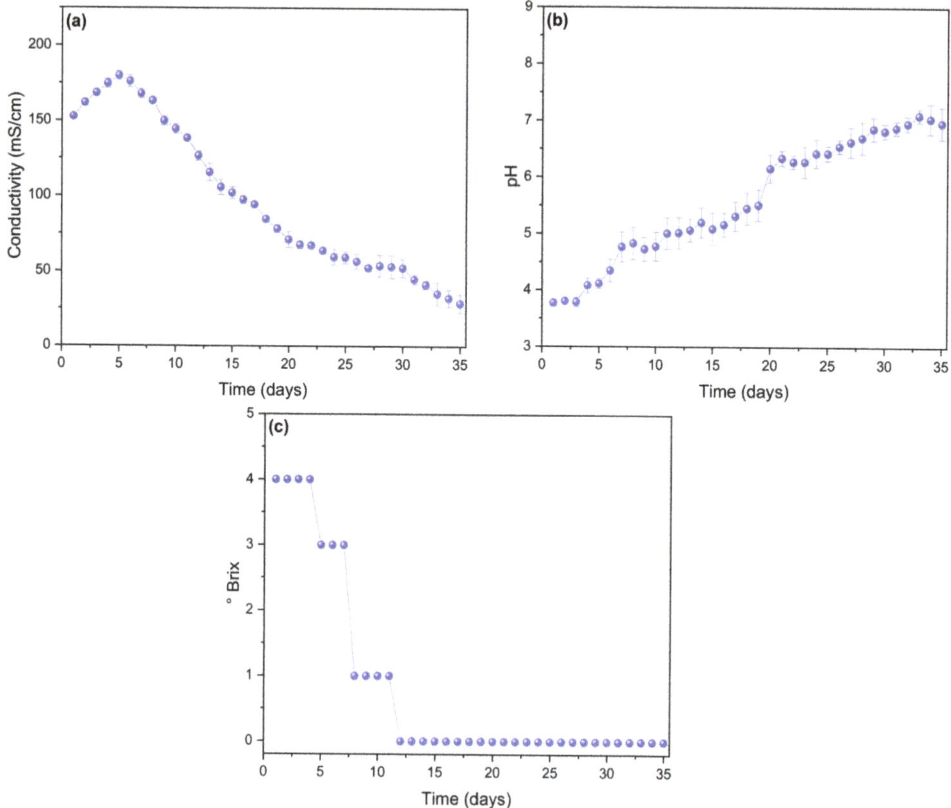

Figure 3. Values of (**a**) conductivity, (**b**) pH, and (**c**) ° Brix of the microbial fuel cells.

Figure 5 shows the values of the transmittance spectrum by FTIR of the onion waste at the initial and final time of operation in the microbial fuel cells. The 3291 cm^{-1} peak belongs to the hydroxyl group (range 3200–3300 cm^{-1}), the stretches close to 2929 cm^{-1} belong to the methylene-CH, the 1630 cm^{-1} peak belongs to the quinone or conjugated ketone (C=C, stretch), the peaks between 1300–1400 cm^{-1} belong to OH bonds and the 1027 cm^{-1} peak to -CC-stretch, ethers [57–59]. The decrease in the intensity of the transmittance peaks is mainly due to the degradation of the compounds in the electrical energy generation process [60].

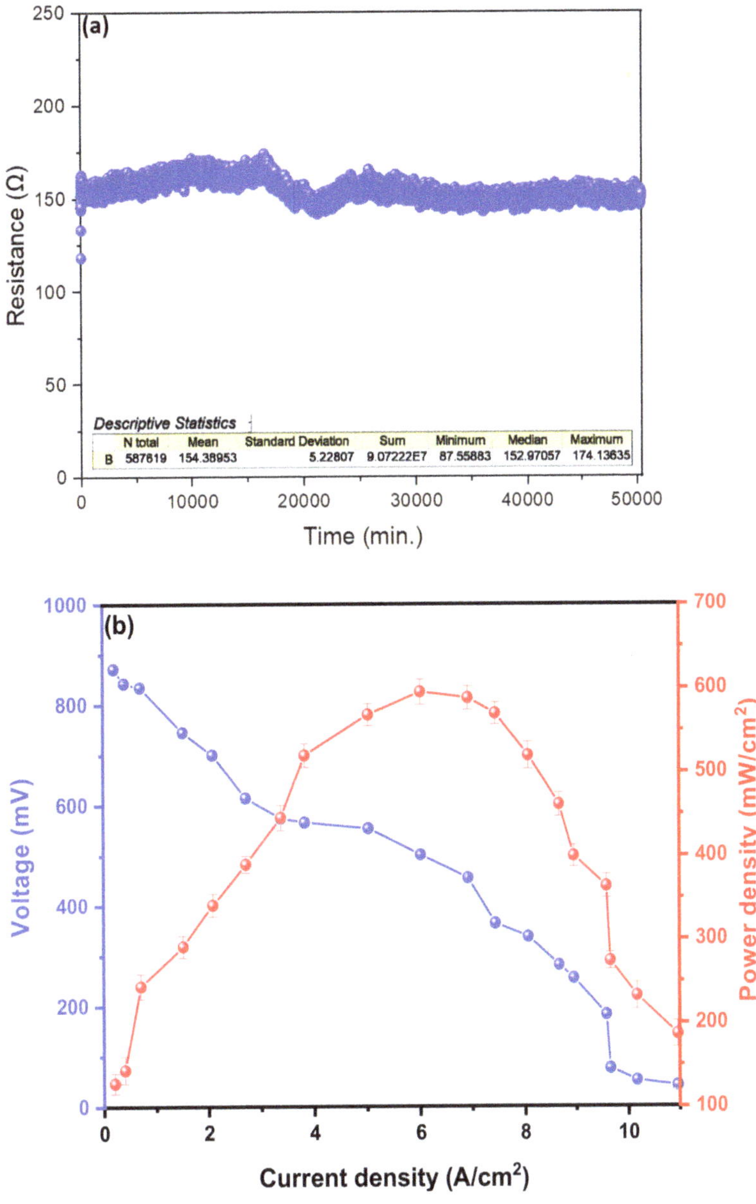

Figure 4. Characterization of (**a**) internal resistance and (**b**) power and voltage density about the current density of the MFCs.

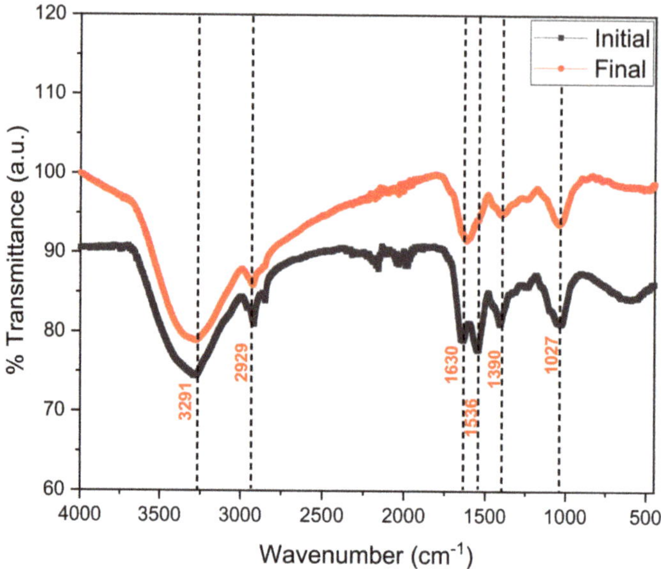

Figure 5. FTIR spectrophotometry of the initial and final onion residues.

The Analysis and Research Center of the "Biodes Laboratorios" laboratory carried out the molecular identification. The genetic material (DNA) was extracted from pure or axenic cultures of bacteria and fungi, isolated from anode plates with onion substrate, using the CTAB extraction method [61]. The PCR products for the 16S rDNA gene (bacteria) and the ITF region (yeast) of each isolate were sequenced in the Macrogen laboratory (USA) [62]. Then the relationship of the sequences was evaluated in the bioinformatics Software MEGA X (Molecular Evolutionary Genetics Analysis) to be later aligned and compared with other sequences in the bioinformatics program BLAST (Basic Local Alignment Search Tool), through which the percentage of identity was obtained for the identification of bacteria and fungi.

A BLAST characterization of the rDNA sequence of the bacteria and yeast isolated from the anode plate of the onion microbial fuel cells was performed, where a 100% identity percentage was obtained corresponding to the Pseudomona eruginosa species, 99.93% to the Acinetobacter bereziniae species, 100%, to the Stenotrophomonas maltophilia species (see Table 1), and 100% to the Yarrowia lipolytica species (see Table 2).

The phylogenetic tree was constructed using the Maximum Likelihood method with Bootstrap phylogeny test with 100 replicates to show differences in general phylogenetic distances. A BLAST characterization of the rDNA sequence of the bacteria isolated from the anode plate of the fuel cells was performed (Figure 6). Onion microbial bacteria, identified as *Pseudomonas aeruginosa* species, is a Gram-negative, facultative aerobic bacterium [62], which can use carbon and nitrogen sources, obtaining energy from the oxidation of sugars; this species is persistent in the environment [63]. Likewise, it is worth mentioning that this species possesses electron mediators such as phenazine-1-carboxylic acid, pyocyanin, pyoverdine, among others, which allows it to survive in anaerobic conditions [64,65].

Table 1. BLAST characterization of the rDNA sequence of bacteria isolated from the MFC anode plate with onion substrate.

BLAST Characterization	Consensus Sequence Length (nt)	% Maximum Identity	Accession Number	Phylogeny
Pseudomona aeruginosa	1442	100.00%	MT633047.1	Cellular organisms; Bacteria; Proteobacteria; Gammaproteobacteria; Pseudomonadales; Pseudomonadaceae; Pseudomonas; Pseudomonas aeruginosa group
Acinetobacter bereziniae	1468	99.93 %	CP018259.1	Cellular organisms; Bacteria; Proteobacteria; Gammaproteobacteria; Pseudomonadales; Moraxellaceae; Acinetobacter
Stenotrophomonas maltophilia	1477	100.00%	NR_041577.1	Cellular organisms; Bacteria; Proteobacteria; Gammaproteobacteria; Xanthomonadales; Xanthomonadaceae; Stenotrophomonas; Stenotrophomonas maltophilia group

Table 2. BLAST characterization of the yeast rDNA sequence isolated from the MFC anode plate with onion substrate.

Caracterización BLAST	Consensus Sequence Length (nt)	% Maximum Identity	Accession Number	Phylogeny
Yarrowia lipolytica	369	100.00%	MN124085.1	Cellular organisms; Eukaryota; Opisthokonta; Fungi; Dikarya; Ascomycota; saccharomyceta; Saccharomycotina; Saccharomycetes; Saccharomycetales; Dipodascaceae; Yarrowia

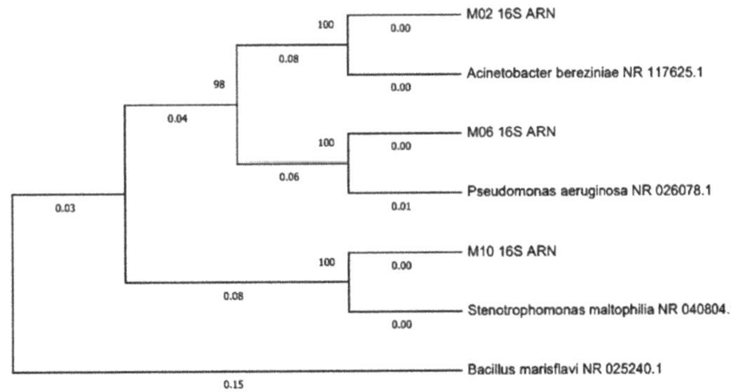

Figure 6. Dendrogram of groups of bacteria isolated from the MFC anode plate with onion substrate.

A study by Ali et al. exposed the efficiency of using the species Pseudomonas aeruginosa, which generated 136 ± 87 mW/m^2 from the use of glucose, followed by fructose and sucrose. This study observed that the cell fed with glucose presented higher bacterial adhesion [66]. On the other hand, *Acinetobacter bereziniae* was identified as a Gram-negative, aerobic, non-fermentative, oxidase-negative, and immobile organism [67]. This microorganism has been detected as a contaminant in human and animal milk and the environment. This bacterium is characterised by surface hydrophobicity, which contributes to its adherence to surfaces [68]. Likewise, this genus was proposed as a model microorganism for environmental microbiological studies, pathogenicity tests, and industrial chemical production [69].

It is worth mentioning that the species Stenotrophomonas maltophilia was also identified, a cosmopolitan and ubiquitous bacterium found in a series of environmental habitats, mainly associated with plants [70]. This bacterium can form biofilms on various surfaces [71]. These bacteria transfer electrons directly to the anode through outer membrane carrier proteins, such as cytochrome c, or membrane appendages called nanowires [72]. Romo et al. 2019 investigated the composition of the microbial community of the cathode of a Microbial Fuel Cell for the reduction of Cr (VI) in tannery effluents, finding four bacterial phyla (Proteobacteria, Actinobacteria, Fimicutes and Bacteroides) by pyrosequencing 454 of the 16S rRNA gene, with the genus Pseudomonas being the most abundant. On the other hand, they stressed the importance of the biofilms formed in the electrodes and the generation of electricity through microbial combustion cells with a salt bridge. In this way, the system's potential is demonstrated together with the microbial communities in the bioremediation processes of contaminated effluents [73]. The phylogenetic tree for yeast was built with the Maximum Likelihood method without Bootstrap phylogeny test, for which ribosomal DNA sequences based on the ITS regions of the *Yarrowia lipolytica* species were used (Figure 7). It is an ascomycete yeast with a high lipolytic and proteolytic capacity. This species has been isolated from meat, fermented dairy, sewage, and water contaminated by hydrocarbons [74]. In Santiago et al. 2020, a *Sacharomyces cereviceae* culture was used as fuel, generating a voltage of 0.761 volts and a PD_{max} and CD_{max} of 8196 mW/cm^2 and 8383 mA/cm^2, respectively, in the Zn-Cu cell, while in the Zn-Zn cell, 5684 mW/cm^2 and 0.238 mA/cm^2 of PD_{max} and CD_{max} were generated, respectively.

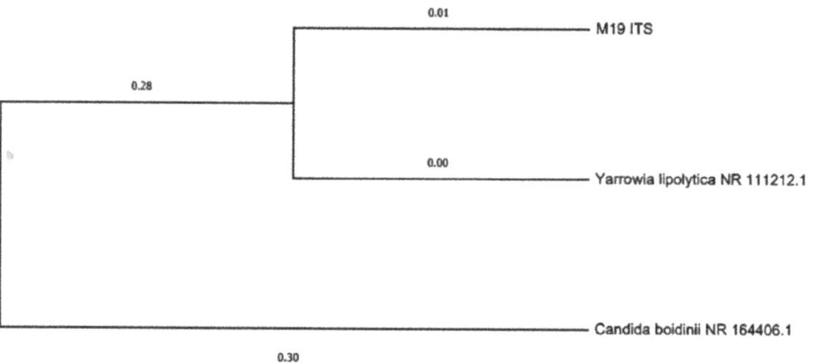

Figure 7. Dendrogram based on the ITS regions of the rDNA regions of a *Yarrowia lipolytica* culture isolated from the anode plate of the MFC with onion substrate.

4. Conclusions and Future Development

Bioelectricity was successfully generated using onion waste using low-cost microbial fuel cells manufactured with zinc and copper electrodes. The maximum voltage and current obtained during the monitoring were 0.991 ± 0.02 V and 4.459 ± 0.0608 mA, respectively, values which, compared to other reports in the literature, are higher and show suitable electrical parameters that did not decrease in their entirety by the end of the monitoring. The optimal pH in the cells was 6.968 ± 0.286, and its conductivity values increased to 179.987 ± 2.859 mS/cm, while the ° Brix rapidly dropped to zero on day 12. The internal resistance of the cells was 154.389 ± 5.228 Ω, which explains why, of its high electrical values, this low resistance value is mainly due to the compounds of onion residues, which decrease due to degradation in the bioelectricity generation process, as shown by the transmittance spectrum of the FTIR. The maximum power density was 595.69 ± 15.05 mW/cm^2 at a current density of 6.02 A/cm^2 with a maximum voltage of 871.92 ± 7.9 mV. Finally, a microbial consortium adhered to the anode was identified by molecular techniques, which obtained a percentage of identity of 100% to the *Pseudomona eruginosa* species, 99.93% to the *Acinetobacter bereziniae* species, 100% to the *Stenotrophomonas maltophilia* species, and 100% to the species *Yarrowia lipolytica*.

In the near future, researchers will face new challenges transitioning from the laboratory to real social environments such as installations of CFMs in houses, industries and food supply centres, the passivation of electrode materials by electromechanical deposition to improve the effects of corrosion of the material, and the connection of the electrodes in the MFCs. Likewise, the effects that can be given by using different nanomaterials for coating the electrodes should be investigated.

Author Contributions: Conceptualization, R.-F.S.; methodology S.M.B.; software, R.N.-N.; validation, N.M.O. formal analysis, R.-F.S. and M.D.L.C.-N.; investigation R.-F.S. data curation, M.D.L.C.-N.; writing—original draft preparation, M.E.; writing—review and editing, R.-F.S. and M.E.; project administration, R.-F.S. and R.N.-N. All authors have read and agreed to the published version of the manuscript.

Funding: This research was funded by Consejo Nacional de Ciencia, Tecnología e Innovación Tecnológica—CONCYTEC/PROCIENCIA according to Project agreement 370-2019.

Institutional Review Board Statement: Not applicable.

Informed Consent Statement: Not applicable.

Data Availability Statement: Not applicable.

Acknowledgments: The authors thank Juan Pastrana, for the help provided in the grammar of the English language.

Conflicts of Interest: The authors declare no conflict of interest.

References

1. Lott, B.E.; Okusanya, B.O.; Anderson, E.J.; Kram, N.A.; Rodriguez, M.; Thomson, C.A.; Rosales, C.; Ehiri, J.E. Interventions to increase uptake of Human Papillomavirus (HPV) vaccination in minority populations: A systematic review. *Prev. Med. Rep.* **2020**, *19*, 101163. [CrossRef] [PubMed]
2. Yaqoob, A.A.; Khatoon, A.; Mohd Setapar, S.H.; Umar, K.; Parveen, T.; Mohamad Ibrahim, M.N.; Ahmad, A.; Rafatullah, M. Outlook on the role of microbial fuel cells in remediation of environmental pollutants with electricity generation. *Catalysts* **2020**, *10*, 819. [CrossRef]
3. Ludlow, J.; Jalil-Vega, F.; Rivera, X.S.; Garrido, R.; Hawkes, A.; Staffell, I.; Balcombe, P. Organic waste to energy: Resource potential and barriers to uptake in Chile. *Sustain. Prod. Consum.* **2021**, *28*, 1522–1537. [CrossRef]
4. Nazario-Naveda, R.; Benites, S.M. Sugar Industry Waste for Bioelectricity Generation. *Environ. Res. Eng. Manag.* **2021**, *77*, 15–22.
5. Kaza, S.; Yao, L.; Bhada-Tata, P.; Van Woerden, F. *What a Waste 2.0: A Global Snapshot of Solid Waste Management to 2050*; World Bank Publications; World Bank: Washington, DC, USA, 2018.
6. Szulc, W.; Rutkowska, B.; Gawroński, S.; Wszelaczyńska, E. Possibilities of using organic waste after biological and physical processing—An overview. *Processes* **2021**, *9*, 1501. [CrossRef]
7. Moya, D.; Aldás, C.; López, G.; Kaparaju, P. Municipal solid waste as a valuable renewable energy resource: A worldwide opportunity of energy recovery by using Waste-To-Energy Technologies. *Energy Procedia* **2017**, *134*, 286–295. [CrossRef]
8. Pellejero, G.; Miglierina, A.; Aschkar, G.; Turcato, M.; Jiménez-Ballesta, R. Effects of the onion residue compost as an organic fertilizer in a vegetable culture in the Lower Valley of the Rio Negro. *Int. J. Recycl. Org. Waste Agric.* **2017**, *6*, 159–166. [CrossRef]
9. Pagliano, G.; Ventorino, V.; Panico, A.; Pepe, O. Integrated systems for biopolymers and bioenergy production from organic waste and by-products: A review of microbial processes. *Biotechnol. Biofuels* **2017**, *10*, 1–24. [CrossRef]
10. Aravindhan, A.; Lingeshwaran, N.; Goutami, K. Bio-Gas production from different organic wastes. *Mater. Today Proc.* **2021**, *47*, 5457–5461.
11. Segundo, R.F.; Renny, N.N.; Moises, G.C.; Daniel, D.N.; Natalia, D.D.; Karen, V.R. Generation of Bioelectricity from Organic Fruit Waste. *Environ. Res. Eng. Manag.* **2021**, *77*, 6–14.
12. Saba, B.; Christy, A.D. Bioelectricity Generation in Algal Microbial Fuel Cells. In *Handbook of Algal Science, Technology and Medicine*; Academic Press: Cambridge, MA, USA, 2020; pp. 377–384.
13. Jatoi, A.S.; Akhter, F.; Mazari, S.A.; Sabzoi, N.; Aziz, S.; Soomro, S.A.; Mubarak, N.M.; Baloch, H.; Memon, A.Q.; Ahmed, S. Advanced microbial fuel cell for wastewater treatment—A review. *Environ. Sci. Pollut. Res.* **2021**, *28*, 5005–5019. [CrossRef]

14. Leung, D.H.L.; Lim, Y.S.; Uma, K.; Pan, G.T.; Lin, J.H.; Chong, S.; Yang, T.C.K. Engineering *S. oneidensis* for performance improvement of microbial fuel cell—A mini-review. *Appl. Biochem. Biotechnol.* **2021**, *193*, 1170–1186. [CrossRef]
15. Raychaudhuri, A.; Behera, M. Review of the process optimization in microbial fuel cells using design of experiment methodology. *J. Hazard. Toxic Radioact. Waste* **2020**, *24*, 04020013. [CrossRef]
16. Cecconet, D.; Molognoni, D.; Callegari, A.; Capodaglio, A.G. Agro-food industry wastewater treatment with microbial fuel cells: Energetic recovery issues. *Int. J. Hydrogen Energy* **2018**, *43*, 500–511. [CrossRef]
17. de Asefi, B.; Li, S.L.; Moreno, H.A.; Sanchez-Torres, V.; Hu, A.; Li, J.; Yu, C.P. Characterization of electricity production and microbial community of food waste-fed microbial fuel cells. *Process Saf. Environ. Prot.* **2019**, *125*, 83–91. [CrossRef]
18. Fogg, A.; Gadhamshetty, V.; Franco, D.; Wilder, J.; Agapi, S.; Komisar, S. Can a microbial fuel cell resist the oxidation of Tomato pomace? *J. Power Sources* **2015**, *279*, 781–790. [CrossRef]
19. Shrestha, N.; Fogg, A.; Wilder, J.; Franco, D.; Komisar, S.; Gadhamshetty, V. Electricity generation from defective tomatoes. *Bioelectrochemistry* **2016**, *112*, 67–76. [CrossRef]
20. Yaqoob, A.A.; Ibrahim, M.N.M.; Umar, K. Biomass-derived composite anode electrode: Synthesis, characterizations, and application in microbial fuel cells (MFCs). *J. Environ. Chem. Eng.* **2021**, *9*, 106111. [CrossRef]
21. Prokopov, T.; Chonova, V.; Slavov, A.; Dessev, T.; Dimitrov, N.; Petkova, N. Effects on the quality and health-enhancing properties of industrial onion waste powder on bread. *J. Food Sci. Technol.* **2018**, *55*, 5091–5097. [CrossRef]
22. Sagar, N.A.; Pareek, S.; Gonzalez-Aguilar, G.A. Quantification of flavonoids, total phenols and antioxidant properties of onion skin: A comparative study of fifteen Indian cultivars. *J. Food Sci. Technol.* **2020**, *57*, 2423–2432. [CrossRef]
23. Liguori, L.; Califano, R.; Albanese, D.; Raimo, F.; Crescitelli, A.; Di Matteo, M. Chemical composition and antioxidant properties of five white onion (*Allium cepa* L.) landraces. *J. Food Qual.* **2017**, *2017*, 6873651. [CrossRef]
24. Bedrníček, J.; Laknerová, I.; Linhartová, Z.; Kadlec, J.; Samková, E.; Bárta, J.; Bártová, V.; Mraz, J.; Pešek, M.; Winterová, R.; et al. Onion waste as a rich source of antioxidants for meat products. *Czech J. Food Sci.* **2019**, *37*, 268–275. [CrossRef]
25. Celano, R.; Docimo, T.; Piccinelli, A.L.; Gazzerro, P.; Tucci, M.; Di Sanzo, R.; Carabetta, S.; Campone, L.; Russo, M.; Rastrelli, L. Onion peel: Turning a food waste into a resource. *Antioxidants* **2021**, *10*, 304. [CrossRef]
26. Velenturf, A.P.; Purnell, P. Principles for a sustainable circular economy. *Sustain. Prod. Consum.* **2021**, *27*, 1437–1457. [CrossRef]
27. Chorolque, A.; Pellejero, G.; Sosa, M.C.; Palacios, J.; Aschkar, G.; García-Delgado, C.; Jiménez-Ballesta, R. Biological control of soil-borne phytopathogenic fungi through onion waste composting: Implications for circular economy perspective. *Int. J. Environ. Sci. Technol.* **2021**, 1–10. [CrossRef]
28. Kumar, M.; Barbhai, M.D.; Hasan, M.; Punia, S.; Dhumal, S.; Rais, N.; Chandran, D.; Pandiselvam, R.; Kothakota, A.; Tomar, M.; et al. Onion (*Allium cepa* L.) peels: A review on bioactive compounds and biomedical activities. *Biomed. Pharmacother.* **2022**, *146*, 112498. [CrossRef]
29. Mihai, F.C.; Gündoğdu, S.; Markley, L.A.; Olivelli, A.; Khan, F.R.; Gwinnett, C.; Gutberlet, J.; Reyna-Bensusan, N.; Llanquileo-Melgarejo, P.; Meidiana, C.; et al. Plastic pollution, waste management issues, and circular economy opportunities in rural communities. *Sustainability* **2022**, *14*, 20. [CrossRef]
30. Provin, A.P.; de Aguiar Dutra, A.R. Circular economy for fashion industry: Use of waste from the food industry for the production of biotextiles. *Technol. Forecast. Soc. Change* **2021**, *169*, 120858. [CrossRef]
31. Stillitano, T.; Spada, E.; Iofrida, N.; Falcone, G.; De Luca, A.I. Sustainable agri-food processes and circular economy pathways in a life cycle perspective: State of the art of applicative research. *Sustainability* **2021**, *13*, 2472. [CrossRef]
32. Sinha, S.; Tripathi, P. Trends and challenges in valorisation of food waste in developing economies: A case study of India. *Case Stud. Chem. Environ. Eng.* **2021**, *4*, 100162. [CrossRef]
33. Rojas-Flores, S.; Noriega, M.D.L.C.; Benites, S.M.; Gonzales, G.A.; Salinas, A.S.; Palacios, F.S. Generation of bioelectricity from fruit waste. *Energy Rep.* **2020**, *6*, 37–42. [CrossRef]
34. Hindatu, Y.; Annuar, M.S.M.; Gumel, A.M. Mini review: Anode modification for improved performance of microbial fuel cell. *Renew. Sustain. Energy Rev.* **2017**, *73*, 236–248. [CrossRef]
35. Santiago, B.; Rojas-Flores, S.; De La Cruz Noriega, M.; Cabanillas-Chirinos, L.; Otiniano, N.M.; Silva-Palacios, F.; Luis, A.S. Bioelectricity from Saccharomyces cerevisiae yeast through low-cost microbial fuel cells. In Proceedings of the 18th LACCEI International Multi-Conference for Engineering, Education, and Technology: Engineering, Integration, and Alliances for a Sustainable Development, Virtual, 27–31 July 2020.
36. Rojas-Flores, S.J.; Benites, S.M.; Agüero Quiñones, R.; Enríquez-León, R.; Angelats Silva, L. Bioelectricity through microbial fuel cells from decomposed fruits using lead and copper electrodes (Bioelectricidad mediante Celdas de Combustible Microbiana a partir de frutas descompuestas usando electrodos de plomo y cobre). In Proceedings of the 18th LACCEI International Multi-Conference for Engineering, Education, and Technology: Engineering, Integration, and Alliances for a Sustainable Development, Virtual, 27–31 July 2020.
37. Rojas Flores, S.; Naveda, R.N.; Paredes, E.A.; Orbegoso, J.A.; Céspedes, T.C.; Salvatierra, A.R.; Rodríguez, M.S. Agricultural Wastes for Electricity Generation Using Microbial Fuel Cells. *Open Biotechnol. J.* **2020**, *14*, 52–58. [CrossRef]
38. Flores, S.J.R.; Benites, S.M.; Rosa, A.L.R.A.L.; Zoilita, A.L.Z.A.L.; Luis, A.S.L. The Using Lime (Citrus× aurantiifolia), Orange (Citrus× sinensis), and Tangerine (*Citrus reticulata*) Waste as a Substrate for Generating Bioelectricity: Using lime (Citrus× aurantiifolia), orange (Citrus× sinensis), and tangerine (*Citrus reticulata*) waste as a substrate for generating bioelectricity. *Environ. Res. Eng. Manag.* **2020**, *76*, 24–34.

39. Din, M.I.; Iqbal, M.; Hussain, Z.; Khalid, R. Bioelectricity generation from waste potatoes using single chambered microbial fuel cell. *Energy Sources Pt. A Recovery Util. Environ. Eff.* **2020**, 1–11. [CrossRef]
40. Moqsud, M.A.; Omine, K.; Yasufuku, N.; Hyodo, M.; Nakata, Y. Microbial fuel cell (MFC) for bioelectricity generation from organic wastes. *Waste Manag.* **2013**, *33*, 2465–2469. [CrossRef]
41. Subha, C.; Kavitha, S.; Abisheka, S.; Tamilarasan, K.; Arulazhagan, P.; Banu, J.R. Bioelectricity generation and effect studies from organic rich chocolaterie wastewater using continuous upflow anaerobic microbial fuel cell. *Fuel* **2019**, *251*, 224–232. [CrossRef]
42. Moharir, P.V.; Tembhurkar, A.R. Effect of recirculation on bioelectricity generation using microbial fuel cell with food waste leachate as substrate. *Int. J. Hydrogen Energy* **2018**, *43*, 10061–10069. [CrossRef]
43. Rahman, W.; Yusup, S.; Mohammad, S.A. Screening of fruit waste as substrate for microbial fuel cell (MFC). *AIP Conf. Proc.* **2021**, *2332*, 020003. [CrossRef]
44. Rossi, R.; Cario, B.P.; Santoro, C.; Yang, W.; Saikaly, P.E.; Logan, B.E. Evaluation of electrode and solution area-based resistances enables quantitative comparisons of factors impacting microbial fuel cell performance. *Environ. Sci. Technol.* **2019**, *53*, 3977–3986. [CrossRef]
45. Stefanova, A.; Angelov, A.; Bratkova, S.; Genova, P.; Nikolova, K. Influence of electrical conductivity and temperature in a microbial fuel cell for treatment of mining wastewater. In *Annals of the Constantin Brâncuși University of Târgu Jiu*; Letters and Social Science Series; University Constantin Brancusi of Targu Jiu: Târgu Jiu, Romania, 2018.
46. Gandu, B.; Rozenfeld, S.; Hirsch, L.O.; Schechter, A.; Cahan, R. Enhancement of Electrochemical Activity in Bioelectrochemical Systems by Using Bacterial Anodes: An Overview. *Bioelectrochem. Syst.* **2020**, *1*, 211–238.
47. Geng, Y.K.; Yuan, L.; Liu, T.; Li, Z.H.; Zheng, X.; Sheng, G.P. Thermal/alkaline pretreatment of waste activated sludge combined with a microbial fuel cell operated at alkaline pH for efficient energy recovery. *Appl. Energy* **2020**, *275*, 115291. [CrossRef]
48. Ren, Y.; Chen, J.; Li, X.; Yang, N.; Wang, X. Enhanced bioelectricity generation of air-cathode buffer-free microbial fuel cells through short-term anolyte pH adjustment. *Bioelectrochemistry* **2018**, *120*, 145–149. [CrossRef]
49. Clark, C.J.; Shaw, M.L.; Wright, K.M.; McCallum, J.A. Quantification of free sugars, fructan, pungency and sweetness indices in onion populations by FT-MIR spectroscopy. *J. Sci. Food Agric.* **2018**, *98*, 5525–5533. [CrossRef]
50. Lawson, K.; Rossi, R.; Regan, J.M.; Logan, B.E. Impact of cathodic electron acceptor on microbial fuel cell internal resistance. *Bioresour. Technol.* **2020**, *316*, 123919. [CrossRef]
51. Arkatkar, A.; Mungray, A.K.; Sharma, P. Effect of microbial growth on internal resistances in MFC: A case study. In *Innovations in Infrastructure*; Springer: Singapore, 2019; pp. 469–479.
52. Rossi, R.; Logan, B.E. Impact of external resistance acclimation on charge transfer and diffusion resistance in bench-scale microbial fuel cells. *Bioresour. Technol.* **2020**, *318*, 123921. [CrossRef]
53. Echeverría, M. Bioelectricity production with organic substrates, nitrates and lead using high Andean soils. *Innov. Res. A Driv. Force Socio-Econo-Technol. Dev.* **2020**, *1277*, 198.
54. Yang, Y.; Lin, E.; Sun, S.; Chen, H.; Chow, A.T. Direct electricity production from subaqueous wetland sediments and banana peels using membrane-less microbial fuel cells. *Ind. Crops Prod.* **2019**, *128*, 70–79. [CrossRef]
55. Asensio, Y.; Fernandez-Marchante, C.M.; Lobato, J.; Cañizares, P.; Rodrigo, M.A. Influence of the ion-exchange membrane on the performance of double-compartment microbial fuel cells. *J. Electroanal. Chem.* **2018**, *808*, 427–432. [CrossRef]
56. Verma, M.; Singh, S.S.J.; Rose, N.M. Phytochemical screening of onion skin (*Allium cepa*) dye extract. *J. Pharmacogn. Phytochem.* **2018**, *7*, 1414–1417.
57. Zzeyani, S.; Mikou, M.; Naja, J.; Bouyazza, L.; Fekkar, G.; Aiboudi, M. Assessment of the waste lubricating oils management with antioxidants vegetables extracts based resources using EPR and FTIR spectroscopy techniques. *Energy* **2019**, *180*, 206–215. [CrossRef]
58. Liu, D.; Wu, F. Biosynthesis of Pd nanoparticle using onion extract for electrochemical determination of carbendazim. *Int. J. Electrochem. Sci* **2017**, *12*, 2125–2134. [CrossRef]
59. Luo, H.; Liu, G.; Zhang, R.; Jin, S. Phenol degradation in microbial fuel cells. *Chem. Eng. J.* **2009**, *147*, 259–264. [CrossRef]
60. Gustincich, S.; Manfioletti, G.; Del Sal, G.; Schneider, C.; Carninci, P. A fast method for high-quality genomic DNA extraction from whole human blood. *Biotechniques* **1991**, *11*, 298–300.
61. Valenzuela-González, F.; Casillas-Hernández, R.; Villalpando, E.; Vargas-Albores, F. El gen ARNr 16S en el estudio de comunidades microbianas marinas (The 16S rRNA gene in the study of marine microbial communities). *Cienc. Mar.* **2015**, *41*, 297–313. [CrossRef]
62. Paz-Zarza, V.M.; Mangwani-Mordani, S.; Martínez-Maldonado, A.; Álvarez-Hernández, D.; Solano-Gálvez, S.G.; Vázquez-López, R. Pseudomonas aeruginosa: Patogenicidad y resistencia antimicrobiana en la infección urinaria (Pseudomonas aeruginosa: Pathogenicity and antimicrobial resistance in urinary tract infection). *Rev. Chilena Infectol.* **2019**, *36*, 180–189. [CrossRef]
63. Strateva, T.; Yordanov, D. Pseudomonas aeruginosa—A phenomenon of bacterial resistance. *J. Med. Microbiol.* **2009**, *58*, 1133–1148. [CrossRef]
64. Yousaf, S.; Anam, M.; Ali, N. Evaluating the production and bio-stimulating effect of 5-methyl 1, hydroxy phenazine on microbial fuel cell performance. *Int. J. Environ. Sci. Technol.* **2017**, *14*, 1439–1450. [CrossRef]
65. Raghavulu, S.V.; Modestra, J.A.; Amulya, K.; Reddy, C.N.; Venkata Mohan, S. Relative effect of bioaugmentation withelectrochemically active and nonactive bacteria on bioelec-trogenesis in microbial fuel cell. *Bioresour. Technol.* **2013**, *146*, 696703. [CrossRef]

66. Ali, N.; Anam, M.; Yousaf, S.; Maleeha, S.; Bangash, Z. Characterization of the Electric Current Generation Potential of the Pseudomonas aeruginosa Using Glucose, Fructose, and Sucrose in Double Chamber Microbial Fuel Cell. *Iran. J. Biotechnol.* **2017**, *15*, 216–223. [CrossRef]
67. Brady, M.F.; Jamal, Z.; Pervin, N. Acinetobacter. In *StatPearls*; StatPearls Publishing: Treasure Island, FL, USA, 2021.
68. Reyes, S.M.; Bolettieri, E.; Allen, D.; Hay, A.G. Genome Sequences of Four Strains of Acinetobacter bereziniae Isolated from Human Milk Pumped with a Personal Breast Pump and Hand-Washed Milk Collection Supplies. *Microbiol. Resour. Announc.* **2020**, *9*, e00770-20. [CrossRef] [PubMed]
69. An, S.; Berg, G. Stenotrophomonas maltophilia. *Trends Microbiol.* **2018**, *26*, 637–638. [CrossRef] [PubMed]
70. Flores-Treviño, H.; Bocanegra-Ibarias, P.; Camacho-Ortiz, A.; Morfín-Otero, R.; Salazar-Sesatty, H.A.; Garza-González, E. Stenotrophomonas maltophilia biofilm: Its role in infectious diseases. *Expert Rev. Anti-Infect. Ther.* **2019**, *17*, 877–893. [CrossRef]
71. Toding, O.S.L.; Virginia, C.; Suhartini, S. Conversion banana and orange peel waste into electricity using microbial fuel cell. In *IOP Conference Series: Earth and Environmental Science*; IOP Publishing: Bristol, UK, 2018; Volume 209, p. 12049.
72. Romo, D.M.; Gutiérrez, N.H.; Pazos, J.O.; Figueroa, L.V.; Ordóñez, L.A. Bacterial diversity in the Cr(VI) reducing biocathode of a Microbial Fuel Cell with salt bridge. *Rev. Argentina Microbiol.* **2019**, *51*, 110–118. [CrossRef] [PubMed]
73. Groenewald, M.; Boekhout, T.; Neuvéglise, C.; Gaillardin, C.; van Dijck, P.W.; Wyss, M. Yarrowia lipolytica: Safety assessment of an oleaginous yeast with a great industrial potential. *Crit. Rev. Microbiol.* **2014**, *40*, 187–206. [CrossRef] [PubMed]
74. Fröhlich-Wyder, M.T.; Arias-Roth, E.; Jakob, E. Cheese yeasts. *Yeast* **2019**, *36*, 129–141. [CrossRef] [PubMed]

 *molecules*

Article

Acid-Catalyzed Liquefaction of Biomasses from Poplar Clones for Short Rotation Coppice Cultivations

Ivo Paulo [1], Luis Costa [1], Abel Rodrigues [2,3], Sofia Orišková [1], Sandro Matos [1,4], Diogo Gonçalves [1], Ana Raquel Gonçalves [1], Luciana Silva [1], Salomé Vieira [1], João Carlos Bordado [1] and Rui Galhano dos Santos [1,*]

[1] CERENA-Centre for Natural Resources and the Environment, Instituto Superior Técnico, Av. Rovisco Pais, 1049-001 Lisboa, Portugal; ivo.paulo@tecnico.ulisboa.pt (I.P.); luis.d.costa@ist.utl.pt (L.C.); sofia.oriskova@tecnico.ulisboa.pt (S.O.); sandro.matos@tecnico.ulisboa.pt (S.M.); diogo.azevedo.goncalves@tecnico.ulisboa.pt (D.G.); raquelpgoncalves95@hotmail.com (A.R.G.); lbsilva@fc.ul.pt (L.S.); salomevieira@tecnico.ulisboa.pt (S.V.); jcbordado@ist.utl.pt (J.C.B.)
[2] INIAV—Instituto Nacional de Investigação Agrária e Veterinária, I.P., Ministry of Agriculture, 2780-159 Oeiras, Portugal; abel.rodrigues@iniav.pt
[3] IDMEC—Instituto de Engenharia Mecânica, Instituto Superior Técnico, Av. Rovisco Pais, 1049-001 Lisboa, Portugal
[4] WOODCHEM SA., Estrada das Moitas Altas, 2401-902 Leiria, Portugal
* Correspondence: rui.galhano@ist.utl.pt

Citation: Paulo, I.; Costa, L.; Rodrigues, A.; Orišková, S.; Matos, S.; Gonçalves, D.; Gonçalves, A.R.; Silva, L.; Vieira, S.; Bordado, J.C.; et al. Acid-Catalyzed Liquefaction of Biomasses from Poplar Clones for Short Rotation Coppice Cultivations. *Molecules* 2022, 27, 304. https://doi.org/10.3390/molecules27010304

Academic Editors: Mohamad Nasir Mohamad Ibrahim, Patricia Graciela Vázquez and Mohd Hazwan Hussin

Received: 22 November 2021
Accepted: 2 January 2022
Published: 4 January 2022

Publisher's Note: MDPI stays neutral with regard to jurisdictional claims in published maps and institutional affiliations.

Abstract: Liquefaction of biomass delivers a liquid bio-oil with relevant chemical and energetic applications. In this study we coupled it with short rotation coppice (SRC) intensively managed poplar cultivations aimed at biomass production while safeguarding environmental principles of soil quality and biodiversity. We carried out acid-catalyzed liquefaction, at 160 °C and atmospheric pressure, with eight poplar clones from SRC cultivations. The bio-oil yields were high, ranging between 70.7 and 81.5%. Average gains of bio-oil, by comparison of raw biomasses, in elementary carbon and hydrogen and high heating, were 25.6, 67, and 74%, respectively. Loss of oxygen and O/C ratios averaged 38 and 51%, respectively. Amounts of elementary carbon, oxygen, and hydrogen in bio-oil were 65, 26, and 8.7%, and HHV averaged 30.5 MJkg^{-1}. Correlation analysis showed the interrelation between elementary carbon with HHV in bio-oil or with oxygen loss. Overall, from 55 correlations, 21 significant and high correlations among a set of 11 variables were found. Among the most relevant ones, the percentage of elementary carbon presented five significant correlations with the percentage of O (-0.980), percentage of C gain (0.902), percentage of O loss (0.973), HHV gain (0.917), and O/C loss (0.943). The amount of carbon is directly correlated with the amount of oxygen, conversely, the decrease in oxygen content increases the elementary carbon and hydrogen concentration, which leads to an improvement in HHV. HHV gain showed a strong positive dependence on the percentage of C (0.917) and percentage of C gain (0.943), while the elementary oxygen (-0.885) and its percentage of O loss (0.978) adversely affect the HHV gain. Consequently, the O/C loss (0.970) increases the HHV positively. van Krevelen's analysis indicated that bio-oils are chemically compatible with liquid fossil fuels. FTIR-ATR evidenced the presence of derivatives of depolymerization of lignin and cellulose in raw biomasses in bio-oil. TGA/DTG confirmed the bio-oil burning aptitude by the high average 53% mass loss of volatiles associated with lowered peaking decomposition temperatures by 100 °C than raw biomasses. Overall, this research shows the potential of bio-oil from liquefaction of SRC biomasses for the contribution of renewable energy and chemical deliverables, and thereby, to a greener global economy.

Keywords: poplar; genotypes; liquefaction; short rotation crops

1. Introduction

As society and scientific knowledge develop, it becomes increasingly important to reduce the dependency on petrochemicals due to their dwindling reserves and their negative impact on the environment inflicted by their exploration. The European Union (EU) has set a goal to increase the budget for R&D into eco-innovations due to the importance that the European Commission has identified in this field [1], with a 294.5 billion € investment in 2020, corresponding to about 2.18% of GDP [2]. With this, the EU wants to take a leading role in the development of policies that aim to propel industries and practices into a more sustainable and environmentally friendly future.

Lignocellulosic biomasses are a renewable, continuous, and sustainable feedstock delivering liquid, gaseous, and solid materials useful for several industries [3,4]. Some significant hurdles prevail in their competitiveness by comparison with petrochemical sources. The costs of harvesting, manufacturing lines, the environmental management, the logistics of transport of light lignocellulosic materials, and their conversion wastes are examples of their drawbacks which tend to be surpassed as the need to accelerate a transition to green and circular economy gains relevance. Challenges are also posed [5] in the domain of socio–cultural–economic impacts of tree harvesting, deforestation, and the destruction of century-old forests. Given the complexity of these issues, each country identifies (within its sphere of direct influence) the most attractive lignocellulosic materials and waste streams to employ as feedstock for valorization [6].

Alternative forms of forest and land management are thus required to control the risks of excessive deforestation and the uncontrolled use of lignocellulosic materials. Short rotation coppicing comes as a possibility for biomass production while safeguarding ecosystem biodiversity and soil quality. These coppices are carbon neutral and intensively managed industrial crops, with productive cycles between two to five years, a plant density between 1200 to 10,000 per hectare, and six to seven productive cycles. The lands are subjected to fallow/rotation after that. In Europe, poplar is the predominant SRC species due to its high biomass productivity (ranging between 12 and 20 $Mgha^{-1}y^{-1}$ or higher) and its great potential for genetic improvement [7–11].

Poplar (*Populus* sp.) is common in SRC cultivations, with well-known fuel aptitudes [12,13] that can be hybridized to improve the bark/wood ratio, resistance to diseases, or calorific power. Besides ash and water, poplar wood contains major extractive organic compounds (ranging between 1.5 and 3%) and biopolymers, such as lignin, cellulose, and hemicellulose (ranging between 17 and 23%, 43 to 46%, and 29 to 36%, respectively) [14].

Recently published work on thermochemical conversion of poplar clones from SRCs includes studies on torrefaction [15,16], pyrolysis [17–20], and hydrothermal liquefaction [21], and solvent liquefaction [6,22–25].

Our research work focused on converting eight SRC poplar clones through acid-catalyzed liquefaction, allowing the conversion of biomass thermochemically through mild temperatures at ambient pressure [26]. A bio-oil with a high heating value greater than 40 $MJkg^{-1}$ was previously obtained through this process [27].

Thermochemical liquefaction of different biomass feedstocks has been studied, including spruce [28], pinewood [22,29,30], eucalyptus [31–33], potato peels [34], cork powder [35], spent coffee beans [36], beech [37], or wheat straw [28,38,39]. The results from such studies demonstrated that different biomasses, with distinct chemical compositions and structures, can be used for the acid-catalyzed liquefaction to produce bio-oils in high yields. The previous studies demonstrated, the produced liquid bio-oil can be further used as an environmentally friendly raw material for the chemical industry or fuels [18,26,40–46].

The acid-catalyzed liquefaction is a thermochemical process that converts the main biopolymers (cellulose, hemicelluloses, and lignin) into low molecular weight compounds, a bio-oil. The properties of the obtained bio-oil are close to those of petroleum except for the oxygen content. For instance, from the soybean's liquefaction, a bio-oil with a higher heating value of 44.22 $MJkg^{-1}$ and a H/C molar ratio of 1.9 was obtained [27].

For this work, we chose eight samples of different SRC poplar genotypes as the feedstock for the acid-catalyzed liquefaction process based on our experience with this species. Poplar proliferates preferably in milder climates, such as those found in southern Europe, even though some plantations produce satisfactory yields in northern Europe as well [47].

Given the potential of SRC for biomass production, under environmental sustainability, the proposed objectives target enlarging the scope of thermochemical conversion for added-value products, e.g., chemicals and fuels. The acid-catalyzed liquefaction was applied to an array of commercial poplar clones, to study their potential to produce bio-oil in high yield. We characterized the bio-oils and solid residues to assess their use as biofuels and chemicals. To the best of our knowledge, the comparison between liquified biomasses from different poplar clones was never disclosed. Within this context, our work supports further development on thermochemical conversion technologies to value this type of biomass feedstock.

2. Materials and Methods

We employed eight different poplar genotypes as biomass feedstocks. Their genotype, origin, parentage, and hemicellulose, cellulose, and lignin content are shown in Table 1. The selected genotypes were *AF8*, *Bakan* (Bak), *Brandaris* (Bra), *Ellert* (Ell), *Grimminge* (Gri), *Hees* (Hee), *Skado* (Ska), and *Wolterson* (Wol). The biomass samples were not pre-treated, except from shredding on a Retsch© SM 2000 mill equipped with a 4 mm sieve to decrease the grain size and thus increase the surface area. We purchased the solvent 2-Ethylhexanol and the catalyst 97% p-Toluenesulfonic acid (PTSA) from Sigma–Aldrich. Technical acetone for washing purposes was acquired locally.

Table 1. Hemicellulose, cellulose, and lignin estimated the content of the poplar genotypes [48].

			Lignocellulosic Content (%)		
Genotype	**Origin**	**Parentage**	**Hemicellulose**	**Cellulose**	**Lignin**
AF8	Portugal	Hybrid P. generosa	23	48	28
Bakan	Belgium	Hybrid P. trichocarpa × P. maximowiczii	19	52	28
Brandaris	Belgium	Species P. nigra	23	47	29
Ellert	Belgium	Hybrid P. canadensis	24	48	26
Grimminge	Belgium	Triple hybrid P. deltoides × (P. trichocarpa × P. deltoides)	24	48	27
Hees	Belgium	Hybrid P. canadensis	23	50	26
Skado	Belgium	Hybrid P. trichocarpa × P. maximowiczii	20	49	30
Wolterson	Belgium	Species P. nigra	24	48	27

2.1. Liquefaction Procedure

The liquefaction, an acid-catalyzed process, was performed at 160 °C, at ambient pressure, for the predetermined reaction time. The biomass samples and the solvent were fed into the LENZ Glass Reactor, in a solvent:biomass ratio of 5:1. 2-Ethylhexanol (2-EH) was used as a solvent and the weight of biomass was based on its dry state.

The mass of catalyst, PTSA, was set at 3% (w/w) of the mass of solvent and biomass samples. After 90 min at 160 °C, the process was quenched to 80 °C. Afterward, the bio-oil crude was filtered to retrieve the solid residues.

Upon process completion, the reactor cooled to room temperature to be further vacuum filtrated. The solid residues were washed with acetone and dried in the oven at 110 ± 3 °C for 24 h. The excess solvent of the bio-oil samples was removed under a vacuum. The process conversion, bio-oil yield, was calculated as per the weight of the solid fraction obtained after filtration, according to Equation (1):

$$\text{Bio-oil yield (\%)} = (1 - m_{s0}/m_{si}) \times 100, \tag{1}$$

where m_{si} is the mass of dry biomass fed to the reactor, in grams, and m_{s0} is the mass of solid residues obtained at the end of the process, in grams.

2.2. Fourier Transformed Infrared (FTIR-ATR) Analysis of Biomass and Bio-Oil

The FTIR-ATR analysis was performed on a Spectrum Two–Perkin Elmer spectrometer. The spectra were captured from 4000 to 600 cm^{-1} and treated in Perkin Elmer–Spectrum IR software.

2.3. Elemental Analysis

The carbon (C), hydrogen (H), and nitrogen (N) content in biomass, solid residues, and bio-oil were assessed by a LECO TruSpec CHN analyzer, whilst a LECO CNS2000 analyzer determined sulfur (S) content.

2.4. Higher Heating Value (HHV) Calculation

Commonly, biomass and its derivatives, i.e., bio-oil and residues, contain up to 97–99% of C, H, O. Additional elements, such as sulfur and nitrogen, are present in negligible amounts, below the detection limit, and thus difficult to measure or quantify [22,49]. We assessed the oxygen content according to Equation (2):

$$O\ (\%) = 100 - C\ (\%) - H\ (\%), \tag{2}$$

According to Rodrigues et al. [15], the elemental analysis of poplar clones vary from 51.5–52.2%, 5.2–5.4%, 42.0–43.3%, and 0.4–0.7% for C, H, O, and N, respectively. The elemental composition of poplar clones in SRC in the Czech Republic was similar to such values [13].

The higher heat value (HHV) of the biomass and solid residues was assessed using the method disclosed by Yin et al. [50] using Equation (3). The HHV of bio-oils was evaluated by Equation (4), which is specifically established for bio-oils [51].

$$HHV\ (MJ/kg) = 0.2949C + 0.8250H, \tag{3}$$

$$HHV\ (MJ/kg) = 0.363302C + 1.087033H - 0.1009920, \tag{4}$$

2.5. Energy Densification Ratio (EDR) Calculation

The energy densification ratio (EDR), a dimensionless indicator, informs on how the HHV was improved thanks to the liquefaction process [52]. We used Equation (5) to calculate the EDR values:

$$EDR = HHV_{bio\text{-}oil} / HHV_{biomass}, \tag{5}$$

where $HHV_{bio\text{-}oil}$ and $HHV_{biomass}$ are the higher heating values of bio-oil and biomass samples, respectively.

2.6. Van Krevelen Diagram

The van Krevelen diagrams are useful to spot variations between different types of kerogen and fuels. This diagram cross-plots the hydrogen and carbon atomic ratio (10H/C) as a function of the oxygen to carbon atomic ratios of carbonaceous compounds. The van Krevelen diagram is suitable for identifying and revealing compositional differences between organic products [45]. Using the data obtained via elemental analysis, we plotted the chemical compositions of biomass, bio-oil, and solid residues in the van Krevelen diagram.

2.7. Thermogravimetric Analysis (TGA)

The thermogravimetric analysis of raw biomass, bio-oils, and solid residues was performed using the Hitachi-STA7200. The evaluation was accomplished between 25 and 600 °C in a nitrogen atmosphere, with a 100 mL/min flow and a heating rate of 5 °C/min.

2.8. Pearson's Correlations

The ultimate analysis and HHV data were used to access correlations between 11 variables using SPSS Statistics software. The analysis was performed to find a correlation pattern within the bio-oil variables and quantify the interactions between them. The C, H, and O content of the bio-oils were correlated with the gain of C, H, H/C, O/C, HHV, and with the loss of O, ash, and moisture content. The number of correlations was assessed according to Equation (6):

$$\text{N Pearson's r} = (n^2 - n)/2, \tag{6}$$

where N is the number of correlations and n is the number of variables.

3. Results and Discussion

Poplars are increasingly used for biofuel production due to their high growth and biomass productivity. They have a significant holocellulose (ranging between 47 and 52% for cellulose and 19 and 24% for hemicelluloses) and a moderate lignin content (ranging between 26 and 30%). This work presents the liquefaction results of eight different poplar genotypes. The experimental conditions, such as solvent and catalyst, were previously optimized in studies for other biomasses [29,32,33], as well as for poplar [26]. In particular, the system 2-ethylhexanol/PTSA has been shown to allow the production of bio-oils in high yields [22,53].

For comparison, results from Rodrigues et al. [15] regarding poplar genotypes' torrefaction were used. The liquefaction assays were conducted in duplicate at 160 °C for 90 min, using 3 wt. % PTSA as the catalyst and 1:5 biomass:solvent ratio. At first glance, bio-oil yields higher than 70% indicated the SRC poplar clones' aptitude for liquefaction. The bio-oil yields are shown in Figure 1. Overall, the process led to a bio-oil yield ranging from 70.7 to 81.5%. The highest bio-oil yields of around 81% were obtained for the *AF8* and *Skado* genotypes, while the lowest conversion was achieved with the *Brandaris* sample. Overall, the conversions are in accordance with the literature concerning similar thermochemical conversion under the same experimental conditions [29,54]. In comparison, the microwave-assisted pyrolysis of poplar, where the bio-oil achieved a maximum of 30.8%, and the conducted liquefaction process led to higher yields [17].

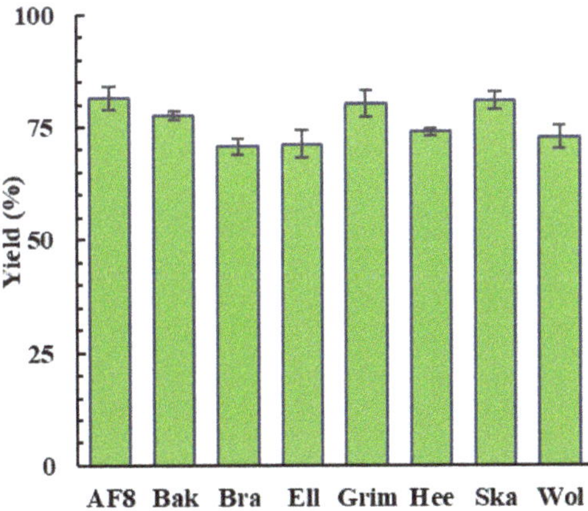

Figure 1. Comparison of the yield of bio-oils obtained via acid-catalyzed liquefaction of poplar clones.

It should be noted that the solid residue fraction can be given by the difference for the complete conversion once the gaseous streams are reduced and can be neglected [29,30]. The solids contain unreacted biomass as well as any decomposition products. In fact, during the liquefaction, solid residues can be produced from the decomposition of lignocellulosic biomass, and are commonly referred to as humins [29,30,33]. Such occurrence is well-explained and is associated with the recondensation of decomposition products of reactions [55–57].

The chemical characterization of biomass and its bio-oil counterparts, as well as the solid residues and torrefied samples, is shown in Table 2. It encompasses the ultimate analysis, moisture, and ash content, calculated HHV, H/C, and O/C ratios. Table 3 presents the values of % C gain, % H gain, % O gain, % ash loss, % moisture loss, H/C gain, and O/C loss of the bio-oils from biomass liquefaction, compared to the untreated biomass samples.

Table 2. Chemical characterization of poplar clones, bio-oils, solid residues, and torrefied biomass.

| | Samples | Chemical Composition [1] (%) | | | | Ash (%) | Moisture (%) | HHV [2] (MJ/kg) | 10H/C | O/C | Empirical Formula |
		C	H	N	O						
Biomass [15]	AF8	51.5	5.2	<0.5	43.3	2.18	9.58	17.20	1.01	0.84	
	Bakan	51.5	5.4	0.6	42.5	1.56	10.80	17.49	1.04	0.83	
	Brandaris	52.0	5.2	0.7	42.1	2.87	8.08	17.58	1.00	0.81	
	Ellert	51.8	5.3	0.6	42.4	2.25	10.10	17.52	1.02	0.82	
	Grimminge	52.2	5.3	0.5	42.0	1.76	9.42	17.75	1.02	0.81	$CH_{1.22}O_{0.61}$
	Hees	51.8	5.2	0.7	42.3	2.39	7.90	17.48	1.01	0.82	
	Skado	51.6	5.3	0.4	42.7	1.47	9.91	17.45	1.03	0.83	
	Wolterson	51.9	5.3	0.7	42.1	2.28	9.73	17.64	1.02	0.81	
	Mean	51.79	5.27	0.59	42.42	2.09	9.44	17.51	1.02	0.82	
Bio-oil	AF8	64.4	8.4	<0.5	27.2	0.4	1.30	29.85	1.31	0.42	
	Bakan	65.6	8.9	<0.5	25.5	0.3	1.13	30.95	1.36	0.39	
	Brandaris	64.4	8.6	<0.5	27.0	0.3	1.49	30.03	1.33	0.42	
	Ellert	66.1	8.8	<0.5	25.1	0.4	1.15	31.06	1.33	0.38	
	Grimminge	64.8	8.9	<0.5	26.3	0.1	0.96	30.62	1.38	0.41	$CH_{1.61}O_{0.30}$
	Hees	65.3	8.8	<0.5	25.9	0.2	1.18	30.67	1.34	0.40	
	Skado	64.5	8.4	<0.5	27.1	0.3	1.37	29.90	1.31	0.42	
	Wolterson	65.3	8.9	<0.5	25.8	0.2	1.42	30.84	1.37	0.39	
	Mean	65.05	8.72	–	26.23	0.28	1.25	30.49	1.34	0.40	
Solid residues	AF8	52.3	5.9	<0.5	41.8	0.3	–	18.44	1.13	0.80	
	Bakan	50.4	5.8	<0.5	43.8	0.3	–	17.36	1.15	0.87	
	Brandaris	50.0	5.0	<0.5	45.0	0.2	–	16.25	1.00	0.90	
	Ellert	49.0	5.7	<0.5	45.3	0.7	–	16.52	1.16	0.93	
	Grimminge	50.6	5.8	<0.5	43.6	0.3	–	17.48	1.15	0.86	$CH_{1.34}O_{0.66}$
	Hees	48.6	5.8	<0.5	45.6	0.8	–	16.48	1.19	0.94	
	Skado	50.9	5.4	<0.5	43.7	0.1	–	17.14	1.06	0.86	
	Wolterson	49.7	5.6	0.5	44.2	0.6	–	16.81	1.12	0.89	
	Mean	50.19	5.62	0.52	44.13	0.41	–	17.06	1.12	0.88	
Torrefied biomass [15]	AF8	66.3	4.9	0.36	28.44	3.46	–	24.2	0.74	0.43	
	Bakan	65.9	4.94	0.66	28.5	2.7	–	24.1	0.75	0.43	
	Brandaris	66.9	4.99	0.88	27.29	4.0	–	24.6	0.75	0.41	
	Ellert	67.8	5.1	0.73	26.37	3.49	–	25.2	0.75	0.39	
	Grimminge	68.3	5.06	0.69	25.94	2.97	–	25.4	0.74	0.38	$CH_{0.89}O_{0.32}$
	Hees	67.3	4.8	0.84	27.06	3.54	–	24.6	0.71	0.40	
	Skado	67.4	5.04	0.54	27.02	2.63	–	24.9	0.75	0.40	
	Wolterson	63.5	4.95	0.7	30.85	2.93	–	22.9	0.74	0.49	
	Mean	66.7	4.97	0.68	27.68	3.22	–	24.47	0.75	0.42	

[1] Dry basis; [2] calculated HHV.

Table 3. Ratios of C, H, HHV H/C gain and of O, Ash, Moisture, and O/C loss of the bio-oils obtained from the liquefaction of poplar clone samples.

Sample	C Gain (%)	H Gain (%)	O Loss (%)	Ash Loss (%)	HHV Gain (%)	Moisture Loss (%)	H/C Gain (%)	O/C Loss (%)
AF8	25.05	61.54	37.18	81.65	73.56	95.82	29.70	50.00
Bakan	27.38	64.81	40.00	80.77	76.98	97.22	30.77	53.01
Brandaris	23.85	65.38	35.87	89.55	70.82	96.29	33.00	48.15
Ellert	27.61	66.04	40.80	82.22	77.26	96.04	30.39	53.66
Grimminge	24.14	67.92	37.38	94.32	72.50	98.94	35.29	49.38
Hees	26.06	69.23	38.77	91.63	75.48	97.47	32.67	51.22
Skado	25.00	58.49	36.53	79.59	71.34	96.97	27.18	49.40
Wolterson	25.82	67.92	38.72	91.23	74.83	97.94	34.31	51.85
Mean	25.61	65.17	38.16	86.37	74.10	97.09	31.67	50.83

The bio-oil elemental analysis is in accordance with those obtained for other biomasses, e.g., pinewood, eucalyptus, and tomato pomace [22,30,32,58]. The results from the chemical analysis of bio-oil proved that the sets of liquefied biomasses were very distinct from the chemical composition of the raw biomasses and the solid residue. The carbon content (%, dry basis) was higher for the bio-oil, ranging from 64.4% (*AF8* and *Brandaris*) to 66.1% (*Ellert*). On the other hand, the elementary carbon content of the residues averaged 50.19%, presenting values between 48.6% (*Hees*) and 52.3% (*AF8*). While the solid residues showed a slightly lower elementary carbon content than the feedstock, the% carbon gain for the bio-oils was, on average, ~25%. An advantage of bio-oil is that its ash content is much lower than that of the torrefied biomass. The liquefaction delivered bio-oils with a very similar average carbon content (~67%) to that from the torrefied biomass (see Figure 2) [15].

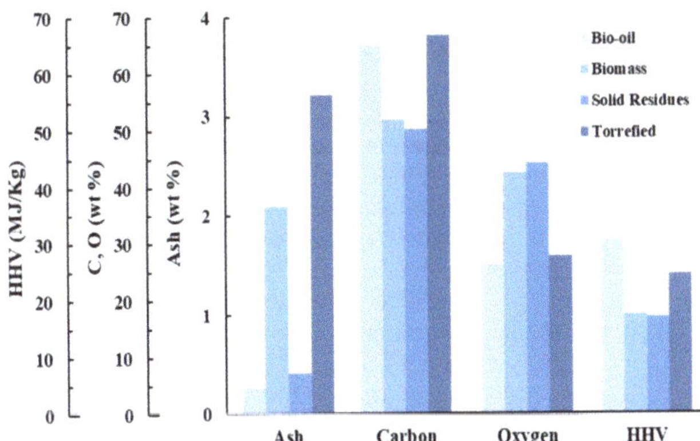

Figure 2. Comparison of the average ash, carbon, oxygen contents, and HHV between bio-oils, biomass, solid residues, and torrefied biomass.

Regarding hydrogen content (%, dry basis), the bio-oils, averaging ~8.7%, presented values ranging from 8.4% (*AF8* and *Skado*) to 8.9% (*Bakan, Grimminge, Wolterson*) (Table 1). Such values demonstrated an average 65% gain in the hydrogen content by comparison with the raw biomass. The solid residues presented a hydrogen content like that from the raw biomass. Additionally, the hydrogen content was up to 75% higher than the torrefied biomass, wherein losses of hydrogen and H/C ratios were detected [15].

As expected, the oxygen content (%, dry basis) was considerably lower for the bio-oils and torrefied biomass compared to their biomass counterparts, with values averaging ~26%, ~28%, and ~42%, respectively (see Figure 2). The highest oxygen content was obtained for the genotype *AF8* genotype samples (27.2%) and the lowest for genotype *Ellert* (25.1%).

The correspondent oxygen amount of the solid residues were in line with those of the raw biomasses. The average % O loss (~38%) was concomitant with increases in C and H contents of around 25.61% and 65.17% (Table 2). The loss of oxygen occurs through water elimination, which is retrieved by distillation during the process [29]. Consequently, the H/C and O/C ratios showed significant gains of around 32 and 51% (Table 2). The ash and moisture content ranged from 0.1 to 0.4% and 0.96 to 1.49%, respectively, for the obtained bio-oils. A significant decrease in ash and moisture content compared with the raw poplar genotypes was obtained upon biomass liquefaction (see Figure 2).

The average ratios O/C and 10H/C of bio-oils from biomass poplar clones were 0.40 and 1.34, respectively (Table 1). The variations of these ranges between clones were small, with the O/C ratio ranging between 0.38 and 0.42 and the 10H/C ratio ranging between 1.31 and 1.38. The O/C of bio-oils was lower than that of raw poplar biomass samples and like the O/C values of torrefied poplar clones. On the other hand, the H/C increased considerably due to the rise in hydrogen in bio-oil samples.

These variations explain the improvement in the calculated HHV since lower oxygen content and O/C ratios lead to higher HHV. On average, the bio-oils presented an HHV of 30.49 MJ/kg. *AF8* presented the lowest value (29.85 MJ/kg), while *Wolterson* presented the highest (30.84 MJ/kg). The HHV of the biomass was on average 17.51 MJ/kg, and that of torrefied biomass was 24.5 MJ/kg, demonstrating that the HHV was remarkably improved with the liquefaction. The presence of residual solvent can also potentiate, although not to the fullest extent, the observed increase on the HHV. However, the solvent was removed, and work from Condeço et al., 2021, showed that liquefaction processes with low conversion led to lower HHV oil, indicating that increases in HHV result from the conversion of the lignocellulosic materials into bio-oil [59]. The values of Ma et al. [27], although higher (>40 MJkg^{-1}) than those obtained in this work, were indicative of the potential of this solvolytic liquefaction for delivering bio-oil with high heating power.

The HHV gain of bio-oil by comparison with raw biomass averaged 74%. Compared with torrefied biomass [15], wherein the increase in HHV was ~40%, bio-oil still presented a high energy densification ratio (74%). Figure 2 highlights the significant increase in the % elemental carbon and the decrease in the % O content, which increases the HHV of the biomass when compared to the raw and torrefied biomass.

The energy densification ratio (EDR) was employed to calculate the effectiveness of the process. The increase in the EDR results from solid mass decrease due to dehydration and decarboxylation reactions [52]. The average bio-oil EDR of 1.74 showed that the liquefaction of poplar biomass led to higher energy densification. On the other hand, the solid residues led to a slight loss in the heating values, which accounted for an EDR of 0.97. The lower average EDR of 1.39 for torrefied biomass reflected the aptitude of liquefaction for delivering a bio-oil product with high calorific potential.

The van Krevelen diagram identified the fuel quality changed with the chemical composition variation. Usually, biomasses with lower O/C and H/C ratios are considered good fuel aptitudes due to lower water vapor, minimum energy loss, and less smoke upon combustion [60].

Overall, the van Krevelen diagram (Figure 3) showed that the bio-oil locations were close to those of liquid fossil fuels (such as diesel or gasoline), demonstrating that liquefaction leads to liquid products similar to fossil fuels. By comparison, torrefaction leads to products similar and compatible with fossil coals/peat. On the other hand, the solid residues were closer to the highly oxidized compounds. In comparison with biomass, the atomic ratios of O/C and H/C of solid residues increased, while for bio-oil, the H/C increased, and O/C decreased. This suggests that bio-oil is a better fuel than raw biomass itself. The decrease in O/C atomic ratios leads to an increase in the high energy bonds (C-C) and a reduction in low energy bonds (O-C) leading to an HHV improvement. González-Arias et al. postulated that such change might be explained by the occurrence of dehydration reactions that leads to hydroxyl groups loss and by the decarboxylation reactions that eliminate the carboxyl and carbonyl groups [52].

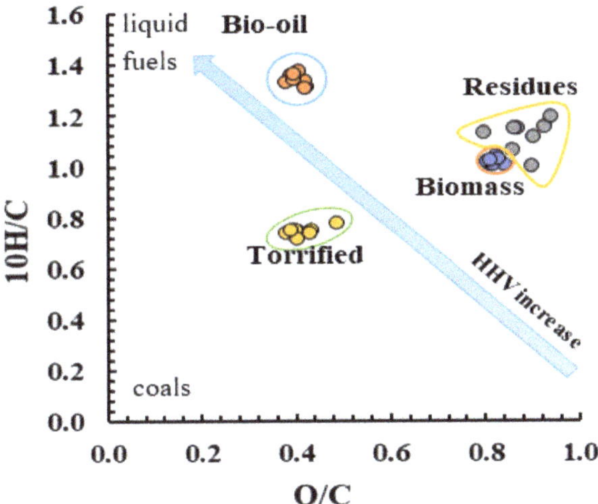

Figure 3. Van Krevelen diagram comparing the H/C and O/C ratios of biomass, solid residues, torrefied biomass, and bio-oils.

The distribution of Pearson correlations among chemical variables and calorific power of bio-oils and raw biomass samples reflected the above-described tendencies (Table 4). From 55 correlations, 21 significant and high correlations (r > 0.7) among a set of 11 variables were found. Among the most relevant, the % elementary carbon presented five significant correlations with % O (-0.980), % C gain (0.902), % O loss (0.973), HHV gain (0.917), and O/C loss (0.943). As expected, the amount of carbon was directly correlated with the amount of oxygen. The decrease in oxygen content increased the elementary carbon and hydrogen concentration, which led to an improvement in HHV. The HHV gain showed a strong positive dependence with the % C (0.917) and % C gain (0.943). Conversely, the elementary oxygen (-0.885) and its % O loss (0.978) adversely affected the HHV gain. Consequently, the O/C loss (0.970) increased the HHV positively.

Table 4. Pearson's correlation (r) from elemental analysis and bio-oil variables.

		C	H	O	C Gain	H Gain	O Loss	Ash Loss	HHV Gain	Moisture Loss	H/C Gain	O/C Loss
Variables (%)	C	1	0.669	-0.980[2]	0.902[2]	0.485	0.973[2]	0.042	0.917[2]	0.049	0.090	0.943
	H	0.669	1	-0.803[1]	0.389	0.814[1]	0.617	0.595	0.561	0.676	0.712	0.537
	O	-0.980[2]	-0.803[1]	1	-0.827[1]	0.605	0.915[2]	-0.192	-0.885[2]	-0.219	-0.261	-0.899
	C Gain	0.902[2]	0.389	-0.827[1]	1	0.150	0.947[2]	-0.311	0.943[2]	-0.201	-0.272	0.972
	H Gain	0.485	0.814[1]	-0.605	0.150	1	0.405	0.650	0.400	0.502	0.868	0.291
	O Loss	0.973[2]	0.617	-0.945[2]	0.947[2]	0.405	1	-0.119	0.978[2]	0.000	0.012	0.984
	Ash Loss	0.042	0.595	-0.192	-0.311	0.650	-0.119	1	-0.207	0.813	0.739	-0.205
	HHV Gain	0.917[2]	0.561	-0.885[2]	0.943[2]	0.400	0.978[2]	-0.207	1	-0.062	-0.001	0.970
	Moisture Loss	0.049	0.676	-0.219	-0.201	0.502	0.000	0.813[1]	-0.062	1	0.643	-0.068
	H/C Gain	0.090	0.712[1]	-0.261	-0.272	0.868[2]	0.012	0.739[1]	-0.001	0.643	1	-0.091
	O/C Loss	0.943[2]	0.537	-0.899[2]	0.972[2]	0.291	0.984[2]	-0.205	0.970[2]	-0.068	-0.091	1

[1] $p < 0.05$; [2] $p < 0.01$.

Figure 4 and Table 5 show the ATR-FTIR spectra and data of raw biomass (Figure 4a), bio-oil (Figure 4b), and solid residues (Figure 4c) for the eight poplar genotypes. No significant differences in the profiles of functional groups were detected among samples of poplar genotypes within each profile. The major spectral differences concerned the absorption intensity in the range between 2800 and 3000 cm^{-1} assigned to C–H stretching vibrations. In this range, the bio-oils had a higher absorption than biomass and solid residues. In the range 1370 and 1730 cm^{-1}, solid residues have visibly lower absorption than bio-oils or biomass. In the range 1100–1200 cm^{-1} we saw lower absorption in bio-oils in comparison with biomass and solid residue samples. Overall, the spectra of all samples displayed a broad band around 3500 cm^{-1}, a characteristic band resulting from OH stretching vibration.

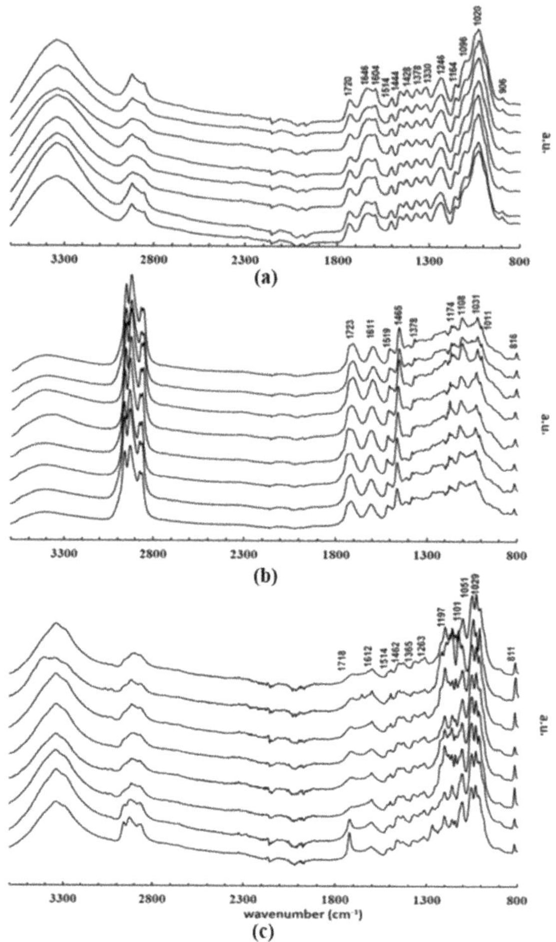

Figure 4. FTIR-ATR spectra of (**a**) biomass; (**b**) bio-oil, and (**c**) solid residue.

The absorption differences between 2800 and 3000 cm^{-1}, assigned to C–H stretching vibrations, point out the presence of derivatives of holocellulose and lignin in bio-oil. In the range from 1370 to 1730 cm^{-1}, biomass spectra showed the peaks at 1604 and 1514 cm^{-1}, generally attributed to the presence of lignin. These same peaks were identified on the bio-oil spectra at 1611 and 1519 cm^{-1}, respectively, and were practically non-existent in the spectra of solid residues. These peaks in bio-oil spectra revealed that lignin was depolymerized, hence its derivatives were present. Moreover, the peaks related to syringyl and guaicyl units at 1378 and 1246 cm^{-1}, respectively, were identified within the bio-oil samples. On the other hand, the peaks related to hemicellulose and cellulose associated with the stretching and vibrations of functional groups (see Table 5) were identified in the biomass as well as the bio-oil samples (peaks at 1465, 1174, 1108, 1031 cm^{-1}). The differences in absorption in the range from 1100 to 1200 cm^{-1} reflected the inherent chemical differences between the profile of bio-oil and the other two. Additionally, at 1723 cm^{-1}, a peak was shown due to the vibrational states of carbonyl functional groups present in aldehydes, ketones, acids, or esters, which resulted from the conversion of cellulose or hemicellulose into levulinic acid, furfural, and related compounds [61]. On the other hand, in the biomass, the correspondent peak profiled in the biomass sample spectra (1720 cm^{-1}) is associated with hemicellulose and lignin [59]. The peak at 1646 cm^{-1}, assigned to the OH bending of water, confirmed the presence of water in the biomass samples. Regarding the solid residues' spectra, peaks related to lignin were identified at 1612, 1514, 1365, and 1263 cm^{-1}, and those concerning holocellulose appeared at 1462, 1197, 1101, and 1029 cm^{-1} (Table 5). These peaks indicated the presence of unreacted biomass in solid residues.

Table 5. FTIR-ATR relevant peaks for biomass, bio-oil, and solid residues.

Peaks (cm^{-1})			Band Assignment		Ref.
Biomass	**Bio-Oil**	**Residues**	**Functional Group**	**Compounds**	
1720	1723	1718	C=O carbonyls in ester groups and acetyl groups in xylan	Ketones, esters, hemicellulose, and carboxylic acids and esters	[62–64]
1646			O-H bending	Water	[65,66]
1604	1611	1612	C=C aromatic ring vibration	Lignin	[62,67]
1514	1519	1514	C=C aromatic ring stretching	Lignin	[32,68]
1444	1465	1462	OCH$_3$-, -CH$_2$-, and C-H stretching	Cellulose, hemicellulose	[69]
1378	1378	1365	Aromatic C-H deformation	Syringyl rings	[63]
1330			C-O syringyl ring	Lignin	[62]
1246	1248	1263	Aromatic ring vibration	Guaicyl lignin	[62]
1164	1174	1197	C-O-C asymmetrical stretching	Cellulose, hemicellulose	[62]
1096	1108	1101	C-O-C stretching	Cellulose, hemicellulose	[64]
1020	1031	1029	C-O, C=C, and C-C-O stretching	Cellulose, hemicellulose, lignin	[62]
906			Glycosidic linkage	Cellulose, hemicellulose	[62,68]
	816	811	C-H out-of-plane	Cellulose, hemicellulose	[70]

The TGA/DTG curves and mass losses of biomass, bio-oils, and solid residues between 0 and 600 °C, are shown in Figure 5 and Table 6. Regarding raw biomass, TGA analysis evidenced four stages of decomposition. During the first stage, at temperatures ranged between 25 °C and 120 °C, a mass loss averaging 7% occurred, concerning volatile components and free water content. The second (120–300 °C) and third stages (300–400 °C) averaged weight losses of 19 and 47%, respectively. The biomass's biopolymolecular structure suffered restructuration at this point, releasing smaller compounds (e.g., H_2O, CO, CO_2, etc.). The cellulose and hemicellulose, alongside lignin, decomposed to form volatiles and low molecular weight compounds during these two stages. While the decomposition of holocellulose led mainly to volatiles; lignin produces primarily carbon. Cellulose, xylan, and lignin contained about 91, 77, and 66% of volatile matter, respectively. The fourth stage, beginning at 400 °C, involved a slower decomposition and significantly lower mass loss of 7% (similar to the first stage) and was associated with the volatilization of carbon via C–C and C–H bonds cleavage [71].

The TGA curves analysis showed that bio-oils from poplar liquefaction were more volatile than the fresh raw material, thus requiring lower peaking temperatures to vaporize and decompose. The maximum temperatures of TGA decomposition were about 325 °C and 225 °C for raw biomasses and bio-oil, respectively. The TGA curves also revealed that the bio-oils decomposed in a three-stage pattern, between about 50 °C and 600 °C (Table 6). The onset temperature of thermal decomposition of bio-oils was about 50 °C. The bio-oil samples presented the first weight loss, ca. 16%, up to 185 °C, due to volatilization of moisture and low molecular weight compounds. From 180–300 °C, the second stage exhibited an average mass loss of ~37%, corresponding to the bio-oil's heavier components that require low temperatures to decompose or volatilize. Seehar et al. hypothesized that the mass loss at these temperatures might denote the presence of chemical structures analogs to those from gasoline, diesel, and jet fuel [72]. The third stage (300–600 °C) showed an average mass loss of 17%, which can be attributed to residual char formation from the sample's slow degradation.

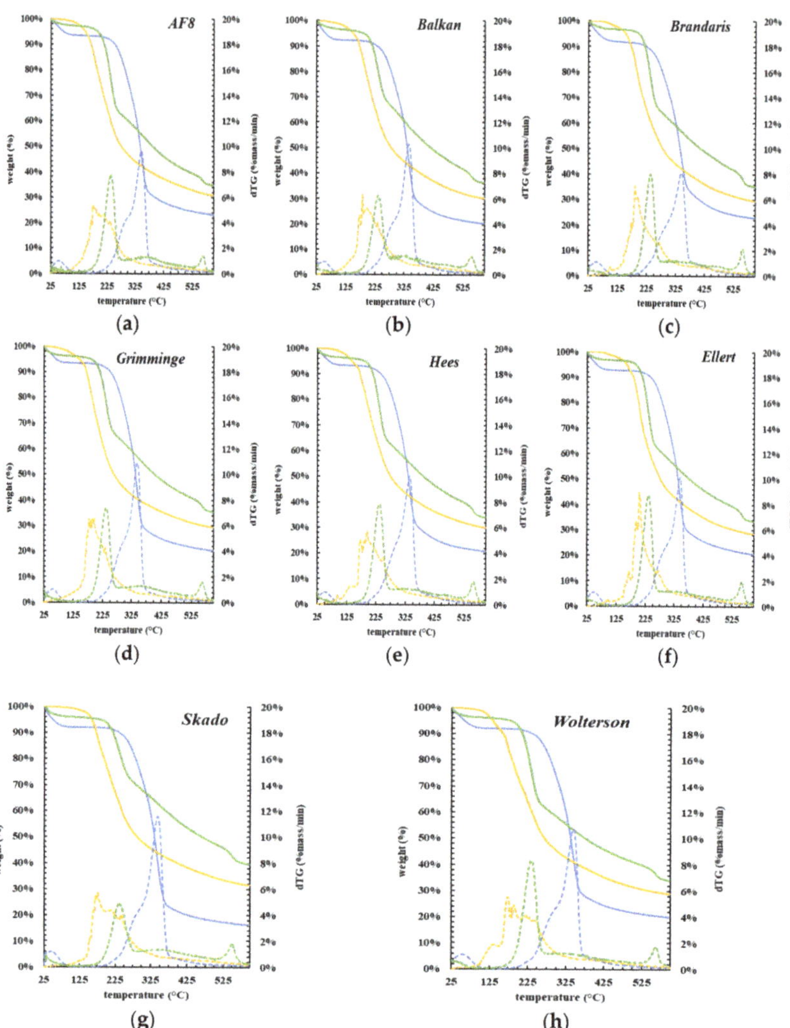

Figure 5. TGA and DTG thermograms of biomass (blue), bio-oil (yellow), and solid residues (green) of all poplar clones: (**a**) *AF8*; (**b**) *Balkan*; (**c**) *Brandaris*; (**d**) *Grimminge*; (**e**) *Hees*; (**f**) *Ellert*; (**g**) *Skado*; (**h**) *Wolterson*. The dashed line is gTG and the solid line is TGA.

Generally, the thermogravimetric curves of bio-oil samples showed an average mass loss of up to 70%. The mass loss as volatiles up to 300 °C, summing up ca. 53%, can denote some gasoline, jet fuel, and diesel segments [72]. Such decomposition profile supports their potential use in combustion applications [73]. The DTG curves showed that the bio-oil dropped weight at lower temperatures, confirming the presence of a lighter product than their biomass counterparts. Most of the mass loss was verified below 230 °C.

Table 6. Mass loss from TGA curves for biomass, bio-oils, and solid residues.

Samples		1st Stage		2nd Stage		3rd Stage		4th Stage	
		Temp. Range (°C)	Mass Loss (%)	Temp. Range (°C)	Mass Loss (%)	Temp. Range (°C)	Mass Loss (%)	Temp. Range (°C)	Mass Loss (%)
AF8	Biomass	<120	6	80–300	18	300–400	46	>400	7
	Bio-oil	50–185	18	185–300	35	300–600	16	–	–
	Residue	<115	3	125–260	32	260–525	26	>525	4
Balkan	Biomass	<120	8	80–300	20	300–400	46	>400	6
	Bio-oil	50–185	15	185–300	37	300–600	18	–	
	Residue	<115	4	125–260	26	260–525	28	>525	6
Brandaris	Biomass	<120	8	80–300	19	300–400	42	>400	8
	Bio-oil	50–185	16	185–300	37	300–600	17	–	–
	Residue	<115	3	125–260	30	260–525	25	>525	7
Ellert	Biomass	<120	7	80–300	20	300–400	46	>400	7
	Bio-oil	50–185	10	185–300	44	300–600	18	–	
	Residue	<115	3	125–260	34	260–525	24	>525	6
Grimminge	Biomass	<120	6	80–300	19	300–400	49	>400	6
	Bio-oil	50–185	17	185–300	38	300–600	16	–	–
	Residue	<115	4	125–260	29	260–525	26	>525	6
Hees	Biomass	<120	7	80–300	20	300–400	46	>400	6
	Bio-oil	50–185	16	185–300	38	300–600	16	–	–
	Residue	<115	4	125–260	31	260–525	25	>525	6
Skado	Biomass	<120	8	80–300	19	300–400	51	>400	6
	Bio-oil	50–185	17	185–300	35	300–600	17	–	–
	Residue	<115	4	125–260	22	260–525	28	>525	6
Wolterson	Biomass	<120	8	80–300	19	300–400	47	>400	6
	Bio-oil	50–185	20	185–300	35	300–600	16	–	–
	Residue	<115	4	125–260	33	260–525	23	>525	6
Mean	Biomass	<120	7	80–300	19	300–400	47	>400	7
	Bio-oil	50–185	16	185–300	37	300–600	17	–	–
	Residue	<115	3	125–260	30	260–525	26	>525	6

The TGA curves of the solid residues showed four decomposition stages. Within the first stage (temperatures up to 115 °C) they showed a low average mass loss of 3%, related to the loss of moisture and other light compounds. The second (125–260 °C) and third (260–525 °C) stage of thermal decomposition of solid residues corresponded to average mass losses of 30 and 26%, respectively. Such stages displayed peaking temperatures of 325 °C and 250 °C, typical of cellulose and hemicellulose, respectively, suggesting that the biomass liquefaction was incomplete [59]. In the fourth stage, corresponding to temperatures higher than 525 °C, a slight mass loss of 6% was attributed to heavy compounds resulting from the condensation of liquefaction products that led to insoluble solids. It is worth noting that, on DTG curves, a slight mass loss above 350 °C, <2%mass/min, led to the peak of the 4th stage. This suggests the presence of heavy compounds by comparison with the DTG of biomass and bio-oils.

4. Conclusions

This study evaluated the aptitude of bio-oils obtained via acid-catalyzed liquefaction of poplar woody biomasses from eight clones from short rotation crops. The laboratory assays were performed under mild conditions of 160 °C and ambient pressure, and the resulting bio-oil yield ranged between 70.7 and 81.5%, within the scope of cited literature. Loss of oxygen and O/C ratios averaged 38 and 51%, respectively. Elementary amounts of carbon, oxygen, and hydrogen in bio-oil were 65, 26, and 8.7%, respectively, and HHV averaged a value of 30.5 MJkg^{-1}. Correlation analysis showed the interconnectedness between, e.g., elementary carbon with HHV in bio-oil or with oxygen loss. The van Krevelen diagram proved that bio-oils are more chemically compatible with liquid fossil fuels, such as diesel or gasoline than the initial biomass. FTIR analysis evidenced the drastic chemical conversion of raw woody biomass through the presence of derivatives of depolymerization of lignin and holocellulose in bio-oil. Results of TGA/DTG in a nitrogen atmosphere confirmed the burning aptitude of bio-oil by the high mass losses of volatiles of 53% and by peaking decomposition temperatures lowered by 100 °C than those of raw biomasses. Overall, the TGA analysis showed that bio-oils from poplar liquefaction were more volatile than the fresh feedstock, thus requiring lower peaking temperatures to vaporize and decompose. Additionally, in comparison with biomass, the bio-oil atomic ratios of H/C increased, and O/C decreased. This reflects the fact that bio-oil is a better fuel than raw biomass. Liquefaction results from this research confirmed the potential of biomasses from SRC cultivations to produce energy and chemicals.

Author Contributions: Conceptualization, R.G.d.S. and A.R.; methodology, R.G.d.S. and A.R.; formal analysis, R.G.d.S., A.R., I.P., D.G., L.S. and S.V.; investigation, R.G.d.S., L.C. and S.M.; resources, R.G.d.S., J.C.B., and A.R.; writing—original draft preparation, R.G.d.S., J.C.B., A.R. and A.R.G.; writing—review and editing, R.G.d.S., A.R. and S.O.; supervision, R.G.d.S.; project administration, R.G.d.S.; funding acquisition, R.G.d.S. and J.C.B. All authors have read and agreed to the published version of the manuscript.

Funding: The authors gratefully acknowledge the funding of the FCT project Clean Forest (PCIF/GVB/0167/2018) to develop this work.

Institutional Review Board Statement: Not applicable.

Informed Consent Statement: Not applicable.

Data Availability Statement: Not applicable.

Acknowledgments: This study was also supported by FCT, through CERENA strategic project (FCT-UIDB/04028/2020) and IDMEC, under LAETA, project FCT-UIDB/50022/2020, which supported the infrastructures. The authors also thank Reinhart Ceulemans from the University of Antwerp for providing biomass from Bakan, Brandaris, Ellert, Grimminge, Hees, and Skado clones, and ACHAR for providing biomass from AF2 and AF8 clones.

Conflicts of Interest: The authors declare no conflict of interest.

Sample Availability: Samples of the compounds are not available from the authors.

References

1. Timmermans, F.; Katainen, J. *Reflection Paper: Towards a Sustainable Europe by 2030*; European Commission: Brussels, Belgium, 2019.
2. Office of the European Commission, P. *R&D and Eco-Innovation Fact Sheet*; European Commission: Brussels, Belgium, 2020.
3. Silvestre, W.P.; Pauletti, G.F.; Godinho, M.; Baldasso, C. Fodder radish seed cake pyrolysis for bio-oil production in a rotary kiln reactor. *Chem. Eng. Process. Process Intensif.* **2018**, *124*, 235–244. [CrossRef]
4. Konwar, L.J.; Mikkola, J.P.; Bordoloi, N.; Saikia, R.; Chutia, R.S.; Kataki, R. Sidestreams from bioenergy and biorefinery complexes as a resource for circular bioeconomy. In *Waste Biorefinery: Potential and Perspectives*; Elsevier B.V: Amsterdam, The Netherlands, 2018; pp. 85–125. ISBN 9780444639929.
5. Searle, S. *Sustainability Challenges of Lignocellulosic Bioenergy Crops*; International Council on Clean Transportation (ICCT): Washington, DC, USA, 2018.
6. Song, C.; Zhang, C.; Zhang, S.; Lin, H.; Kim, Y.; Ramakrishnan, M.; Du, Y.; Zhang, Y.; Zheng, H.; Barceló, D. Thermochemical liquefaction of agricultural and forestry wastes into biofuels and chemicals from circular economy perspectives. *Sci. Total Environ.* **2020**, *749*, 141972. [CrossRef] [PubMed]
7. Vanbeveren, S.P.P.; Ceulemans, R. Biodiversity in short-rotation coppice. *Renew. Sustain. Energy Rev.* **2019**, *111*, 34–43. [CrossRef]
8. Research, F. Resources Short Rotation Coppice. Available online: https://www.forestresearch.gov.uk/tools-and-resources/fthr/biomass-energy-resources/fuel/energy-crops/short-rotation-coppice/ (accessed on 13 September 2021).
9. Rodrigues, A.; Vanbeveren, S.P.P.; Costa, M.; Ceulemans, R. Relationship between soil chemical composition and potential fuel quality of biomass from poplar short rotation coppices in Portugal and Belgium. *Biomass Bioenergy* **2017**, *105*, 66–72. [CrossRef]

10. Rodrigues, A.M.; Costa, M.M.G.; Nunes, L.J.R. Short rotation woody coppices for biomass production: An integrated analysis of the potential as an energy alternative. *Curr. Sustain. Energy Rep.* **2021**, *8*, 70–89. [CrossRef]

11. Sixto, H.; Hernandez, M.J.; Barrio, M.; Carrasco, J.; Cañellas, I. Populus genus for the biomass production for energy use: A review. *Investig. Agrar. Sist. Recur. For.* **2007**, *16*, 277. [CrossRef]

12. Robbins, M.P.; Evans, G.; Valentine, J.; Donnison, I.S.; Allison, G.G. New opportunities for the exploitation of energy crops by thermochemical conversion in northern Europe and the UK. *Prog. Energy Combust. Sci.* **2012**, *38*, 138–155. [CrossRef]

13. Štochlová, P.; Novotná, K.; Costa, M.; Rodrigues, A. Biomass production of poplar short rotation coppice over five and six rotations and its aptitude as a fuel. *Biomass Bioenergy* **2019**, *122*, 183–192. [CrossRef]

14. Sjöström, E.; Alen, R.; Timell, T.E. *Analytical Methods in Wood Chemistry, Pulping and Papermaking, Springer Series in Wood Science*; Springer: Berlin/Heidelberg, Germany, 1999.

15. Rodrigues, A.; Loureiro, L.; Nunes, L.J.R. Torrefaction of woody biomasses from poplar SRC and Portuguese roundwood: Properties of torrefied products. *Biomass Bioenergy* **2018**, *108*, 55–65. [CrossRef]

16. Álvarez, A.; Migoya, S.; Menéndez, R.; Gutiérrez, G.; Pizarro, C.; Bueno, J.L. Torrefaction of Short Rotation Coppice Willow. Characterization, hydrophobicity assessment and kinetics of the process. *Fuel* **2021**, *295*, 2021. [CrossRef]

17. Bartoli, M.; Rosi, L.; Giovannelli, A.; Frediani, P.; Frediani, M. Bio-oil from residues of short rotation coppice of poplar using a microwave assisted pyrolysis. *J. Anal. Appl. Pyrolysis* **2016**, *119*, 224–232. [CrossRef]

18. Yogalakshmi, K.N.; Sivashanmugam, T.P.D.P.; Kavitha, S.; Kannah, R.Y.; Varjani, S.; AdishKumar, S.; Kumar, G.; Banu, J.R. Lignocellulosic biomass-based pyrolysis: A comprehensive review. *Chemosphere* **2022**, *286*.

19. Soares Dias, A.P.; Rego, F.; Fonseca, F.; Casquilho, M.; Rosa, F.; Rodrigues, A. Catalyzed pyrolysis of SRC poplar biomass. Alkaline carbonates and zeolites catalysts. *Energy* **2019**, *183*, 1114–1122. [CrossRef]

20. Rego, F.; Soares Dias, A.P.; Casquilho, M.; Rosa, F.C.; Rodrigues, A. Pyrolysis kinetics of short rotation coppice poplar biomass. *Energy* **2020**, *207*, 2020. [CrossRef]

21. Wu, X.F.; Zhou, Q.; Li, M.F.; Li, S.X.; Bian, J.; Peng, F. Conversion of poplar into bio-oil via subcritical hydrothermal liquefaction: Structure and antioxidant capacity. *Bioresour. Technol.* **2018**, *270*, 216–222. [CrossRef]

22. Amado, M.; Bastos, D.; Gaspar, D.; Matos, S.; Vieira, S.; Bordado, J.M.; Galhano dos Santos, R. Thermochemical liquefaction of pinewood shaves—Evaluating the performance of cleaner and sustainable alternative solvents. *J. Clean. Prod.* **2021**, *304*, 127088. [CrossRef]

23. Rohde, V.; Hahn, T.; Wagner, M.; Böringer, S.; Tübke, B.; Brosse, N.; Dahmen, N.; Schmiedl, D. Potential of a short rotation coppice poplar as a feedstock for platform chemicals and lignin-based building blocks. *Ind. Crops Prod.* **2018**, *123*, 698–706. [CrossRef]

24. Kormin, S.; Rus, A.Z.M. Preparation and Characterization of Biopolyol from Liquefied Oil Palm Fruit Waste: Part 1. *Mater. Sci. Forum* **2017**, *882*, 108–112. [CrossRef]

25. Kormin, S.; Zafiah Rus, A.M.; Onn Malaysia, H.; Raja, P.; Pahat, B. *Preparation and Characterization of Biopolyol From Liquefied Oil Palm Fruit Waste: Part 2*; Trans Tech Publications, Ltd.: Stafa-Zurich, Switzerland.

26. Mateus, M.M.; do Vale, M.; Rodrigues, A.; Bordado, J.C.J.C.; Galhano dos Santos, R. Is biomass liquefaction an option for the viability of poplar short rotation coppices? A preliminary experimental approach. *Energy* **2017**, *124*, 40–45. [CrossRef]

27. Ma, L.; Wang, T.; Liu, Q.; Zhang, X.; Ma, W.; Zhang, Q. A review of thermal–chemical conversion of lignocellulosic biomass in China. *Biotechnol. Adv.* **2012**, *30*, 859–873. [CrossRef]

28. Kunaver, M.; Jasiukaitytė, E.; Čuk, N. Ultrasonically assisted liquefaction of lignocellulosic materials. *Bioresour. Technol.* **2012**, *103*, 360–366. [CrossRef]

29. Braz, A.; Mateus, M.M.M.M.; dos Santos, R.G.D.R.G.; Machado, R.; Bordado, J.M.J.M.; Correia, M.J.N.J.N. Modelling of pine wood sawdust thermochemical liquefaction. *Biomass Bioenergy* **2019**, *120*, 200–210. [CrossRef]

30. Goncalves, D.; Orišková, S.; Matos, S.; Machado, H.; Vieira, S.; Bastos, D.; Gaspar, D.; Paiva, R.; Bordado, J.C.; Rodrigues, A.; et al. Thermochemical Liquefaction as a Cleaner and Efficient Route for Valuing Pinewood Residues from Forest Fires. *Molecules* **2021**, *26*, 7156. [CrossRef]

31. Zhang, H.; Pang, H.; Shi, J.; Fu, T.; Liao, B. Investigation of liquefied wood residues based on cellulose, hemicellulose, and lignin. *J. Appl. Polym. Sci.* **2012**, *123*, 850–856. [CrossRef]

32. Fernandes, F.; Matos, S.; Gaspar, D.; Silva, L.; Paulo, I.; Vieira, S.; Pinto, P.C.R.; Bordado, J.; dos Santos, R.G. Boosting the Higher Heating Value of Eucalyptus globulus via Thermochemical Liquefaction. *Sustainability* **2021**, *13*, 3717. [CrossRef]

33. Mateus, M.M.M.M.; Guerreiro, D.; Ferreira, O.; Bordado, J.C.J.C.; dos Santos, R.G. Heuristic analysis of Eucalyptus globulus bark depolymerization via acid-liquefaction. *Cellulose* **2017**, *24*, 659–668. [CrossRef]

34. Dos Santos, R.G.; Ventura, P.; Bordado, J.C.; Mateus, M.M. Valorizing potato peel waste: An overview of the latest publications. *Rev. Environ. Sci. Biotechnol.* **2016**, *15*, 585–592. [CrossRef]

35. Soares, B.; Gama, N.; Freire, C.; Barros-Timmons, A.; Brandao, I.; Silva, R.; Neto, C.P.; Ferreira, A.; Pascoal Neto, C.; Ferreira, A. Ecopolyol Production from Industrial Cork Powder via Acid Liquefaction Using Polyhydric Alcohols. *ACS Sustain. Chem. Eng.* **2014**, *2*, 846–854. [CrossRef]

36. Soares, B.; Gama, N.; Freire, C.S.R.; Barros-Timmons, A.; Brandão, I.; Silva, R.; Neto, C.P.; Ferreira, A. Spent coffee grounds as a renewable source for ecopolyols production. *J. Chem. Technol. Biotechnol.* **2015**, *90*, 1480–1488. [CrossRef]

37. Daneshvar, S.; Behrooz, R.; Najafi, S.K.; Mir, G.; Sadeghi, M. Preparation of Polyurethane Adhesive from Wood Sawdust polyol: Application of Response Surface Methodology for Optimization of Catalyst and Glycerol. *Biointerface Res. Appl. Chem.* **2022**, *12*, 1870–1883. [CrossRef]
38. Wang, H.; Chen, H.-Z. A novel method of utilizing the biomass resource: Rapid liquefaction of wheat straw and preparation of biodegradable polyurethane foam (PUF). *J. Chin. Inst. Chem. Eng.* **2007**, *38*, 95–102. [CrossRef]
39. Liang, L.; Mao, Z.; Li, Y.; Wan, C.; Wang, T.; Zhang, L.; Zhang, L. Liquefaction of crop residues for polyol production. *Bioresources* **2006**, *1*, 248–256. [CrossRef]
40. Patel, A.; Shah, A.R. Integrated lignocellulosic biorefinery: Gateway for production of second generation ethanol and value added products. *J. Bioresour. Bioprod.* **2021**, *6*, 108–128. [CrossRef]
41. Dos Santos, R.G.; Carvalho, R.; Silva, E.R.; Bordado, J.C.; Cardoso, A.C.; Do Rosário Costa, M.; Mateus, M.M. Natural polymeric water-based adhesive from cork liquefaction. *Ind. Crops Prod.* **2016**, *84*, 314–319. [CrossRef]
42. Vale, M.; Mateus, M.M.; dos Santos, R.G.; Nieto de Castro, C.; de Schrijver, A.; Bordado, J.C.; Marques, A.C. Replacement of petroleum-derived diols by sustainable biopolyols in one component polyurethane foams. *J. Clean. Prod.* **2019**, *212*, 1036–1043. [CrossRef]
43. Mohan, D.; Pittman, C.U.; Steele, P.H. Pyrolysis of wood/biomass for bio-oil: A critical review. *Energy Fuels* **2006**, *20*, 848–889. [CrossRef]
44. Arvindnarayan, S.; Prabhu, K.K.S.; Shobana, S.; Kumar, G.; Dharmaraja, J. Upgrading of micro algal derived bio-fuels in thermochemical liquefaction path and its perspectives: A review. *Int. Biodeterior. Biodegrad.* **2017**, *119*, 260–272. [CrossRef]
45. dos Santos, R.G.; Bordado, J.C.; Mateus, M.M. Potential biofuels from liquefied industrial wastes—Preliminary evaluation of heats of combustion and van Krevelen correlations. *J. Clean. Prod.* **2016**, *137*, 195–199. [CrossRef]
46. Lee, S.-H.; Teramoto, Y.; Shiraishi, N. Biodegradable polyurethane foam from liquefied waste paper and its thermal stability, biodegradability, and genotoxicity. *J. Appl. Polym. Sci.* **2002**, *83*, 1482–1489. [CrossRef]
47. Dimitriou, I.; Rutz, D. *Sustainable Short Rotation Coppice. A Handbook*; y WIP Renewable Energies: Munich, Germany, 2015; ISBN 978-3-936338-36-2.
48. Rego, F.; Soares Dias, A.P.; Casquilho, M.; Rosa, F.C.; Rodrigues, A. Fast determination of lignocellulosic composition of poplar biomass by thermogravimetry. *Biomass Bioenergy* **2019**, *122*, 375–380. [CrossRef]
49. Sheng, C.; Azevedo, J.L.T. Estimating the higher heating value of biomass fuels from basic analysis data. *Biomass Bioenergy* **2005**, *28*, 499–507. [CrossRef]
50. Yin, C.-Y. Prediction of higher heating values of biomass from proximate and ultimate analyses. *Fuel* **2011**, *90*, 1128–1132. [CrossRef]
51. Mateus, M.M.; Bordado, J.M.; dos Santos, R.G. Estimation of higher heating value (HHV) of bio-oils from thermochemical liquefaction by linear correlation. *Fuel* **2021**, *302*, 121149. [CrossRef]
52. González-Arias, J.; Sánchez, M.E.; Martínez, E.J.; Covalski, C.; Alonso-Simón, A.; González, R.; Cara-Jiménez, J. Hydrothermal carbonization of olive tree pruning as a sustainableway for improving biomass energy potential: Effect of reaction parameters on fuel properties. *Processes* **2020**, *8*, 1201. [CrossRef]
53. Mateus, M.M.M.M.; Carvalho, R.; Bordado, J.C.J.C.; dos Santos, R.G. Biomass acid-catalyzed liquefaction—Catalysts performance and polyhydric alcohol influence. *Data Br.* **2015**, *5*, 736–738. [CrossRef]
54. dos Santos, R.G.; Acero, N.F.; Matos, S.; Carvalho, R.; Vale, M.; Marques, A.C.; Bordado, J.C.; Mateus, M.M. One-Component Spray Polyurethane Foam from Liquefied Pinewood Polyols: Pursuing Eco-Friendly Materials. *J. Polym. Environ.* **2018**, *26*, 91–100. [CrossRef]
55. Hassan, E.M.; Shukry, N. Polyhydric alcohol liquefaction of some lignocellulosic agricultural residues. *Ind. Crops Prod.* **2008**, *27*, 33–38. [CrossRef]
56. Pan, H.; Zheng, Z.; Hse, C.Y. Microwave-assisted liquefaction of wood with polyhydric alcohols and its application in preparation of polyurethane (PU) foams. *Eur. J. Wood Wood Prod.* **2012**, *70*, 461–470. [CrossRef]
57. dos Santos, R.G.; Bordado, J.C.; Mateus, M.M. Microwave-assisted Liquefaction of Cork—From an Industrial Waste to Sustainable Chemicals. *Ind Eng Manag.* **2015**, *4*, 173–177. [CrossRef]
58. Vidal, M.; Bastos, D.; Silva, L.; Gaspar, D.; Paulo, I.; Matos, S.; Vieira, S.; Bordado, J.M.; dos Santos, R.G. Up-cycling tomato pomace by thermochemical liquefaction—A response surface methodology assessment. *Biomass Bioenergy* **2022**, *156*, 106324. [CrossRef]
59. Condeço, J.A.D.; Hariharakrishnan, S.; Ofili, O.M.; Mateus, M.M.; Bordado, J.M.; Correia, M.J.N. Energetic valorisation of agricultural residues by solvent-based liquefaction. *Biomass Bioenergy* **2021**, *147*, 106003. [CrossRef]
60. Jadhav, A.; Ahmed, I.; Baloch, A.G.; Jadhav, H.; Nizamuddin, S.; Siddiqui, M.T.H.; Baloch, H.A.; Qureshi, S.S.; Mubarak, N.M. Utilization of oil palm fronds for bio-oil and bio-char production using hydrothermal liquefaction technology. *Biomass Convers. Biorefinery* **2019**, 1–9. [CrossRef]
61. Grilc, M.; Likozar, B.; Levec, J. Kinetic model of homogeneous lignocellulosic biomass solvolysis in glycerol and imidazolium-based ionic liquids with subsequent heterogeneous hydrodeoxygenation over $NiMo/Al_2O_3$ catalyst. *Catal. Today* **2015**, *256*, 302–314. [CrossRef]
62. Xu, F.; Yu, J.; Tesso, T.; Dowell, F.; Wang, D. Qualitative and quantitative analysis of lignocellulosic biomass using infrared techniques: A mini-review. *Appl. Energy* **2013**, *104*, 801–809. [CrossRef]

63. Mateus, M.M.; Ventura, P.; Rego, A.; Mota, C.; Castanheira, I.; Bordado, J.M.; dos Santos, R.G. Acid liquefaction of potato (*Solanum tuberosum*) and sweet potato (*Ipomoea batatas*) cultivars peels—Pre-screening of antioxidant activity/total phenolic and sugar contents. *BioResources* **2017**, *12*, 1463–1478. [CrossRef]
64. Traoré, M.; Kaal, J.; Martínez Cortizas, A. Differentiation between pine woods according to species and growing location using FTIR-ATR. *Wood Sci. Technol.* **2017**, *52*, 487–504. [CrossRef]
65. Zohdi, V.; Whelan, D.R.; Wood, B.R.; Pearson, J.T.; Bambery, K.R.; Black, M.J. Importance of Tissue Preparation Methods in FTIR Micro-Spectroscopical Analysis of Biological Tissues: 'Traps for New Users'. *PLoS ONE* **2015**, *10*, e0116491. [CrossRef]
66. Zhuang, J.; Li, M.; Pu, Y.; Ragauskas, A.J.; Yoo, C.G. Observation of potential contaminants in processed biomass using fourier transform infrared spectroscopy. *Appl. Sci.* **2020**, *10*, 4345. [CrossRef]
67. Bui, N.Q.; Fongarland, P.; Rataboul, F.; Dartiguelongue, C.; Charon, N.; Vallée, C.; Essayem, N. FTIR as a simple tool to quantify unconverted lignin from chars in biomass liquefaction process: Application to SC ethanol liquefaction of pine wood. *Fuel Process. Technol.* **2015**, *134*, 378–386. [CrossRef]
68. Yona, A.M.C.; Budija, F.; Kričej, B.; Kutnar, A.; Pavlič, M.; Pori, P.; Tavzes, Č.; Petrič, M.; Kricej, B.; Kutnar, A.; et al. Production of biomaterials from cork: Liquefaction in polyhydric alcohols at moderate temperatures. *Ind. Crops Prod.* **2014**, *54*, 296–301. [CrossRef]
69. Popescu, C.M.; Popescu, M.C.; Singurel, G.; Vasile, C.; Argyropoulos, D.S.; Willfor, S. Spectral characterization of eucalyptus wood. *Appl. Spectrosc.* **2007**, *61*, 1168–1177. [CrossRef] [PubMed]
70. Mateus, M.M.; Gaspar, D.; Matos, S.; Rego, A.; Motta, C.; Castanheira, I.; Bordado, J.M.; Dos Santos, R.G. Converting a residue from an edible source (Ceratonia siliqua L.) into a bio-oil. *J. Environ. Chem. Eng.* **2019**, *7*. [CrossRef]
71. Zhang, Y.; Liu, Z.; Hui, L.; Wang, H. Diols as solvent media for liquefaction of corn stalk at ambient pressure. *BioResources* **2019**, *13*, 6818–6836. [CrossRef]
72. Seehar, T.H.; Toor, S.S.; Shah, A.A.; Pedersen, T.H.; Rosendahl, L.A. Biocrude production from wheat straw at sub and supercritical hydrothermal liquefaction. *Energies* **2020**, *13*, 3114. [CrossRef]
73. Shawal, N.N.; Murtala, A.M.; Adilah, A.K.; Hamza, U.D. Identification of Functional Groups of Sustainable Bio-Oil Substrate and its Potential for Specialty Chemicals Source. *Adv. Mater. Res.* **2012**, *557–559*, 1179–1185. [CrossRef]

MDPI

St. Alban-Anlage 66

4052 Basel

Switzerland

Tel. +41 61 683 77 34

Fax +41 61 302 89 18

www.mdpi.com

Molecules Editorial Office

E-mail: molecules@mdpi.com

www.mdpi.com/journal/molecules

9 783036 564333